Soilless Crop Production

Sanjay Misra started his career as a computer professional first in **NIIT** where he served for four years and was bestowed with the **Award of Excellence**. He also worked as senior faculty at **APTECH** for sometime and then joined **Hartron Workstation** and stayed there for more than three years. Afterwards, he served **ZED Career Academy & GENX Tech.** for more than two years and then he joined **Aldite Corporation** and served there for two years. During this period, he had also worked on various **computer applications**, **web-sites** and **component programming** in C++, **JAVA**, **JAVA Servlet**, **Beans**, etc. To his credit he has two more books, *viz.* 'Bits & Bytes of C++' and 'CCC-Course on Computer Concepts'.

Though Sanjay Misra is a biology graduate from Delhi University but is well trained in Computer Applications. Out of curiosity, as an amateur gardener he used to grow horticultural plants and fishes by working on projects of hydroponics and aquaponics continuously at his home by sparing time from his computer profession. It was in 2011, when he started a commercial hydroponics and aquaponics farm in Allahabad along with his brother Dr. Sanyat Misra, under the guidance of his father Dr. R. L. Misra, a retired Project Coordinator (Floriculture) *cum* Principal Scientist from IARI, New Delhi, and who had been Officer-in-Charge of Indo-Israel Project on Horticultural Research & Development Farm at IARI, New Delhi where many horticultural crops were being grown under protected cultivation through soilless culture. Dr. R.L. Misra is author of several books on **Horticulture**.

Dr. Sanyat Misra is a forestry graduate from University of Agicultural Sciences, Bengaluru (Karnataka), did his master's degree in Horticulture from G.B. Pant University of Agriculture & Technology, Pantnagar, Udham Singh Nagar (Uttarakhand) and doctoral degree from Dr. Yashwant Singh Chauhan University of Horticulture & Forestry, Nauni, Solan (H.P.), and now he is serving as Assistant Professor cum Junior Scientist for the last 12 years in the Department of Horticulture, Birsa Agricultural University, Kanke, Ranchi (Jharkhand). To his credit, he has more than five books of international repute related to Floriculture and Computer Science, apart from many scientific papers and technical articles in floriculture.

Dr. R.L. Misra served Indian Agricultural Research Institute, New Delhi for 40 years in various capacities, serving temperate regions of Shimla and Kullu (both under Himachal Pradesh) for more than 16 years and New Delhi for about 24 years, exclusively on horticultural crops with main emphasis being on flowers. He had been an excellent teacher and guided 22 research scholars for their M.Sc. and Ph.D. degrees, and had been teaching all the eight floricultural courses in IARI. Together, he had also been an outstanding researcher and developed 30 gladiolus varieties. He had been Officer-in-Charge, Indo-Israel Project on R&D Farms, IARI, New Delhi where protected cultivation of vegetables and flowers in soilless culture was being done apart from open cultivation of fruits, vegetables and flowers. He retired as Principal Scientist cum Project Coordinator, All India Coordinated Research Project on Floriculture (ICAR), stationed at IARI, New Delhi. To his credit, he has more than one dozen books out of which three books are of international repute, and some 500 publications including research papers, review articles, book chapters, bulletins, symposium papers and technical articles. Still he is continuing as Adjunct Faculty in Floriculture at IARI, New Delhi. He had been a member of the expert committee in the selection from Assistant Professor to Associate Professor, to Professor and Scientists in some 24 universities and other scientific organizations and still he is continuing. He had been and is still continuing as an examiner in some 23 universities and organizations. After retirement, he is busy in literature build up, and three floricultural books of international scope are almost ready for going to the press.

Soilless Crop Production

Sanjay Misra
Sanyat Misra
R.L. Misra

2017
Daya Publishing House®
A Division of
Astral International Pvt. Ltd.
New Delhi – 110 002

Cataloging in Publication Data--DK
Courtesy: D.K. Agencies (P) Ltd. <docinfo@dkagencies.com>
Misra, Sanjay, author.
Soilless crop production / authors, Sanjay Misra, Sanyat Misra, R.L. Misra.
pages cm
Includes bibliographical references and index.
ISBN 978-93-86071-47-7 (International Edition)

1. Hydroponics.   I. Misra, Sanyat, 1974- author. II. Misra, R. L., 1946- author. III. Title.
SB126.5.M57 2017   DDC 631.585   23

Published by   : **Daya Publishing House®**
A Division of
Astral International Pvt. Ltd.
– ISO 9001:2015 Certified Company –
4736/23, Ansari Road, Darya Ganj
New Delhi-110 002
Ph. 011-43549197, 23278134
E-mail: info@astralint.com
Website: www.astralint.com

Dedication

We would like to dedicate this book to **Mrs. Prema Devi Misra,** wife of Dr. R.L. Misra and mother of Sanjay Misra and Dr. Sanyat Misra because she always blessed us and celebrated when any of our publications was out. We also dedicate to Tanu (daughter of Mr. Sanjeev Misra-eldest son of Dr. R.L. Misra) who after death of her mother Vibha, was taken over by her maternal grandparents, and because of the pending Court Cases in U.P., still we have not been able to recover her from them. She was taken over in September 2001when she was hardly four years old, on the pretension of keeping her for only two months but soon they filed a court case claiming her possession on the ground that she is deserted by her father. Court case is still going on, and she is attaining 18 years of age now. We also feel too much about her step brother **Aayush** who was also one day taken away by his mother, as just to stay only in Mumbai when Mr. Sanjeev was transferred to Chennai, she did not accompany her husband, so for him also the Court Case in Mumbai is going on, and this way life of Mr. **Sanjeev Misra**, father of both the kids, is really disturbed. So to Sanjeev, Tanu and Aayush we really dedicate this book. Both the authors, *viz.* Sanjay Misra and Dr. Sanyat Misra did not meet Sanjeev Misra for a quite long time as all the three are at faraway places and are awfully occupied with their own professions at their own places of work.

Acknowledgements

The two authors, Sanjay Misra and Dr. Sanyat Misra are highly thankful to their cousin Mr. **Rajeev Ranjan Mishra** and Mrs. **Sandhya Mishra** (*bhabhi*) who have always inspired them to do something different and better. Further they owe their gratefulness to Mrs. Smita Misra, beloved wife of Mr. Sanjay Misra and Mrs. Rashmi Misra, beloved wife of Dr. Sanyat Misra, as both worked in creating congenial home atmosphere and inspired them continuing their task uninterrupted.

We are also thankful to Mr. **Anil Maurya** who very well managed the business of Mr. Sanjay Misra as he, his father Dr. R.L. Misra and brother Dr. Sanyat Misra were busy in preparing this book.

Preface

In order to understand the art of soilless cultivation of crops, gardeners must be aware of the both, *i.e.* the foundation of gardening and an understanding of the key factors critical to achieving success in soilless environment. We wrote this book to provide a framework for learning these necessary factors in a way that emphasizes not only the importance of different techniques of soilless cultivation of crops, but also the techniques of integrated farming where crops and fish can be grown together in the same system.

This book is intended to give enough grounding to the readers, so that they can grow vegetables, other food and ornamental crops, along with fishes and other aquatic animals in a small space in their houses and to help them start commercial production of many crops in a small area.

This book not only covers hydroponics techniques, its nutrient solution requirements, disorders of plants (nutrient deficiency symptoms, diseases and other insect-pests), media, root and air environmental requirements, but also includes detailed discussion of aquaculture and aquaponics systems and their requirements along with the concepts of controlled environment of agriculture techniques such as greenhouses and growrooms. Through environmentally-controlled greenhouses, we can grow anything, anywhere and in any season for fresh supply of eatables, especially vegetables and fruits, *vis-à-vis* cut flowers in the market to generate good dividend.

Sanjay Misra
Sanyat Misra
R.L. Misra

Contents

Introduction

Today, arable land for crop production is declining drastically due to severe global population influx, *vis-a-vis* urbanization of rural areas. The demand for food and nutrition is inexorably outstripping the supply. In addition, soils may also pose serious limitations for plant's growth due to the presence of disease-causing organisms and nematodes, unsuitable soil reaction, unfavourable soil compaction, poor drainage, degradation due to erosion, *etc*. Further, continuous cultivation of crops has resulted in poor soil fertility, which in turn has reduced the opportunities for natural soil fertility build up by microbes. This situation has led to poor yield and quality. Therefore, the need to grow crops in the nutrient-rich soilless environments such as houses, the roof tops, the balcony, the greenhouses, etc. has also increased.

The soilless crop production will help people to grow healthy vegetables and food crops for their consumption in their houses without requirement of any additional space, and they can even start commercial production of many crops in a small area. Not only food crops including fishes, but ornamental plants can also be grown without the use of soil. In fact, the Jainism faith does not tolerate violence (himsa) in any form and, therefore, this does not permit even to plough the field for fear of killing microscopic and otherwise soil organisms, a case of total non-violence (ahimsa) so here the soilless culture is a best alternative to protect their religious faith and to strengthen their concept.

Although the soil acts as a mineral nutrient reservoir and is the most available growing medium for plants, the soil itself is not essential to plant's growth. Plant roots absorb the mineral nutrients such as calcium, magnesium, potassium, phosphorus, iron, *etc*. present in the soil after being dissolved in water. Most of the terrestrial plants can be grown with their roots in the mineral nutrient solution or in an inert medium, such as gravel, mineral wool, or coconut husk, using soilless culture techniques.

Most of the countries have adopted the soilless techniques for producing vegetables and certain other precious crops because it allows a more accurate control of the root system environment by using optimal water, nutrients, and temperatures to get crop yield of high quality and quantity. It also saves energy and cost that would have been required in soil disinfection. Of course, the construction investments have to be borne initially, along with their further maintenance.

The water uptake of a plant depends on the level of transpiration, inside and outside climatic factors of the greenhouse, and the plant's characteristics such as habitat, growth stages and leaf area. Similarly, the nutrient uptake of a plant is also influenced by the climatic factors, plant species and growth stages. Today, computer-controlled systems for fulfilling the water, nutrient, temperature and humidity requirements of the plants have also been developed to fully automate the system.

Hydroponics and **aquaponics** are two major techniques of soilless crop production though through **aquaponics** even the fish cultivation is also effected. In **hydroponics**, the plants are generally fed with inorganic nutrients that are mixed in water to form nutrient solution. Hydroponic crop production has significantly increased in recent years worldwide, as it allows a more efficient use of water and fertilizers, as well as a better control of climate and pest factors. Furthermore, hydroponic production increases crop quality and productivity, which results in higher competitiveness and economic incomes. In **aquaponics**, the plants and fishes are grown in the same system, and fish excretion (organic matter) is converted to plant's nutrients through microorganism activities. These nutrients are absorbed by the plants and the purified water flows back to fish container. In aquaponics, the inorganic mineral nutrients are almost completely eliminated.

Historical Background

The growing of plants without soil but in the nutrient solution is known as hydroponics which has been in existence since ancient civilization. The Indians,

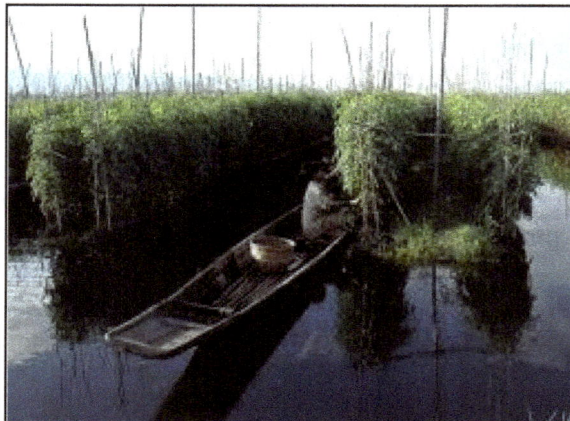

Floating Gardens.

Babylonians, Aztecs and Chinese are a few of the historic cultures where they used some form of hydroponic gardening. About 600 BC, the Hanging Gardens of Babylon (one of the 'Seven Wonders of the World') is the first known example of soilless cultivation. At AD 1100, Aztec Indians created floating gardens, the 'chinampas', as is also being practiced in Kashmir (India), especially in Dal Lake. They used layers of mud and vegetation to suspend crops over freshwater lakes in Central America. In Japan, it is highly widespread. During 1275-1292, Marco Polo wrote of floating gardens he discovered in China. In 1792, an English scientist Joseph Priestley discovered that plants absorb carbon dioxide and give off oxygen. During 1856-1865, two German scientists, Julius von Sachs and W. Knop, standardized a nutrient solution making it possible to grow plants in water only, with no medium holding roots. In 1920, Dennis R. Hoagland developed the 'Hoagland's solution' creating a nutrient formula that is still the basis of the present day practice. During 1920's to 30's, Dr. William F. Gericke, Professor of Botany at the University of California at Davis created the term 'hydroponics' to refer to growing plants in water without soil. Dr. Gericke produced 7.5 metre tall tomato plants through his method of soilless gardening. In 1952, over 3,600,000 kg of fresh produce was grown to meet up the military demand, according to the US Army's special hydroponics branch. During 1960-65, the Nutrient Film Technique (Alan Cooper, UK) and Drip Irrigation Systems (Cornell University, USA) were invented. In 1970, an Italian Dr. Franco Massantini pioneered the Aeroponic method in which roots are suspended in a mist spray. Dutch researchers use rockwool slabs to secure plants in 'ebb and flow' and 'drip systems'. In 1983, Richard Spooner invented 'aeroponics' in which roots of the plants are contained in a chamber which is sprayed continuously or intermittently with a fine mist of water containing nutrients and in this method the roots of the plants hang in the air. The advantage of this system is that the roots have good access to oxygen. During 1986-1988, an Israeli, Dr. Hillel Soffer, senior researcher at the Volcani Institute at Ein Gedi, developed the aero-hydroponic method in which partially submerged roots were sprayed with an oxygen-rich nutrient solution. In 1996, in the Netherlands, Canna introduced Canna Coco, a medium for roots made from coconut husk fibres, a renewable and organic alternative to rockwool. In 1998, Jorge Cervantes unveiled the hydro-organic growing as an alternative by using chemical nutrient salts in 'hydro-organic ~ the natural approach to hydroponics'. In 1999, Kushman wrote the 1st annual hydro report, beginning a tradition of highlighting the best hydro systems and related products.

Problems with Large Scale Soilless Cultivation

Although the soilless cultivation of food crops proved to be the best method for growing vegetables at home or in small scale commercial systems, the large scale commercial setups may have to deal with certain problems such as

☆ The prices of the culture setups for large scale commercial systems are too high.

☆ The farmers have to worry about the occurrence of diseases in the roots of crops. In water culture, nutrient solution is circulated in the entire field. Thus, once root rots occur due to pathogens such as *Phytophthora*, *Pythium*,

Fusarium spp., it spreads over all the fields in a short period of time, and in an extreme case no harvest will be obtained.

☆ There are some problems regarding cultivation techniques too.

❑ The mineral concentration, composition and pH of nutrient solution would often change during cultivation. Growers usually control their nutrient solution by a pH meter and EC (electric conductivity) meter, but this method does not indicate the changes in the composition of the solution.

❑ Most farmers change half or all of the nutrient solution with new solution once every month or every other month to keep the composition uniform, but this is an economically disadvantageous practice.

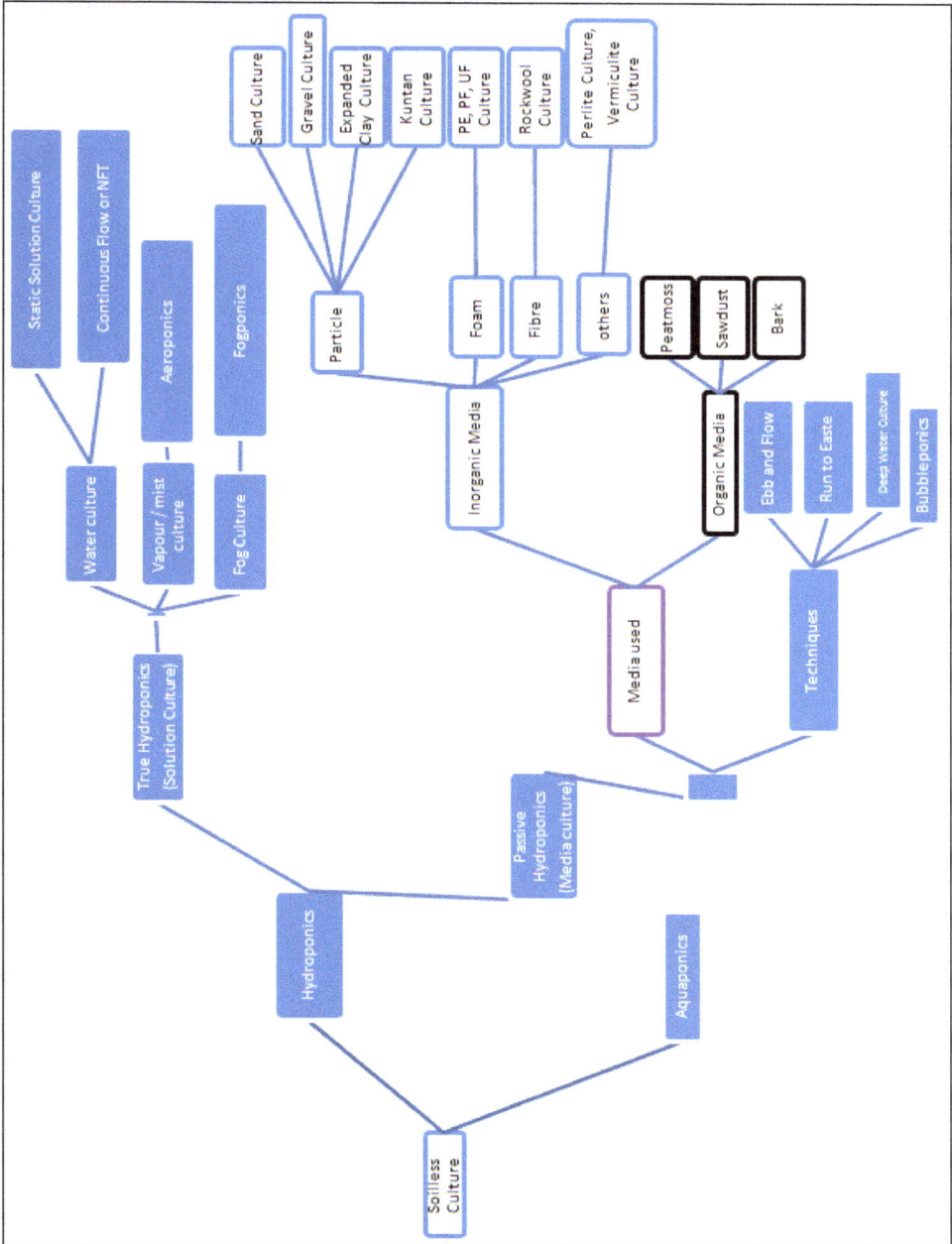

① Hydroponics

Dr. Gericke named the science of soilless gardening as hydroponics though, in real sense, the hydroponics is cultivation of plants in water by dissolving required nutrients. Hydroponics is combination of two Greek words *hydro* for water and *ponos* for labour or work, literally meaning 'water-working'. It is a soilless culture where plants are grown in controlled environment using mineral nutrient solutions. It allows the cultivator to easily, precisely and directly target the root system. Using hydroponics, growers can precisely manipulate crop growth. It is possible to accelerate or decelerate crop harvesting through interventions possible only through hydroponics. In hydroponics, soil is completely withdrawn and the plants are grown either without any growing medium or with some other soil-substitutes called substrates. Some common hydroponics substrates are cocopeat, vermiculite, perlite, and rockwool. The suitably processed substrate is placed in trays, troughs or individual seedling containers. 'Grow bags' and trays are factory-prepared combinations of substrate and container. This removes the need for the substrate to be placed in the container by the cultivator. The seed is placed in the substrate either by hand or by using seedling placement machines. Typical hydroponic crops include capsicum, cucumber, lettuce, strawberry, tomato, a few cut flower crops and certain herbs. Almost all hydroponic systems in temperate regions of the world are enclosed in greenhouse-type structures to provide temperature control, reduce evaporative water loss, and to reduce disease infection and pest infestation.

1.1 Hydroponic Nutrients

The nutrients necessary for plants are mixed in water to form the nutrient solution which is the source of food and water for the plants in hydroponic systems.

The nutrient requirement (particularly nitrogen, phosphorus and potassium) of different plants may be different which depends on the following factors:

☆ Species and variety of plants

☆ Growth stage of plants

☆ Weather

☆ Season of the year

Nutrients appropriate for the plant are mixed in the proper proportion in the pure irrigation water (water is usually passed through a RO-filter). This system of combining the supply of water and nutrients is known as fertigation. A fertigation system is used to distribute the nutrient solution to the plants in any one of many ways such as drip and micro-sprinkler methods. The distribution of the nutrient solution is done according to a schedule known as the fertigation schedule. Fertigation schedule is of critical importance to the health and well-being of the seedlings or plants.

In total there are 17 essential elements required for the growth of plants, of which carbon, hydrogen and oxygen are absorbed from the atmosphere and the remaining 14 elements called as mineral nutrients are classified as:

☆ **Macronutrients** such as nitrogen, phosphorus, potassium, calcium, magnesium, and sulphur. These are required by the plants in large quantities.

☆ **Micronutrients** such as chlorine, boron, iron, copper, manganese, zinc, molybdenum and nickel. These are required by the plants in small quantities.

The nutrient requirements, particularly of **nitrogen, phosphorus** and **potassium** (N-P-K) vary with plant type and their growth stages. For example, tomatoes require higher levels of phosphorus, calcium and potassium whereas leafy vegetables such as lettuce need higher levels of nitrogen.

These nutrients usually are available in powder, liquid, resin or tablet form, which are added to water to make nutrient solution or to the aggregate or reservoir. With the resin form the food is slowly released over long periods of time. Nutrients are available in many different NPK (N=nitrogen, P=phosphorus, K=potassium) formulations expressed as percentages. By selecting different NPK formulations for the plants, the growth, development and flowering can be controlled to a certain degree. The product label for the fertilizer shows the list of ingredients (percentage of each ingredient) available in the fertilizer. The N-P-K (nitrogen, phosphorus and potassium) is always listed first.

Example: If in the list is specified **24-8-16,** it has 24 per cent nitrogen, 8 per cent phosphorus and 16 per cent soluble potash as the macro elements of this fertilizer. The minor elements may be listed as 0.02 per cent boron, 0.07 per cent copper, 0.15 per cent iron, 0.05 per cent manganese, 0.0005 per cent molybdenum and 0.06 per cent zinc.

Many variations of the concentrated nutrient solutions can be purchased from commercial nutrient manufacturers. Although their nutrient solution for a plant may have different chemical combinations, the final compositions would be nearly the same. Commonly used chemicals for macronutrients are potassium nitrate, calcium nitrate, potassium phosphate, and magnesium sulphate. The micronutrients that are added to nutrient solution include iron, manganese, copper, boron, zinc, chlorine and molybdenum. Manufacturers sometimes may add chelating agents [to keep Fe (iron) soluble] and humic acids (to increase nutrient uptake) to the nutrient solution.

1.2 Hydroponic Systems

Hydroponic systems can be 'closed hydroponics systems' or 'open hydroponics systems'. In **closed hydroponic system**, the surplus nutrient solution is recovered, replenished and recycled, *i.e.* the same nutrient solution is re-circulated and the nutrient concentrations are monitored and adjusted according to the requirements. Closed hydroponics can be both simple hydroponic systems (such as 'deep water culture') as well as sophisticated hydroponics (such as NFT). Hydroponic systems that do not involve growing media are usually closed systems, while systems involving growing media may be both closed and open systems.

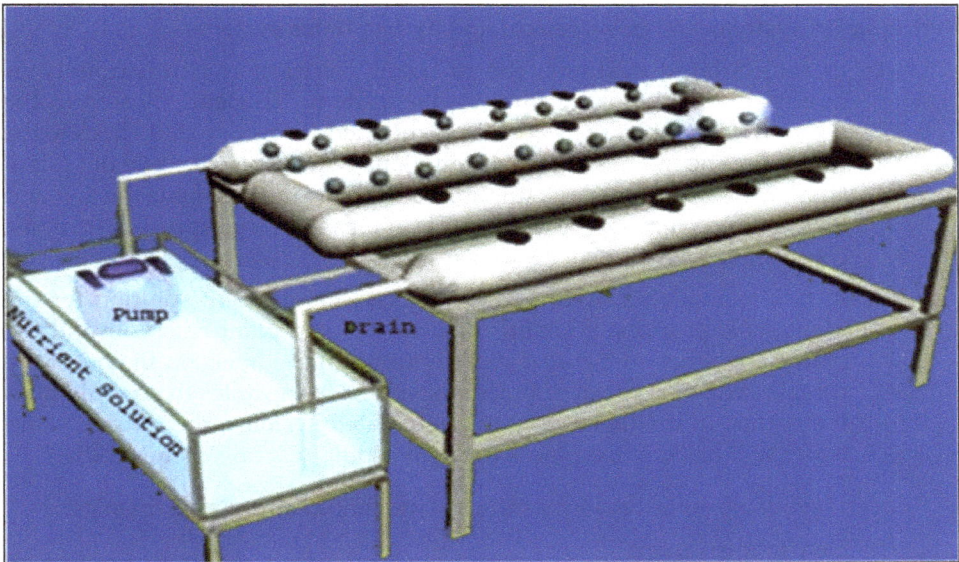

A Closed Hydroponics System.

In **open hydroponic systems** (non-circulating method) the nutrient solution that has been delivered to the plant roots is not reused, *i.e.* a fresh nutrient solution is introduced for each irrigation cycle. The nutrient solution is usually delivered to the plants using a drip system. In open hydroponic systems an adequate run-off must be maintained in order to keep nutrient balance in the root zone.

An Open Hydroponics System.

Closed hydroponics systems can be further classified as **recirculated hydroponics systems** and **non-recirculated hydroponics systems**.

☆ In recirculated systems, the nutrient solution is continuously or intermittently circulated through the system. Nutrient solution in the reservoir is pumped to the growbed and then allowed to drain back to the reservoir. The main issue of system management for this type of system is the increase in nutrient strength and pH as the nutrients are fed and then returned through the system. The increase is more apparent in media based systems then water based systems.

☆ In non-recirculated hydroponics systems, the nutrient solution for the plants is kept in the container which acts as a plant reservoir and the plants are either suspended directly through the holes on the lid of the container or, nets or net pots are suspended in the holes that would hold the plants. In media-based non-recirculated systems, the nutrients and water will reach to the plants through capillary action of the media.

Given below are some reasons as to why we should adopt hydroponics?

☆ Easy to implement

☆ Production with high yield

☆ Healthy and hygienic crops in a limited area with little effort, due to well controlled nutrition levels

☆ Predictable and off season production

☆ Almost free from pests, weeds and soil-borne diseases

☆ No pesticide damage thus saving cost of soil preparation, insecticides, fungicides, *etc*.

☆ No ploughing therefore saving cost of this activity

☆ No need to replace soil

☆ Traceability is possible

☆ Produce has high nutrition value

☆ Quick turnaround between crops

☆ Excellent ergonomics

☆ Requires least labour and water (water can be recycled)

☆ All macro- and micro-nutrients can be used efficiently

☆ Modern way to grow and earn good money

The hydroponics (open or closed systems) can be implemented as:

☆ Solution culture hydroponics (simply called as hydroponics or true hydroponics or active hydroponics): where only the nutrient solution is required and no solid medium is used for the roots.

☆ Medium culture hydroponics (also called as passive hydroponics or passive sub-irrigation or semi-hydroponics): where plants are grown in an absorbent aggregate of porous medium (such as expanded clay pebbles, gravel, mineral wool, or coconut husk) and nutrients.

☆ Aeroponics systems: where roots are suspended in the air and are misted regularly with nutrient solution.

☆ Membrane system hydroponics: capable of delivering nutrients and water to the plant roots under microgravity (*e.g.* space flight) conditions.

☆ Folkewall: specially designed hydroponics growing system with the dual functions of growing plants vertically and purifying wastewater.

Note: Although the soil plants can be used in hydroponic systems, it will be a good idea to start any type of indoor hydroponic system with cubes of rock wool that can be used to germinate the seeds. Once the seeds are germinated in cubes of rock wool, they can be carefully transplanted into the hydroponic system.

1.3 Support Structures

Unlike soils, the growing medium in hydroponics does not provide enough anchorage. Further, the planting medium is almost completely eliminated in liquid cultures. Therefore, growers must provide artificial supporting structures and train plants along those structures, especially when tall growing indeterminate type crop varieties (tomatoes, cucumber, etc.) or crops bearing relatively heavy fruits (bell pepper, egg plant, etc.) are to be grown in hydroponics.

1.4 Hydroponics Equipments

Depending on the requirements, technique and design for building a hydroponics system at home, one or more of the following equipments can be used.

☆ **Growing tray:** It is made up of various different materials and is used to anchor the root system. There are endless designs for the growing tray.

It can be designed according to the requirement of the grower and the plants to be grown. It can be a simple pot for a single plant, a bucket or even larger container for larger plants.

☐ **PVC Pipes:** Type 400 or class 4 - 100 mm PVC pipes can be used in circulating hydroponics as the channels. PVC pipes with thinner walls will sag and may reduce the flow rate of nutrient solution. The result will be lack of oxygen supply for the plant roots. UV resistant pipelines are preferable. Painting these pipes white will prevent the increase of nutrient solution temperature. The flow rate required in hydroponics is very small ranging from 1 to 3 litres per minute. Therefore, an overflow pipe may have to be fitted to adjust the flow rate.

PVC Pipe System

☆ **Pots or net pots:** Although the growing media can be directly added into the growing tray, it may be convenient to add the growing media into pots or net-pots and then these pots are placed in the 'grow tray'.

☆ **Reservoir:** It holds the nutrient solution that is the source of food and water for the plants and can be made out of just about anything, for example, a large tank or bucket.

Net Pot

☆ **Water Pumps:** The water pump must be made up of materials that are non-reactive with nutrient salt solution. Stainless steel shaft, polycarbonate or stainless steel impeller, pump-house and water seal must be there in the pump to be used in hydroponics. One does not require a pump with very high head for circulating the nutrient solution in hydroponics systems. Therefore, a domestic water pump with 0.5 HP will suffice. A safety device must be used with the water pump, as the nutrient solution more effectively conducts electricity compared to water. Therefore, a very sensitive trip switch must be used to disconnect electricity supply whenever the need arises. Water pump can be a submersible pump that can be placed in the nutrient solution tank or it can be an external pump.

☐ **Submersible pump:** It is used to pump the nutrient solution to the plant roots.

Submersible Water Pump   Submersible Water Pump
for Small Systems   for Large Systems

☆ **Delivery system:** It is a path through which nutrient solution in the reservoir is transferred to the plant roots through a pump and then back to the reservoir for reuse. It can be vinyl tubing or regular PVC pipe.

☆ **Simple timer:** It is used to turn 'on and off' the pump, as well as the lights (if lights are being used).

☆ **Air pump** and **air stone:** These are used to oxygenate the nutrient solution and can be the same kind used in fish tanks.

Air Pump

☆ **Timer and oxygen detection sensor:** When the plants are small, their oxygen requirement is low. Therefore, the nutrient solution circulating time period can be limited. Limiting the circulating time periods can also reduce the electricity consumption. For this purpose a timer may be used to set the circulating time manually or an oxygen concentration detection sensor may be included in the system, so that the sensor can activate the pump whenever the oxygen concentration of the nutrient solution level goes down.

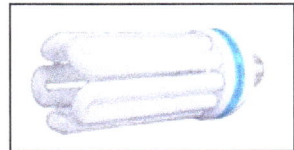

125 watt CFL (Compact Fluorescent Bulb)

☆ **Lighting:** Many different lighting systems, such as compact fluorescent lighting (CFL) or expensive lighting systems from the hydroponics store can be used.

☆ **Heater:** It is used to maintain the temperature of the nutrient solution. An aquarium heater can be used for small systems.

☆ **Proper plant support structure** (such as trellises): It is used to support the heavy plants such as tomatoes, cucumber or squash.

Cucumber Vines Trained on Trellises

✩ **Blowers:** These devices help send airflow through the plants so that plants shake and pollens are distributed to facilitate pollination inside protected structures.

✩ **Pollinators:** These simple electrical devices are used to vibrate the individual plants so that pollination is facilitated inside the protected structures.

Blower

✩ **pH meter:** It is used to test the pH of nutrient solution.

✩ **EC meter:** It is used to test the electrical conductivity of the nutrient solution.

✩ **Nurtimeter:** In addition to the EC, this meter helps in measuring the nutrient contents of the solution.

pH Meter   **EC Meter**   **Nutrimeter**

1.5 Success of Hydroponics Systems

The secret to success with hydroponics systems are

✩ **Right nutrient levels:** The growers should always try to specialise in one or two particular plant varieties so that they can adjust their nutrient mix to suit that one particular plant. It will be better if growers find an all round mix which works for just about everything, by just adjusting the strength.

✩ **Selecting the right plants:** The hydroponics growers should select the right plants for the climate and season, if they are growing plants in an outdoor hydroponics system. However they can grow almost any plant in a glasshouse or even inside under lights.

✩ **Excess of nutrients:** Too much nutrients appear to be toxic and are often worse as compared to not enough nutrients. Fewer nutrients will not only cost less, but the growers don't really have to panic if they are going away for a weekend. Even when their heavy feeder plants are coming into full fruit, fewer nutrients fortnightly will just slow them down and it generally won't kill them.

✩ **Shading:** Different plants require varying degrees of shading, for example, too much sun in the hottest months can lead to 'blossom end rot' in tomatoes.

☆ **Other factors to be considered are:**

❏ Water quality

❏ pH

❏ Temperature

❏ Humidity

❏ Disease and insect-pest control

❏ CO_2

❏ Dissolved oxygen in nutrient solution

1.6 Solution Culture Hydroponics

Hydroponics (also called **active hydroponic)** is a soilless culture where plants are grown in water using mineral nutrient solutions only. The required mineral nutrients are artificially added into plant's water supply from where these are absorbed by the plant roots, and therefore soil is no longer needed for the plants to grow or flourish. Almost any terrestrial plant can be grown with hydroponics.

*Note: The container (reservoir) of nutrient solution should exclude light to prevent algae growth in the nutrient solution. Many types of soilless culture do not use the mineral nutrient solutions required for hydroponics. For example, in **aquaponics system**.*

Solution culture systems provide consistent and immediate control of the root zone environment. The system that is used to supply nutrient solution to the roots should be able to maintain adequate aeration in the root zone and should also be able to provide nutrient solution at a known rate and concentration of nutrients. The system should also be able to maintain the integrity of different nutrient treatments.

NASA Researcher Checking Hydroponic Onions with Bibb Lettuce to his Left and Radishes to the Right.

1.6.1 Advantages

☆ It allows a more efficient use of water and fertilizers.

☆ It allows a better control of climate and pest factors

☆ No soil is needed for hydroponics, therefore it is almost free from soil-borne diseases and pests.

☆ No pesticide damage thus saving cost of soil preparation, insecticides, fungicides, *etc.*

☆ No digging or weeding.

☆ Out of season cropping makes hydroponics a most attractive proposition (extended seasons).

☆ Continuous cropping; crop after crop can be produced with consistently excellent results (no crop rotation).

☆ Lower water requirement as it stays in the system and can be reused. There is no wastage of water due to run-off or evaporation.

☆ Production of high yield healthy crops in a small area due to well controlled nutrition levels.

☆ More equal supply of nutrient solution can be achieved and so more homogeneous crops can be obtained.

☆ It takes a long time to grow seedlings in traditional agriculture, but soilless culture can reduce the time substantially as this process can be performed in a concentrated manner.

☆ Easy to implement.

☆ High productivity with little effort.

☆ It is better for consumption.

☆ Can be used in a small area and also where in-ground agriculture or gardening is not possible.

1.6.2 Disadvantages

☆ The initial set up may be expensive, depending on requirements.

☆ Failure to the hydroponic system leads to rapid plant death.

☆ The high moisture levels and poor aeration in the root zone may result in root rot.

☆ Many hydroponic plants require different fertilizers and containment systems.

Main types of solution culture hydroponics are static solution culture, continuous-flow solution culture, aeroponics, and fogponics.

1.6.3 Some Important Points to be Noted

☆ A well-buffered nutrient should be used which promotes pH stability in the nutrient solution.

- ☐ The usual pH range should be 5.5 - 6.5
- ☐ For vegetative growth, the optimal pH range is 6.0-6.3
- ☐ For flowering and fruiting, the optimal pH range is 5.7-5.9

☆ The nutrient solution's temperature should be maintained between 62°F and 80 F (17-27 C).

☆ The temperature above 72 F (22 C) may cause the dissolved oxygen (DO) dipping too low. In such case use of proper aeration is necessary.

- ☐ At temperature below 60 F (15.6 C), plants tend to slow their metabolism due to lower temperature.
- ☐ Some growers use this to their advantage and reduce nutrient temperature towards the end of the flowering stage to aid in ripening.

☆ Do not allow the stem to become submerged.

☆ In most of the ready-made systems that are available with the stores, the moisture level can be manipulated in the root zone.

- ☐ A dryer root zone can increase essential oil production in aromatic crops such as basil and mint.
- ☐ A wetter root zone can cause plants to focus on vegetative production, particularly large fan-leaved plants, which in turn speeds transpiration and photosynthetic potential.

☆ Dirty growing conditions, warm water, and high nutrient strengths may result in root diseases such as *Pythium, Fusarium, etc.*

1.6.4 Static Solution Culture (Root Dipping Technique)

Static solution hydroponics system is a non-circulating technique where nutrient solution for the plants is kept in the containers (aquarium tank, concrete tank, house appliances such as glass or plastic jars, buckets, tubs, *etc.*) which acts as a plant reservoir. Holes for each plant are made on the lid of the container. Either the plants are suspended directly through the holes or, nets or net pots are suspended in the holes that would hold the plants. Aquarium pumps, valves and tube pipes can be used to provide aeration to the nutrient solution so that the roots get enough oxygen. The depth of the container must be enough (atleast 25–30 cm) to provide enough solution as well as enough space above the solution for oxygen absorbing roots.

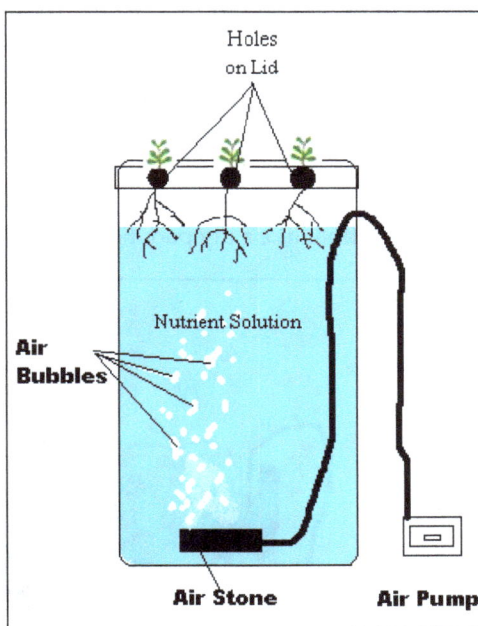

Static Solution Culture Hydroponics

Important: If nutrient solution is not aerated, then the level of the nutrient solution should be kept low so that enough roots are above the solution. This way roots will get enough oxygen. This is in contrast to 'deep water culture' (DWC) where the plant roots are always completely submerged in the highly oxygenated nutrient solution. The **deep water culture** can be built as medium culture hydroponics system where a porous inert medium (that holds the plant roots) and oxygenated nutrient solution are used, or as solution culture hydroponics which uses highly oxygenated nutrient solution only.

Clear containers (transparent containers) should be painted black or covered with black plastic or other materials such as aluminium foil so that the light does not reach inside the container. This is to prevent algal growth.

Although the plants can grow well at room temperature in summer under temperate conditions or during winter in tropical conditions, in some cases, an aquarium heater can be used to adjust the temperature of the nutrient solution.

In static solution culture, as the plant uses the nutrients, the nutrient concentration usually declines with time. When the concentration of the nutrient solution drops below a certain level, the nutrient solution should be changed or more nutrients should be added. The concentration level can be checked by an electrical conductivity (EC) meter. If electrical conductivity meter is not available, then the nutrient solution should be changed periodically depending on the type and progress of the plants.

The level of the solution should not go below a certain level (depending on the type of plants). A float valve can be used to automatically maintain the solution level. If the solution gets below a certain level, water or fresh nutrient solution should be added. More technically, plants or seedlings can be placed into holes in a platform or raft that is generally made of a sheet of buoyant plastic (such as **Styrofoam**) that is floated on the surface of the nutrient solution (floating raft solution culture), so that the level of the solution never drops below the roots. Roots from the plants grow into the nutrient solution as the raft floats directly on top of it.

In its simplest form, the plants are placed on top of nutrient solution and are allowed to float like a lily pad. For heavy plants, a plant support structure or floating

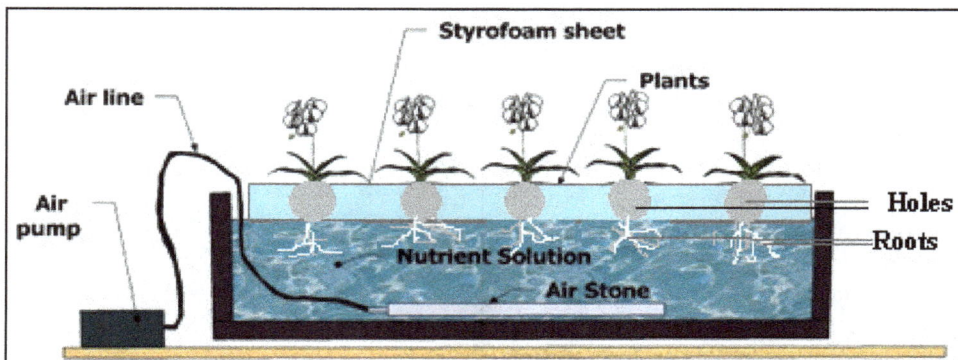

Static Solution Culture (Floating Raft System)

device should be used so that the plants do not sink to the bottom of the reservoir. In this type of system, since the water is stagnate and does not flow like in most other hydroponic systems, the chances of algae growth is much high. The rotten roots should also be timely removed.

In static solution culture, it may seem to be quite difficult to provide identical aeration rates, pH, electrical conductivity, and the nutrient levels in each container, especially as plants become large in relation to the culture container.

1.6.5 Continuous-Flow Solution Culture and NFT-Nutrient Film Technique

In continuous-flow solution culture, the nutrient solution is made to flow constantly through the roots. It can be implemented in a large storage tank of desired size where thousands of plants can be grown. A common variation is **NFT (nutrient film technique),** where a shallow stream of nutrient solution is re-circulated through the bare roots of plants in a watertight channel. The thin film of the nutrient solution ensures that the upper parts of the roots remain moist but above the solution, thus ensuring that the roots are watered and fed but not completely soaked. This way the roots would have better access to oxygen in the air.

In this system, tubes or channels can be used as an inexpensive grow tray for plants. A round tube or PVC pipe (cheap and easily available) with usually equally spaced holes (at the top of the channels) drilled to fit the net pots and seedlings, can be used. The disadvantage of using PVC as a grow tray is that the roots in the middle would have access to a deeper depth of the solution, while roots that are close to the edges would only have a shallow depth, thus causing uneven growth. This problem can be eliminated by using a flat-bottomed channel.

In NFT, the channel (grow tray) is placed above the nutrient solution reservoir and a pump is used to deliver nutrient solution to the high end of the grow tray and a drain pipe brings back the unused solution to the reservoir using gravity. This feed cycle can be intermittent (regulated by a timer and repeated many times per hour) or continuous. The NFT with intermittent feed cycle is sometimes called as Pulsed NFT. Pulsed NFT usually achieves better aeration of roots because roots are exposed to air between each feed. A 'continuous' feed cycle runs non-stop and is usually used in commercial operations. An air stone can be placed in the reservoir that can be connected to an air pump outside, to oxygenate the water.

The grow tray should be placed at an angle to allow the water to flow down towards the drain pipe. At lower end of the channels nutrient solution gets collected and flows to the nutrient solution tank. The solution is monitored for salt concentration before recycling. The flow rate of the nutrient solution should be adjusted to 2-3 litres per minute depending on the length of the channel.

A properly designed NFT system has the right channel slope, the right flow rate, the right channel length, and of course the right composition of nutrient solution.

Plants for this type of hydroponic system are usually started in rockwool cubes and then transferred into the holes of the channels at a certain age, where they grow to maturity.

No growing medium is required for this hydroponic system other than the rockwool cubes where the seedlings were started in.

Design Parameters for NFT

	Recommended	Most Effective	Remarks
Channel slope	1:30 to 1:40	1:100, but difficult to implement	Even with these slopes, water logging may occur
Flow rate	1 to 1.5 litres per minute	1 litre per minute	At planting, rates may be half of this (maximum flow rate should not exceed 2 litres per minute).
Channel length	8 to 12 metres	8 to 10 metres	For rapidly growing crops, oxygen levels remain adequate but nitrogen may be depleted over the length of the channel. Therefore, for channels with length greater than 12 metres, another nutrient feed should be placed halfway along the channel and flow rate should be reduced to 1 litre per minute through each outlet.Plants at the head of long channels are often healthier than those at the outlet end.Also, the channel length depends on the type of plant being grown and the nutrient strength.
Channel depth and width	The width and depth of the channel should be enough to comfortably accommodate plant's mature root system. Channels that are not of proper size may get choked with roots leading to damming, overflowing or create stagnant areas that fail to drain fully. Also, the last plant should be positioned well above the drain's outlet so that roots do not block the drain		

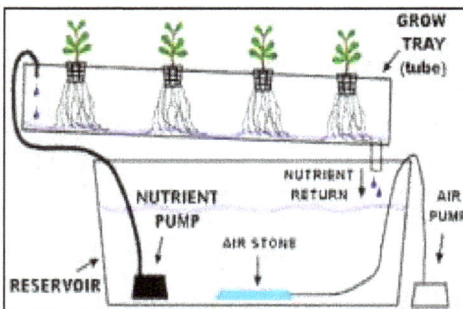

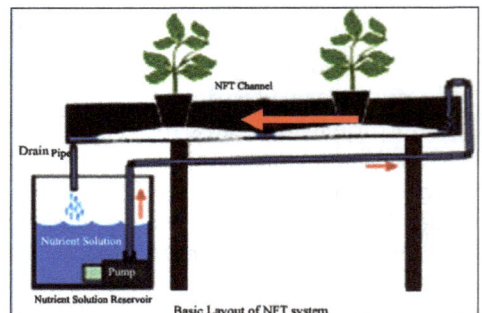

Basic Layout of NFT System

Nutrient solution	Balanced nutrient solution depending on the type and age of plants.

NFT is best suited for light weight (small to medium) and fast growing plants that do not require a lot of support and can be quickly harvested. For heavy plants such as tomatoes or squash, proper support systems such as trellises should be used to support the plants.

NFT Hydroponic System

1.6.5.1 Advantages of NFT

☆ It is easy to maintain the temperature and nutrient concentrations since there is a large storage tank separate from the plants.

☆ In NFT system, plant roots are exposed to sufficient supplies of water, oxygen, and nutrients resulting in healthy plant growth. Thus higher yields of high-quality produce are obtained over an extended period of cropping with lower water and nutrient consumption.

☆ In NFT, it is easy to inspect roots for signs of disease because there is no growing media.

☆ The flood and drain mechanism of NFT helps in preventing localized salt build-up in the root zone and also helps in maintaining uniform root zone pH and conductivity.

☆ NFT is environmentally friendly as there is no or minimum groundwater contamination.

1.6.5.2 Disadvantages of NFT

☆ In case of power failure, pump failure, or blockage in tubes, the system may get crashed resulting in the death of the plants. Therefore, backup electricity and timely check of pump and fill tubes is necessary. To slightly overcome the risk of power failure, two reservoirs can be used. One can be placed above the high end of the channel that feeds the solutions to the plant roots using gravity. The other reservoir is placed at the lowest point to collect the used solution. A pump is used to circulate the nutrient solution from the lower channel to the channel at the high end. If there is a power or pump failure, the grower will at least have enough time as it will take for the reservoir to get empty. Also, the solution from the lower reservoir can be manually placed to the higher reservoir or both the reservoirs can be interchanged periodically. This technique is feasible only for small NFT systems only. For large system, special care should be taken to recover from power and pump failure (such as inverters, generators, and a backup pump).

✰ It is not suitable for plants with large tap-root systems such as carrots.

✰ While the reduced amount of water needed to fill the system may seem like a benefit, it can mean greater temperature and water quality fluctuations in a short period of time, especially if a small volume of storage tank or reservoir is being used.

1.6.6 Deep Flow Technique (DFT)–Pipe System

In DFT–pipe system, 2-3 cm deep nutrient solution flows through 10 cm diameter PVC pipes to which plastic net pots with plants are fitted. The plastic pots contain planting materials and their bottoms touch the nutrient solution that flows in the pipes. The PVC pipes may be arranged in one plane or in zigzag shape depending on the types of crops grown. The figure below shows the main features of a DFT–pipe system.

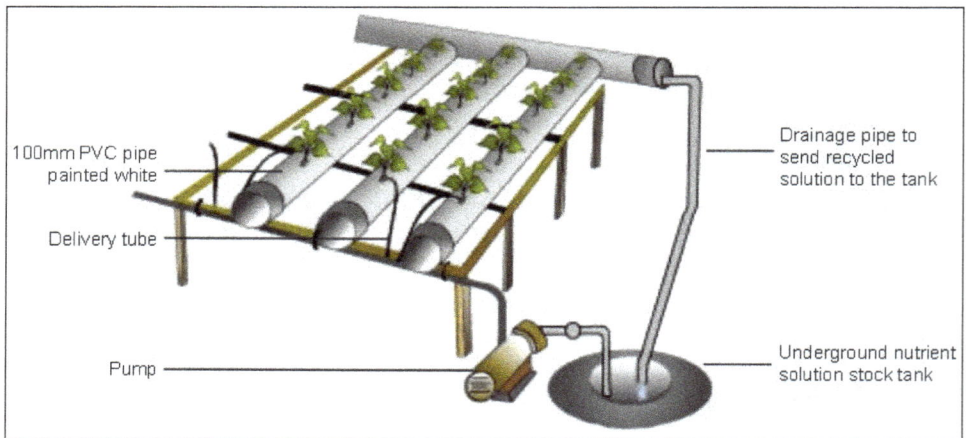

DFT Systems
Source: http://ruaf-asia.iwmi.org/Data/Sites/6/PDFs/H_Eng.pdf[1]

Plants are established in plastic net pots and fixed to the holes made in the PVC pipes. Old coir dust or carbonised rice husk or mixture of both may be used as planting material to fill the net pots. A small piece of net can be placed as a lining in the net pots to prevent the planting material falling into the nutrient solution. Small plastic cups with holes on the sides and bottom may be used instead of net pots. When the recycled solution falls into the solution in the stock tank (reservoir), the nutrient solution gets aerated. The PVC pipes must have a slope of drop of 1:30–1:40 to facilitate the flow of nutrient solution. Painting the PVC pipes white will help reduce the heating up of nutrient solution. This system can be established in the open space or in protected structures.

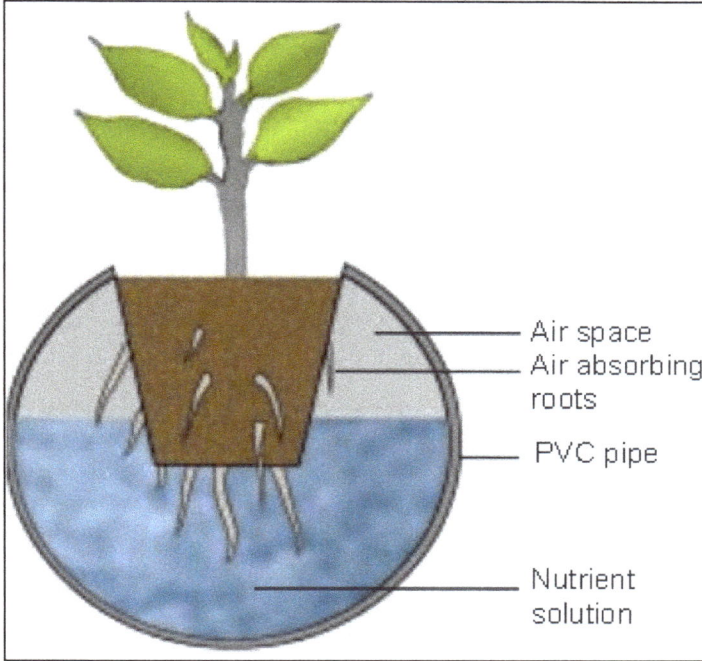

Cross Section of a PVC Pipe System DFT
***Source*: http://ruaf-asia.iwmi.org/Data/Sites/6/PDFs/H_Eng.pdf** [2]

Sources

1-2 http://ruaf-asia.iwmi.org/Data/Sites/6/PDFs/H_Eng.pdf

② Aeroponics

2.1 Aeroponics

Aeroponic is derived from the Greek *aero* (air) and *ponos* (labour). An aeroponics system is a closed or semi-closed system where the plants are grown with their roots suspended in a deep air or mist environment without the use of soil or an aggregate medium. Plants are supported through holes in a panel cover made of expanded polystyrene, PVC, or other material. The roots are periodically wetted with a fine mist of atomized nutrients. The advantage here is that the fine mist sprayers also collect lots of oxygen and direct it to the roots making the plants grow extremely fast. The nutrient solution is then re-circulated to the reservoir.

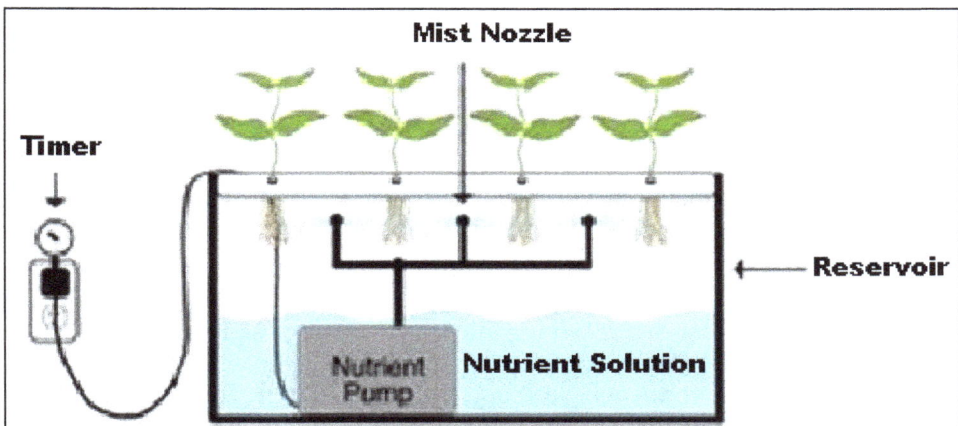

An Aeroponics System Displaying Roots Suspended in Root Chamber.

Lettuce Grown in an Aeroponics Apparatus

Roots Suspended in Aeroponics Chamber

In aeroponics system plants are grown in a closed or semi-closed environment by spraying lower stem of plants and hanging roots with an atomized or sprayed nutrient-rich water solution. In this method plants are grown with their roots suspended in a deep air or growth chamber. The leaves and crown (canopy) extend above, and the roots of the plants are separated using the plant support structure. For larger plants, trellising should be used to support the weight of the plant.

For healthy plant development, the aeroponics environment should be fully controlled to keep away the pests and diseases. There is rarely any disease transmission in an aeroponics system because of reduced plant-to-plant contact and each spray pulse can be sterile. Also, any diseased plant can be immediately removed from the aeroponics apparatus without affecting other plants.

The nutrient cycle in an aeroponics system is of main consideration. The plant roots should never be more than slightly moist or excessively dry. A typical feed/pause cycle is 1.5 to 2 seconds on, followed by nearly 1.5 to 2 minute pause. However, when an accumulator system is incorporated, cycle times can be further reduced to nearly less than 1 second on and nearly 1 minute pause. Another point of consideration in an aeroponics growth of the plants is the size of water droplets. The larger the water droplet, the lesser oxygen is available to the plant roots. The ultrasonic mister generates too fine water droplets that help in producing excessive root-hair without developing a lateral root system. This is required for the sustained growth of plants in an aeroponics system.

Failure of aeroponics apparatus may result in unhealthy and uneven growth of plants, and if the aeroponics system is not recovered for a long time, the plants may even die. To overcome this problem, aeroponics system should be combined with conventional hydroponics which can be used as a backup nutrition and water supply.

Aeroponics is widely used in laboratory studies of plant physiology and pathology. This technique of growing plants is now being used commercially and has proved to be very successful for propagation, seed germination, production of leaf crops, tomato production, seed potato production, *etc.* Today, many (small, medium, and large) aeroponics apparatus are available in the market to be used as in-house crop production as well as for commercial purpose. Plants in a true aeroponics apparatus should have full access to all the available CO_2 in the air for photosynthesis and the plant roots should have full access to the O_2 (oxygen) available in the air. To deliver nutrients to the plant roots, an aeroponics equipment uses sprayers, misters, foggers or other devices. These devices create a fine mist of solution to deliver nutrients to plant roots.

For example,

☆ To cover large areas of roots, a hydro-atomizing spray at 360° can be used.

☆ Ultrasonic foggers can also be used to mist nutrient solutions in low-pressure aeroponics devices.

2.1.1 True Aeroponics v/s Droplet Aeroponics

An apparatus that produces fine mist of nutrient solution is a **true aeroponics system**. This fine mist of solution is generated using a special pump. Whereas the system that produces droplets of nutrient solution is named as **droplet aeroponics system**. Although specially designed droplet aeroponics growing systems are also available, but for small scale production at home, a timer controlled low pressure pump that delivers nutrient solution *via* jets can be used.

2.1.2 Nutrient Cycle

Maintaining the nutrient cycle for aeroponics crop production at home could be a hectic task if a properly designed aeroponics apparatus is not being used. Special care should be taken to maintain the nutrient cycle. Too long feed cycle or the too short pause cycle, discourages both lateral root growth and root hair development. Ideally, roots should never be more than slightly moist or excessively dry. A typical feed/pause cycle is 1.5 to 2 seconds on, followed by nearly 1.5 to 2 minute pause. However, when an accumulator system is incorporated, cycle times can be further reduced to nearly 1 second on and nearly 1 minute pause.

2.1.3 Advantages of the System

☆ The fine mist sprayers also collect lots of oxygen and direct it to the roots making the plants growing extremely fast.

☆ There is a minimal contact between a plant support structure and the plant. The stem, leaves and roots of aeroponics plants are nearly 99 per cent in the air and have full access of the available oxygen (for plant roots) and carbon dioxide (leaves use for photosynthesis).

☆ The aeroponics environment can be fully controlled. Therefore, aeroponics plants can be made free from pests and diseases. As a result many plants can be grown at higher density.

☆ Many plant species can be grown in a true aeroponics system.

☆ Aeroponically grown plants never get waterlogged.

☆ Aeroponically grown plants use approximately 65 per cent less water as compared to **Static Solution Culture** or **NFT** hydroponic plants.

☆ Aeroponically grown plants have approximately 80 per cent more essential minerals and these plants require about   the nutrient input.

☆ Aeroponically grown plants would rarely suffer transplant shock when they are transplanted to soil.

2.1.4 Disadvantages of the System

☆ Some fine mist nozzles get clogged over time with deposits and then do not function properly.

☆ Failure of aeroponics equipment may damage the plants.

☆ Maintaining the nutrient cycle for crop production at home could be a hectic task. Special care should be taken to maintain the nutrient cycle.

2.1.5 Types of Aeroponics Systems

2.1.5.1 Low-Pressure Aeroponics System

The low-pressure aeroponics system is usually suitable for bench top growing and for demonstrating the principles of aeroponics. In this system, the plant roots are suspended above the nutrient solution reservoir or inside the channel connected to a reservoir. A low-pressure pump is used to deliver the nutrient solution *via* jets or ultrasonic transducers. The nutrient solution then drips or drains back into the reservoir. In these systems, as the plants grow to maturity, the root systems of the plants are likely to suffer from dry sections. This may result in an inadequate nutrient uptake. Also, these units are unable to purify the nutrient solution by removing debris and unwanted pathogens.

2.1.5.2 High-Pressure Aeroponics System

The high-pressure aeroponics systems use high-pressure pumps to generate the mist of nutrient solution. These units are little costlier than the low-pressure units and are mostly used in the cultivation of high value crops and plant specimens. These units are equipped with the technologies for air and water purification, nutrient sterilization, low-mass polymers and pressurized nutrient delivery systems.

2.1.5.3 Commercial Aeroponics Systems

Commercial aeroponics systems consist of high-pressure devices and biological systems matrix. The high-pressure devices are used for the cultivation of high value crops and can achieve multiple crop rotations. The biological systems matrix enhances the extended plant life and crop maturation. Since these systems are almost free from pathogens, they are considered to be most reliable and require very low maintenance. The subsystems of commercial aeroponics systems include effluent control systems, precision timing and nutrient solution pressurization, heating and cooling sensors, thermal control of solutions, efficient photon-flux light arrays, spectrum filtration spanning, fail-safe sensors and protection. The more advanced commercial systems include data gathering, monitoring, analytical feedback, and internet connections to various subsystems.

2.1.6 Aeroponics and Tissue Culture

Aeroponics has significantly advanced the tissue culture technology. It clones plants in less time and reduces numerous labour steps associated with tissue culture techniques. With the use of aeroponics, growers clone and transplant air-rooted plants directly into field soil. Aeroponics roots are not susceptible to wilting and leaf loss, or loss due to transplant shock.

Aeroponics could eliminate stage I and stage II plantings into soil. In stage I, tissue culture plants are planted in a sterile media, and in stage II these plants are expanded out for eventual transfer into sterile soil. When plants become strong enough they are transplanted directly to field soil. The entire process of tissue culture is labour intensive and prone to disease infection, and failure.

2.2 Fogponics

Fogponics can be considered as an advanced form of aeroponics. In fogponics, nutrient solution is converted to vapour or fog having very minute particulate size for better absorption by the roots. The smaller particulate size gets absorbed faster by the roots. This vaporized nutrient solution transfers nutrients and oxygen to the enclosed suspended plant roots. Also,

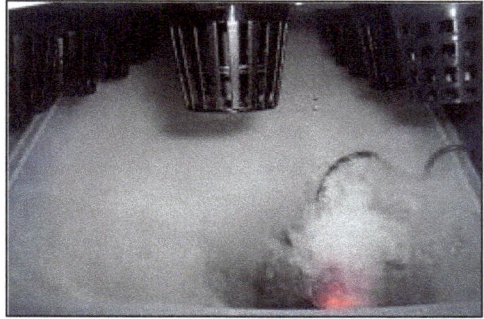

Fogponics Apparatus

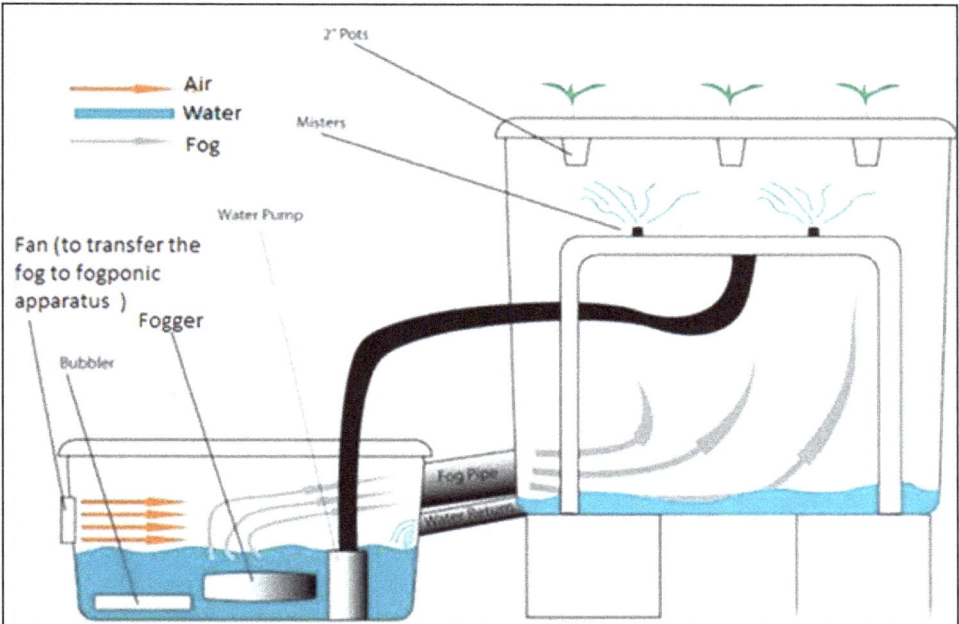

Fogponics System

in fogponics system, the plants require less energy in root growth and mass, and are still able to sustain a large plant. The fogponics system uses nearly the same technique as of aeroponics, except that fogponics utilize approximately 5-30 µm mist within the rooting chamber. Plants best absorb particles from the 1-25 µm range. The water and energy use is reduced as compared with traditional aeroponics and hydroponics systems. Also, the crop yield in fogponics system is higher than that of aeroponics or hydroponics system.

③
Passive Hydroponics (Medium Culture Hydroponics)

Passive hydroponics, or passive sub-irrigation or semi-hydroponics, is a method of growing plants in an absorbent aggregate of porous medium (expanded clay pellets, coconut husk chips, perlite, vermiculite, diatomite, charcoal, rockwool, *etc.*) and nutrients, without the use of soil, peat moss, or bark. Water and fertilizer to the roots are transferred by capillary action of inert porous medium. Water and fertilizer are kept in a reservoir and transported to the roots as necessary thus providing a constant supply of water to the roots.

The hydroponic media contain more air space and thus deliver increased oxygen to the roots. This helps in preventing root rot in epiphytic plants such as orchids and bromeliads. Salt accumulation in the system should be removed from time to time by washing the system.

In its simplest form, the pot (containing porous medium and plants) is kept in a shallow solution of fertilizer and water or on a capillary mat saturated with nutrient solution.

3.1 Advantages

☆ Easy maintenance as watering and feeding involve just topping up the reservoir of growing solution.

☆ The water requirement for the plant can be easily seen.

☆ Mostly hydroponic media are resistant to some types of soil-borne insects.

☆ The aggregate is open therefore allowing air to circulate around the roots.

☆ The hydroponic media contain more air space and thus deliver increased oxygen to the roots.

☆ Reduction of root rot.

☆ The absorbent aggregate maintains humidity around the plant.

☆ Can reduce the labour required to maintain a large collection of plants.

3.2 Disadvantages

☆ Plants that require drying between watering and a dry dormant period may fail to thrive under the constant moisture.

☆ The practice of over-irrigation coupled with a well-draining medium prevents growing media from drying out. However, growing media that are constantly wet have little to no air porosity. Such media may subject roots to oxygen deprivation as the oxygen diffusion rate through water is four orders of magnitude slower than through air.

☆ Salt accumulation in the system should be removed from time to time by washing the system.

There are two main variations for **medium culture hydroponics - sub-irrigation** and **top-irrigation**. For all techniques, most hydroponics reservoirs are now built of plastic or other materials such as concrete, glass, metal, vegetable solids, and wood. The containers should exclude light to prevent algae growth in the nutrient solution.

Some techniques of passive hydroponics are explained below.

3.3 'Ebb and Flow' or 'Flood and Drain' Sub-irrigation

Ebb and Flow (also known as **ebb and flood** and **flood drain**) is an inexpensive, simple and reliable hydroponics system that is best suited for smaller crops such as lettuce or other leafy vegetables. Larger plants can also be grown in this type of hydroponic system but it would just need to set up things a bit different for

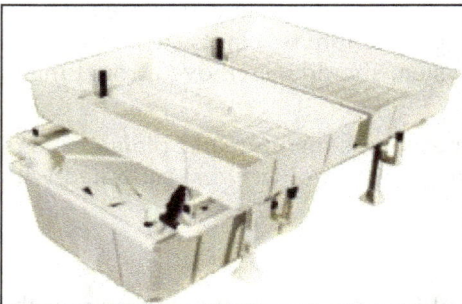

Two Simple Designs of 'Ebb and Flow' System.

the larger plants to deal with the harvesting and replacing of the plants. On a commercial level, 'ebb and flow' hydroponics systems are usually used to grow smaller size plants so as not to have to deal with the added work involved with taller plants. In this system, the nutrient solution reservoir is kept below the water-tight grow tray. Either the grow tray is filled with an inert medium as the rooting medium (such as gravel or coarse sand) or pots of medium stand in the tray. The inert medium is used to anchor the roots and is also a source of temporary reserve of water and nutrients. The nutrient solution is periodically flooded into the grow tray of the system and is allowed to ebb away. Sometimes the flow is constantly on and sometimes it goes on and off with a timer depending on what is being grown. A simple timer-operated pump (*e.g.* pumps used in aquarium, koi ponds, *etc.*) can be used to periodically flood the growing bed (for a short period of 5 to 10 minutes or until the grow tray gets filled) with a nutrient solution from the reservoir. This nutrient solution is drained back from the grow tray to the reservoir by gravity. The ebb system requires a proper drainage system to prevent stagnation of water which may result in fungal growth and root rot. One way to achieve better drainage is by slightly tilting the grow tray. The frequency of flooding depends on the type and stage of the plant being grown and on the water holding capacity of the rooting medium. The media with very high water holding capacity may require watering only once or twice a day. Some media may require two to six or even more times of flooding each day. The aerated nutrient solution provides both nutrition and oxygen to the plant roots. The aeration is accomplished through thin-filming and positive displacement of de-oxygenated air as it is forced out of the root zone by water. Since the solution is not left in constant contact with the plant roots, therefore, the solution does not need to be oxygenated in the reservoir. When the solution is drained back, the roots get re-exposed to air. The root system provides passive oxygenation at a high level and thus to a great extent, suppresses the pathogen growth. 'Ebb and Flow' hydroponic systems are great for hobby growers and for indoor growing as they are easy to build and can be made in any number of sizes.

The repeated use of nutrient solution may result in root diseases and nutrient element insufficiencies. Also, the nutrient solution will require reconstitution, filtering, sterilization, and may even require to be replaced within the growing period. Above all, the rooting medium should be properly sterilized and washed to remove root debris. To prevent the pathogen growth, the temperature of the nutrient solution can be made slightly below the temperature at which the pathogen growth begins, but should not be made so cool that the activities of the roots get suppressed. If there is a suspicion that the anaerobic *Pythium* root rot mould started to proliferate on the surface of the root, hydrogen peroxide (H_2O_2) should be added to nutrient solution. The oxygen liberated from the hydrogen peroxide destroys single-celled organisms. The dosage of hydrogen peroxide may vary with the concentration of peroxide.

If the plants are being fed several times in a day, lower ppm (parts per million) nutrients (600-800) are sufficient and the higher ppm nutrient may burn up the plants from inside. The ppm concentration may increase due to water evaporation and absorption by the plant roots. For the rooting media such as lava rocks the number of feedings can be increased to every 50 to 60 minutes during lights on, and nutrient ppm should be kept between 600 and 800. In lava rocks, the solution drains quickly and a small amount of oxygen and nutrients get trapped in its rough texture.

An E and F system can be designed in many ways:

☆ A single, two-directional path of the solution flow tube can be used to flood the grow bed and to drain back the solution to the reservoir. When the solution is flooded in the tray and the roots get briefly submerged, the pump is made inactive using a switch (timer operated) and the solution is drained back to the reservoir using the same tube. Pumping is usually done through the bottom of the grow bed and then when the pump turns off, the nutrient solution is drained back down through the pump.

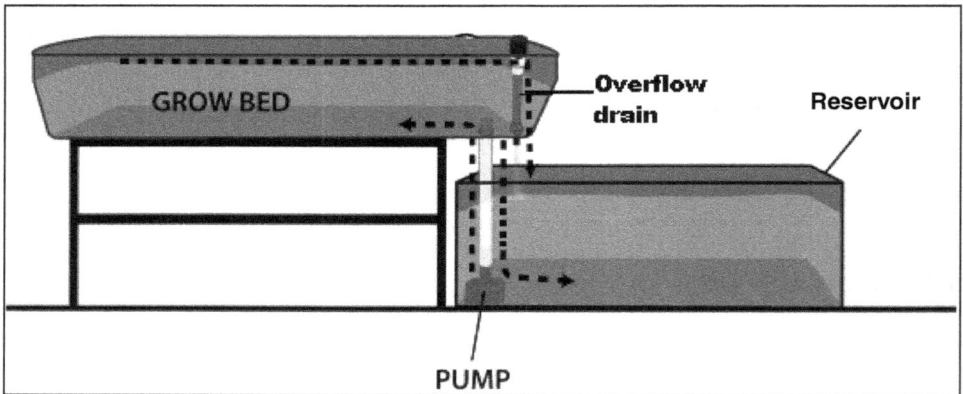

Ebb and Flow System with Single Two-Directional Path for Both, Flood and Drain.

☆ The grow bed space is allowed to flood and then drain either by the use of a pump on a timer to fill the bed and then allow it to drain while the pump is off or by the use of a siphon or other intermittent outflow device

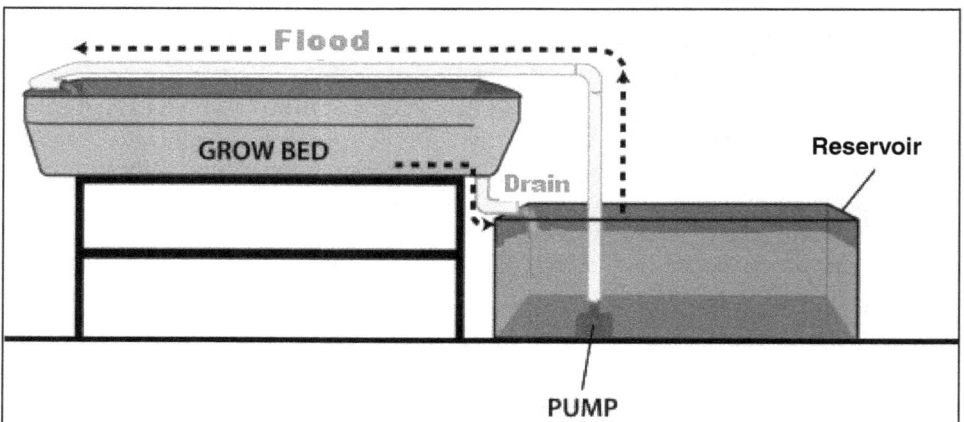

'Ebb and Flow' System with different Paths for Flooding and Draining.

where the bed is constantly filling and then the siphon will drain the bed quickly.

Most experts believe that the flood and drain media based systems using a Bell Siphon to drain are more reliable and simple methods of hydroponics, especially for beginners. It has a better temperature control as compared to the plastic pipe system (NFT). Also, it can be constructed using a wide range of different containers.

If the top 5.0 to 6.25 cm layer of the media is kept dry, it helps in preventing weeds, fungus and harmful bacteria from growing. The ideal depth of media beds is 30 to 45 cm or more depending on the type of plants to be grown in them. The grow bed shell should extend 5 cm above the media bed with an overflow pipe to return water to the reservoir.

3.3.1 Water Cycling

When using media filled grow bed, the media must be periodically flooded and drained. There are several methods by which this can be accomplished. A proper flow is essential for the delivery of oxygen to the roots. This cycling of water from the reservoir to grow bed and back to the reservoir can be performed in several ways. Some of these methods are

☆ Using a siphon

☆ Using a spill over

☆ Using a toilet valve

☆ Using a pump set on a timer

The key is to maintain a flow rate that will cycle the water through the system. The slow flooding, and then rapid draining of the grow beds, provides for excellent access to nutrients for the plants, and high oxygenation for the plant roots. The rapid draining draws oxygen down fully into the roots and this is vital for good growth.

3.3.1.1 Siphon

A siphon is a mechanism for moving water from one reservoir to another, lower reservoir. It is capable of raising water over a barrier. In flood and drain methods, siphons can be installed to quickly empty a grow bed into the reservoir using a simple mechanical method with no moving or electrical parts. A siphon is an elegant way to drain the grow bed in a flood and drain hydroponics system without using a timer to turn the pump on and off.

An auto-siphon can be used that can start and stop itself in response to changing water levels. Two commonly used siphons are **Loop Siphon** or **U-Bend Siphon** and **Bell Siphon**. With any growbed and siphon set up, there must be some way of getting the water through the growbed. This is usually done by cutting a hole through the side or bottom of the tank and installing a drain pipe.

Ebb and Flow Hydroponics with Auto-siphon.

3.3.2 Advantages

★ **Oxygenation**. There is no need to oxygenate the solution because when the solution is drained back, the roots get re-exposed to air. Many plants like some dry time especially if the water is not super aerated.

★ The root system provides passive oxygenation at a high level and thus to a great extent, suppresses the pathogen growth.

★ Less power consumption.

★ For a small system, the grow beds can be manually flooded, one or more times a day depending on the type of inert media and growth stage of plants.

★ It is simple to design, as excessive plumbing is not required.

★ It is a compact cultivation and can be used in a small area.

3.3.3 Disadvantages

★ The repeated use of nutrient solution may result in root diseases and nutrient element insufficiencies. The nutrient solution should be timely checked for ppm (parts per million) concentration of all the essential elements.

☆ If the plants are being fed many times each day, the higher ppm nutrient may burn up the plants from inside.

☆ The nutrient solution will require reconstitution, filtering, sterilization, and may even require to be replaced within the growing period. This could be a hectic task for large systems.

☆ Proper control requires routine checking and replacing with fresh nutrients nearly five to six days to avoid toxicity.

☆ The rooting media should be properly sterilized (using hydrogen peroxide or chlorine solution) and washed (to remove root debris or other things).

☆ Over a period of time, the pH of the nutrient solution may fluctuate to a range which is unhealthy for the plant. Usually the pH should be adjusted each day during the flowering period.

☆ Removal of harvested or damaged plants can be quite difficult for the plants having well developed root system.

☆ Poor drainage may expose the dense roots to stagnant water resulting in root rot and fungal growth.

3.4 Deep Water Culture or Direct Water Culture (DWC)

Deep Water Culture is a method of plant production by means of suspending the plant roots in a solution of nutrient-rich, oxygenated water. Traditional method of **DWC** uses plastic buckets and large containers with the plant contained in a net pot suspended from the centre of the lid. The roots are suspended in the oxygenated nutrient rich solution. An air pump combined

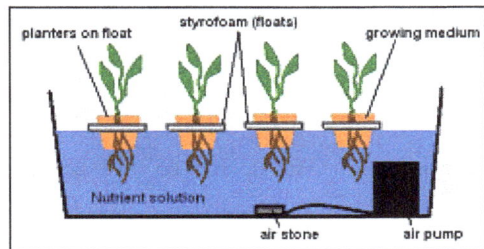

DWC

with porous stones oxygenates the solution. This highly oxygenated nutrient rich solution nourishes the plants for their fast growth. Usually, plants absorb more oxygen directly from the air than from the oxygen dissolved in water but, in deep water culture plant roots are able to absorb large quantities of oxygen and nutrients from the nutrient solution. This leads to rapid growth throughout the life of the plant.

*Note: Deep Water Culture (DWC) is very much similar to **Static Solution Culture** method of hydroponics except that a DWC system can be implemented as both **solution culture** and **medium culture** hydroponics systems and the roots are almost always submerged in a highly oxygenated solution, whereas a **static solution culture** is termed to those hydroponics techniques that have no media and use static nutrient solution only and usually the level of the nutrient solution is kept low so that enough roots are above the solution allowing the roots to get enough oxygen.*

To create a simple DWC system, we need a container (around 10-15 litres capacity container) to hold the nutrient solution and air stones (generates oxygenated bubbles), a lid that can support at least a single plant growing in a net pot, a pump to oxygenate the nutrient solution and a net pot (with or without porous medium). Roots grow out of the net pot into the oxygenated nutrient solution in the container below.

In media based DWC, a porous growing media such as expanded clay pellets is used to absorb the nutrients and water from the underlying reservoir and supply to the roots at initial stage of plant development. The base of the net pot containing porous media and a plant must be slightly submerged in the solution so that the porous media is able to absorb the nutrient solution for the roots. When a large root system gets developed, it will immerse in the nutrient solution.

Maintaining the level of nutrient solution at the same level as of the base of the pots is not necessary because the results will come with a lower level of nutrient solution. Once the plants are ready to flower, the level of the nutrient solution is gradually reduced to expose the roots to the air. The nutrient solution should be replaced or new half strength nutrient solution can be added with approximately once a week or two weeks (It all depends on the plant size, the volume of the nutrient solution in the reservoir, and the fluctuations in pH). Also, the container should be properly washed once a week to remove any algae, mould and salt deposits. Timely measurement and maintenance of pH, especially when the nutrient solution gets replaced is necessary to ensure that its value is between 5.5 and 6.8. If the container is transparent, it should be wrapped with black cloth or plastic to prevent the algae growth.

Note: The DWC suits best when it comes to planting and growing. It will be a good idea to germinate the seeds with cubes of rock wool. The germinated seeds can then be placed into the pots previously filled with expanded clay pellets.

Some Important Points to be Noted

☆ Use a well-buffered nutrient that promotes **pH** stability in the nutrient solution.

☐ The usual pH range should be 5.5 - 6.5.

☐ For vegetative growth, the optimal pH range is 6.0-6.3.

☐ For flowering and fruiting, the optimal pH range is 5.7-5.9.

☆ The nutrient solution's temperature should be maintained between 62°F and 68 F (17-20 C).

☐ The temperature above 22 C may cause the dissolved oxygen (DO) dips too low.

☐ At temperature below 15.50 C, plants tend to slow their metabolism due to lower temperature.

☐ Some growers use this to their advantage and reduce nutrient temperature towards the end of the flowering stage to aid in ripening.

☆ Do not allow the stem to become submerged.

☆ In most DWC systems that are available with the stores, the moisture level can be manipulated in the root zone.

❐ A dryer root zone can increase essential oil production in aromatic crops such as basil and mint.

❐ A wetter root zone can cause plants to focus on vegetative production, particularly large fan leaved plants, which in turn speed transpiration and photosynthetic potential.

☆ Dirty growing conditions, warm water, and high nutrient strengths may result in root diseases such as *Pythium*, *Fusarium*, *etc.*

3.5 Recirculating Deep Water Culture (RDWC) or Modular Deep Water Culture (MDWC)

Recirculating Deep Water Culture systems (also known as RDWC) uses a reservoir that is large enough to provide nutrient solution for multiple interconnected containers. It is an active system where the nutrient solution cycles from the central reservoir around all the containers and back to the reservoir. As the water re-enters the bucket it is broken up

Recirculating Deep Water Culture

and aerated with the use of spray nozzles or similar device. Continuous recirculation oxygenates the water and ensures a good mix of nutrients, **CF** and stabilizes **pH** throughout the entire system. Therefore, testing is required only at the reservoir and not for all the connected DWC containers. This is major advantage of RDWC over single stand-alone DWC.

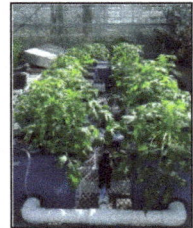

The major drawback of RDWC as compared to single stand-alone DWC is that the root diseases can spread from plant to plant very quickly.

In RDWC, the nutrient solution is highly oxygenated using an air pump combined with porous stones and therefore the plants may grow faster and healthier because of higher amounts of oxygen that the roots receive.

3.6 Run to Waste

In a **run to waste** system, nutrient solution is periodically applied to the surface of the inert porous growing media (such as rock wool, perlite, coco fibre, or sand) in a container, either manually one or more times per day for simple systems with highly porous media or using an automated timer controlled delivery pump and irrigation tubing in a slightly complex system. The delivery frequency for the nutrient solution is usually governed by the key parameters of climate, substrate, substrate conductivity, pH, plant size and growing stage, climate, and nutrient solution concentration. Most of the commercial run to waste systems uses computers to govern the watering frequency.

The nutrient rich waste can be collected, filtered, processed and reused many times.

3.7 Bubbleponics

Bubbleponics is a kind of top-fed deep water culture where highly oxygenated nutrient solution is directly constantly delivered from the reservoir to the plant's root zone using a pump and then it is allowed to run back into the reservoir kept below. Similar to deep water culture, in bubbleponics too there is an airstone in the reservoir that pumps air into the water *via* a hose from outside the reservoir to oxygenate the nutrient solution. Both the airstone and the water pump are allowed to run continuously 24 hours a day.

Note: Bubbleponics can be implemented as medium culture hydroponics or as solution culture hydroponics. If it is being used as a medium culture hydroponics system, the nutrient solution can be delivered periodically using timer controlled pump, otherwise the nutrient solution should be constantly delivered to the plant's roots.

The growth of plants during the first few weeks is greater in bubbleponics systems as compared with the DWC. This is because, in bubbleponics system the plant's roots get easy access to water and nutrients from the beginning and will quickly grow to reach the oxygenated nutrient solution in the reservoir. When the roots get immersed in the nutrient solution plant's growth rate in both, DWC and bubbleponics is almost same.

Sometimes the term, "static solution sulture", "deep water culture" and "bubbleponics" may be used interchangeably.

3.8 Drip Hydroponics System (Top Feed or Top Irrigation)

Drip Hydroponics systems are very flexible systems as they can be used in the wide range of plants (small, medium and large) that can be grown. It can be used to grow many different varieties and sizes of plants. Larger pots and larger grow bags can be used with this type of hydroponic system to grow large cucumber and tomato plants as well as nursery shrubs and even trees. In **drip hydroponics system,** a timer controlled submersible pump is used to drip the nutrient solution to the base of each plant using small drip lines. These drip lines are attached to the main supply line of the submersible pump as shown below.

Note: A drip hydroponics system and most other systems where a pump is used to transfer the nutrient solution to the plant roots can be made as **nutrient**

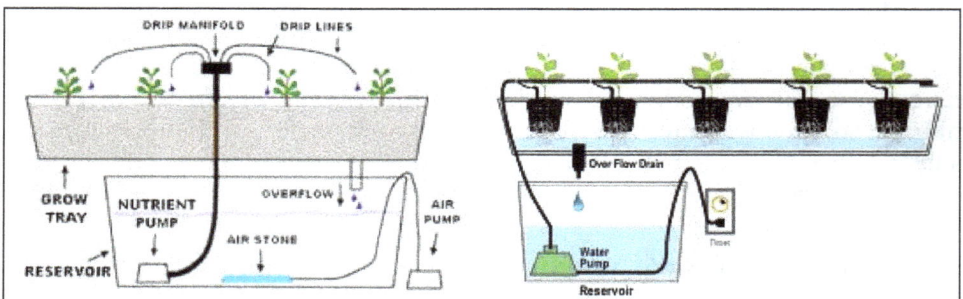

Working Model of Drip Hydroponic System

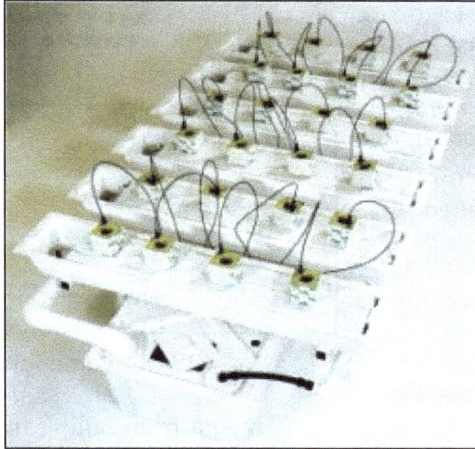

Designs for Drip Hydroponics System

solution recovery system (closed loop system) or as **nutrient solution non-recovery system (open loop system)**.

In a **recovery drip system** the excess nutrient solution that runs off, is collected back in the reservoir using a drain line, for re-use. In this type of system a simple inexpensive timer can be used, as it does not require precise control of the watering cycles. The drawback of this recovery system is that the pH and nutrient strength require periodic checking and adjustments because the pH and nutrient concentration of the nutrient solution gets changed as the plants use up the nutrients in the nutrient solution.

In a **non-recovery drip system**, a precise timer should be used so that watering cycles can be adjusted to make sure that the plants get enough nutrient solution, and to keep the runoff to minimum. This system requires less maintenance as the excess nutrient solution is not recycled back into the reservoir, hence, the nutrient strength and pH of the reservoir will not vary. Thus the care for the nutrient strength and pH has to be taken only when more nutrient solution has to be added in the reservoir.

Larger plants are usually placed on a gravel base on the ground and the supplied nutrient solution is lost into the ground and not re-circulated (non recovery drip system). Usually smaller plants that can be spaced closer together use the re-circulating hydroponic system (recovery drip system). The small feed hoses are usually staked inside the pots or if hydroton is used water drip rings are just set on top of the growing medium.

3.9 Wick System

The Wick system is one of the simplest types of hydroponic system that is very easy and inexpensive to set up. It is a passive system, meaning they have no moving part. In a Wick System: one, two, or more wicks can be used to deliver nutrient solution from the reservoir to the roots by capillary action. Cotton or nylon ropes can be used as wicks. A cotton rope is not as durable as nylon rope.

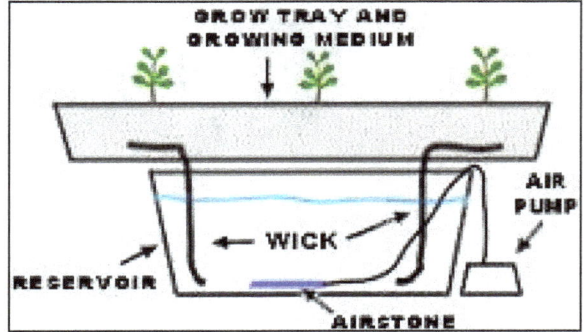

Wick Hydroponic System

The wicks are inserted into the grow tray filled with growing media (such as perlite, vermiculture and coconut fibre that are able to effectively utilize the capillary action of the wicks) through small holes in the grow tray. These holes should be small enough to prevent the growing media from falling through the holes and should be large enough to allow the wicks to pass through them. The grow tray is entirely filled with growing media and the seedlings are placed directly into it. The number of wicks in a system depends on the size of the system, plants to grow, growing media in the grow tray, and the wicking ability of the wicks being used. To attain the maximum growth result, add one wick for each plant making sure that the wicks are placed near the plant roots. Also, for plants that consume more water, more than one wick per plant can be added.

Note: The wicking property of a rope can be easily checked by dipping one end of the rope into a glass or cup of coloured liquid and noticing its sucking ability. This can be performed by comparing the wicking ability with washed and unwashed ropes.

The reservoir below the grow tray is filled with nutrient solution that is supplied to the plant roots through the capillary action of the wick whose one end is placed in the grow tray and the other end is submerged in the nutrient solution in the reservoir. The nutrient solution in the reservoir must be refreshed every week (or two weeks depending on the size of the system), because the strength of the nutrients will get declined as the plants absorb them. Also pH should be regularly checked and maintained.

The nutrient solution in the reservoir can be aerated by placing the air stone (similar to those found in home aquariums) in the reservoir and a pump. The air stone is placed in the nutrient solution in the reservoir and connected to an air pump outside the reservoir. The pump pushes air through the stone, which blows out tiny bubbles to distribute oxygen through the water.

The main drawback of this system is that larger plants or plants that use large amounts of water may use up the nutrient solution faster than the wicks can supply it. The plants that are grown in this type of system are usually lettuce or other leafy vegetables. The herbs, such as rosemary is also a better choice for the wick system as it does not require a lot of water. Plants like tomatoes that consume more water may not show good result.

3.10 Rotary Hydroponics Garden

A rotary hydroponics garden can be created within a circular frame which rotates continuously during the entire growth cycle of the plants being grown. A high intensity grow light is placed in the centre of the rotary garden to simulate sunlight. Most often, this grow light is assisted by a mechanized timer. A complete rotation of the rotary hydroponics system depends on the system's specifics. In most of the systems, the rotation is set to once per hour so that a plant can have 24 rotations within the circle in 24 hour period. The plants are periodically fed with the nutrient solution. In this type of hydroponics system, the plants usually grow much faster as compared to the plants that are grown in soil or other types of hydroponics systems. This is due to the plant's continuous fight against gravity.

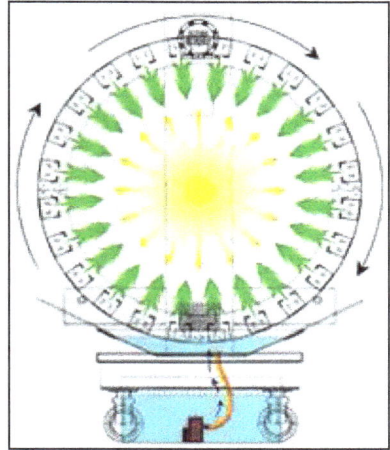

Rotary Hydroponics System

3.11 Membrane Systems

With the rapid technological advancements in hydroponics, a membrane hydroponic system has been developed which is capable of delivering nutrients and water to the plant roots under microgravity (*e.g.* space flight) conditions while still containing the nutrient solution within the system.

The capillary properties of microporous, hydrophilic membranes or tubes in these systems are used to supply the nutrient solution to the plant roots. Due to this capillary property of the system, the nutrient solution will not flow through the membrane when the solution is maintained at a small negative pressure relative to the atmosphere, usually in the range of -0.5 to -1.0 kPa.

3.12 Hybrid Hydroponics

Hydroponics systems can be designed in several ways by combining two or more of the above methods. The following figure demonstrates the combination of **NFT** and **Flood and Drain** methods.

The solution is pumped from the nutrient reservoir to the topmost NFT, from where it is drained back by gravity to another NFT below. Again from here the nutrient solution is drained by gravity to the two **Flood and Drain systems** that have bell siphons which are used to drain back the nutrient solution back to the nutrient solution reservoir (See Figure on next page).

3.13 Growbag

Growbags are large plastic bags filled with a growing medium and various nutrients that are sufficient for one season's growing. Only planting and watering are required by the grower. These bags come in different sizes, colours and formulations

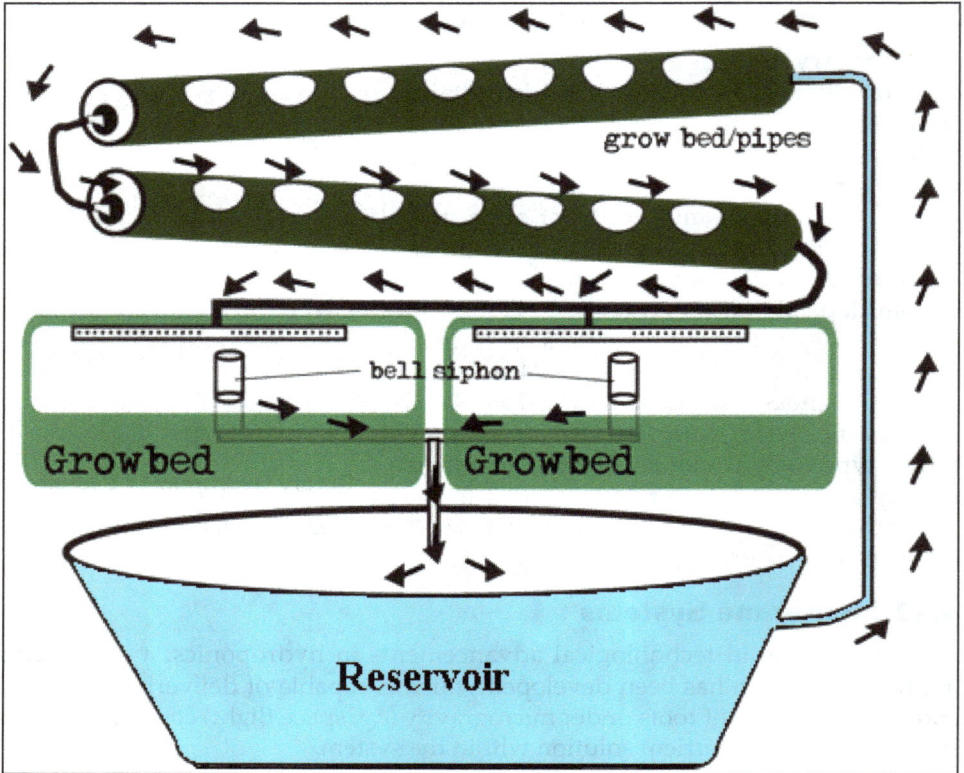

Hydroponics System with Two NFTs and One 'Ebb and Flow'.

suited to specific crops. The growing medium is usually based on a soilless organic material such as, coir, composted green waste, composted bark or composted wood chips, or a mixture of these. Planting is undertaken by first laying the bag flat on the floor or bench of the growing area, then cutting holes for plants in the uppermost surface, into which the plants are inserted.

The standard commercial growbags are usually UV resistant polythene bags that are about 1–2 m long and mostly black from inside. These bags are about 6 cm in height and 18 cm wide. These bags are placed end to end horizontally in rows on the floor with walking space in between. The bags may be placed in paired rows depending on the crop to grow.

Small holes are to be made on the upper surface of the bags from where seeds or other planting materials established in net pots can be placed into the planting

Tomato Plant in Growbag

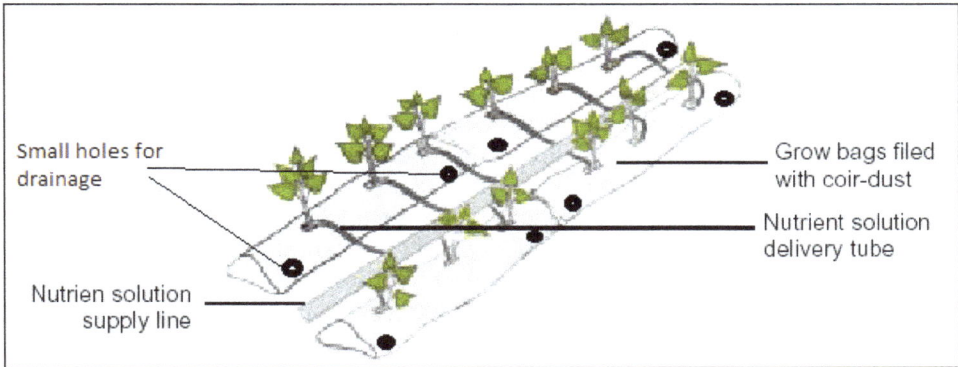

Plants in Growbags

medium (*e.g.* coir dust). Two to four plants can be established per bag. Two small holes should be made low on each side of the bags for drainage.

Fertigation can be practiced with capillary tube leading from main supply line to each plant. The nutrient solution and water may also be added manually to these bags. Depending on the stage of crop growth and the prevailing weather conditions, the amount of water can be varied. Make sure that the growing medium is not completely saturated with water or nutrient solution, as it prevents the oxygen supply to plant roots.

The entire floor should be covered with white UV resistant polythene before placing the bags. This white polythene reflects the sunlight to the plants. It also reduces the relative humidity in between plants and incidence of fungal diseases. When tall growing plants are established, supporting structures will be necessary.

3.13.1 Steps for Planting Tomatoes in Growbags

☆ The pots of young tomato plants should be soaked in a tray of water for an hour to ensure that compost is fully moistened. This will help reduce root damage during transplanting.

☆ Lay bag of compost flat in a sunny position in the garden, on a balcony or in the greenhouse. Shake compost inside the bag and cut a slot out of the bag to expose the compost for planting them into.

☆ Loosen the compost in the bag with a hand fork.

☆ Push some of the compost into the corners of the bag so that it maintains a good shape.

☆ Carefully remove the tomato plants from their pots.

☆ Allow two plants for a 60 litre bag of compost, or three plants for a 75 litre bag.

☆ Make a hole in the compost and place the young tomato plant in it. Set the plant a little deeper than it was growing in the pot because tomato plants are able to produce roots from the stem.

☆ Water the plants in the bag.

3.14 Hanging Bag Technique

Hanging bag technique, also known as 'verti-grow' technique generally uses 1 m long cylinder-shaped, white (interior-black) UV-treated, thick polythene bags, filled with sterilized coconut fibre. These bags are sealed at the bottom end and tied to small PVC pipe at the top, and are suspended vertically from an overhead support above a nutrient solution-collecting channel. Seedlings or other planting materials established in net pots are squeezed into holes on the sides of the hanging bags. The nutrient solution is pumped to top of each hanging bag through a micro-sprinkler attached inside the hanging bags at the top. This micro-sprinkler evenly distributes the nutrient solution inside the hanging bag. Nutrient solution drips down wetting the coconut fibre and plant roots. Excess solution gets collected in the channel below through holes made at the bottom of the hanging bags and flows back to the nutrient solution stock tank.

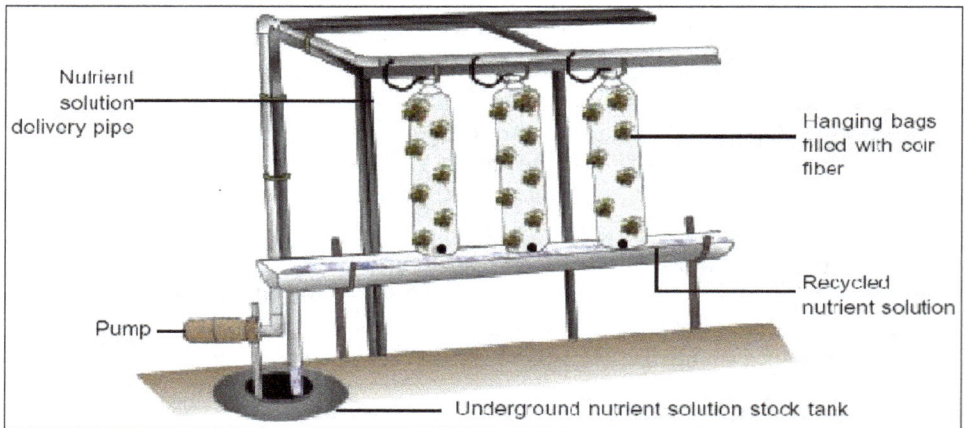

Hanging Bag Technique

This system can be established in the open space or in protected structures. In protected structures, the hanging bags in the rows and amongst the rows must be spaced in such a way that adequate sunlight falls on the bags in the inner rows. The bags are not heavy as they are filled with coconut fibre and can be used for about two years. The number of plants per bag varies depending on the plants. About 20 lettuce plants can be established per bag. This system is suitable for leafy vegetables, strawberry, and small flower plants. Black colour tubes will have to be used for nutrient solution delivery to prevent mould growth inside.

④

Environmental Conditions for Plants

Different species of plants require different environmental conditions to grow and flourish. The growth, development, and productivity of plants are closely connected with the environment. The factors that greatly affect the plant's growth include temperature, light, humidity, oxygen, carbon dioxide, water, wind, rain, nutrients, pH, *etc.* The two types of environments that the growers have to maintain for hydroponics systems are air and root environments. The root environment in hydroponics is usually maintained by nutrient solution and media (for media based systems). These are discussed in chapter 5 and chapter 6.

Hydroponics culture is probably the most intensive method of crop production in today's agricultural industry. Most commercial hydroponics systems are combined with greenhouses and protective covers to make it controlled environment agriculture. With the possibility of adjusting air and root temperature, light, water, plant nutrition, and adverse climate, this combination can be made highly productive, conservative of water and land, and protective of the environment.

4.1 Greenhouses

A **greenhouse** is a structure or building where plants are grown in a controlled environment and may enable certain crops (flowers, vegetables, fruits, and transplants) to be grown throughout the year. Significant inputs of heat and light may be required; particularly with winter production of warm-weather crops. Commercial glass greenhouses are often high tech production facilities for vegetables

Giant Greenhouses in the Netherlands

or flowers. These greenhouses are filled with equipment like screening installations, heating, cooling, lighting, and may be automatically controlled by a computer. Most greenhouses use sprinklers or drip lines for irrigation.

The greenhouse protects plants (soil plants or soilless plants) from heat, wind, cold and rain and helps to increase overall productivity of crops. Large commercial hydroponics systems are usually built or kept under controlled environmental conditions in greenhouses. The size of a greenhouse depends on several factors such as capital investment, training, environmental requirements, market, labour requirements and personal preferences. It is also important to choose a good building site, such as drainage, accessibility, available utilities and amount of sun exposure.

The design of a greenhouse must provide protection from wind, rain, heat, cold and pathogens. The key to success in cooler climates is to build a greenhouse structure that can maintain an interior temperature between 4.5 and 40 C. It is far easier to warm a building a few degrees than it is to cool the interior temperature from outside temperatures of over 40 C down to the required "**comfort zone"** for the plants.

Also, it should be able to do the following:

☆ Permit maximum light transmission to the crop.

☆ Transmit the visible light portion of the solar radiation spectrum which is the only portion utilized by plants for photosynthesis.

☆ Absorb the small amount of ultraviolet radiation in the spectrum and cause some of it to fluoresce into visible light, useful to plants.

☆ Protect the greenhouse interiors from overheating by reflecting or absorbing infrared radiation.

Different types of covering materials for a greenhouse are:

Glass Cover

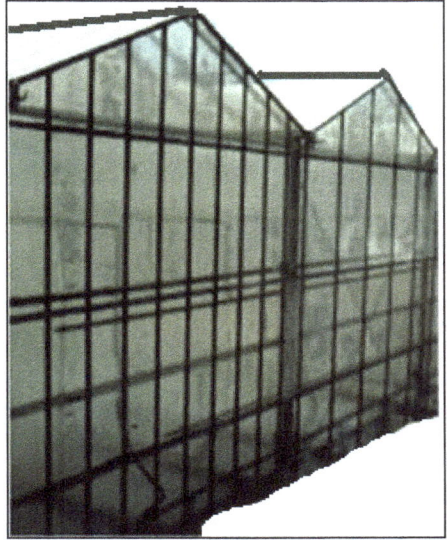
Greenhouse with Glass Cover

☆ It has a very high emissivity for long wave radiation which creates slightly higher air temperatures in the greenhouse during the night.

☆ It is an inflexible, heavy and expensive.

PE (polyethylene) Sheet Film

☆ It is easy to work with PE because it is very flexible and light in weight.

☆ Quite cheap as compared with other films.

☆ Less durable.

☆ It is the most widely used greenhouse cover in the world.

The Air-Inflated Blanket of PE

☆ Approximately, 30-40 per cent heat can be saved during winter by using an air-inflated blanket where two sheets of PE are separated by air pressure maintained by a small continuously running fan.

☆ This double-layer, air-inflated roof also provides stability against high winds.

PVC (Polyvinylchloride)

☆ It has a very high emissivity for long wave radiation (similar to glass), which creates slightly higher air temperatures in the greenhouse during the night.

☆ Less flexible if compared to PE.

☆ The major disadvantage of PVC is its narrow width as compared to PE.

☆ These can be heat welded together to form a large sheet.

☆ PVC film is not photodegradable, as is PE.

☆ PVC film is not suitable for air-inflated roofs because the air pressure may stretch the film and reduce its structural strength.

Double-Skinned Panels made of Polycarbonate and Acrylic

☆ These are very expensive.

4.1.1 Light

Since the photosynthesis plays an important role in the plant growth, therefore, the greenhouse should allow maximum light transmission to the crop. The shading from outside buildings and trees or the taller plants in the greenhouse may reduce the amount of light reaching the crops. The greenhouses and the rows of plants in the greenhouse should be oriented north and south so that the light is evenly distributed across each plant. White floor coverings or paint can also be used to reflect back the light into the crop. Clean white paint is more reflective than metallic or foil, although there is some indication that foil tends to confuse insects and slightly decrease insect pest damage.

4.1.1.1 Ideal Light Spectrum for Plant Growth

Ideally plants will grow in the range of 400-700nm. This covers a range of the visible light spectrum from blue to red. Different types of artificial lights available in the market ranges from fluorescent to 'high intensity discharge (HID)' lighting and each has their merits. Usually HID lighting is used more widely than fluorescent lighting, particularly if the growing area is large. Plants do not see light the same way as we do. Light available to plants is commonly referred to as PAR Watts *i.e.* photo-synthetically active radiation. As the name suggests, this corresponds to light that most encourages photosynthesis. Maximum photosynthesis occurs at two peak wavelengths, 435nm (blue) and 675nm (red). As human beings, we judge light by its brightness, or lumen intensity but the brightest lamp is not necessarily the best for plant growth. The lumen output does, however, provide a guide as to the effectiveness of the lamp.

4.1.1.2 Kelvin and CRI

Kelvin is a measure of colour temperature. Light output in the range of 380-500nm is considered the daylight spectrum. It is translated as 6500K. This colour temperature is the closest to natural sunlight. CRI (colour rendition/rendering index) is a percentage value of the lamps proximity to natural sunlight. For example, a lamp with a CRI of 65 means that it is 65 per cent close to sunlight. Metal Halide lamps are generally in this range of CRI. High Pressure Sodium Lamps (HPS Lamps) generally have a CRI of 25 to 28, meaning that they are only 25 to 28 per cent as close to sunlight.

4.1.1.3 Types of Grow Light

☆ **Fluorescent lights:** These are quite cool and can be placed very close to the tops of the plants (as low as just inches away in some hydroponic systems) with no damage to the plants, thus intensifying the amount of light lumens produced by the bulbs. The fluorescent grow lights do not produce the amount of lights as metal halide and high pressure sodium grow lights do, but they provide very good colour spectrum wavelengths. These grow lights use little power thus saves electric bill to a great extent as compared to metal halide and high pressure sodium grow lights. The **T5 lamps** and **CFL(compact fluorescent lamp)** are rated in fluorescent watts but are usually converted into incandescent watts, hence it can be known what is to be purchased. These grow lights proved to be successful for lettuce and herbs. They can be used on seedlings and for starting larger plants like tomatoes.

☐ **T5** technology offers 400w MH light output with only half the amount of wattage (approx 216w). Lamps are supplied for either growing or flowering crops *i.e.* 6400K spectrum and 2700K spectrum. T5 are more efficient and offer greater life span than standard fluorescent lamps.

The high output T8 bulbs can also be used but, for smaller plants like lettuce, herbs and small ornamentals as they don't put out the lumens nearly as well as the T5 lamps.

☐ **CFL:** There are a number of compact fluorescent lamps in the market. These lamps do not need ballast. They plug into a standard 10 amp power point. These CFL bulbs come in much higher wattages, much higher lumen ratings, different wavelengths to choose from and these bulbs last 8000 hours and more.

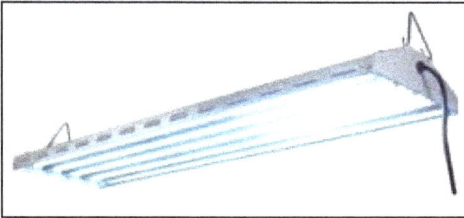

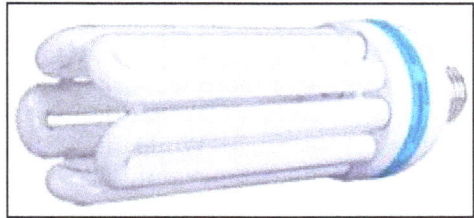

T5 lamps   **125 watt CFL (Compact Fluorescent Bulb)**

☆ **HID (high intensity discharge) systems** normally use high wattage bulbs, either in metal halide for the **blue** light (plants need for growth) or high pressure sodium bulbs for the red lights (plants need for flowering) and sometimes both are used at the same time. All HID systems put out intense light that travels for many metres down to the plants but they produce so much heat that if not placed at proper distance, may burn the plants. HID lamps were most commonly used for street and warehouse lighting but they have been adapted for horticultural use, particularly in environments where natural sunlight is insufficient.

☐ **Metal halide and high pressure sodium grow lights:** These grow lights cannot be placed very close to plants as they produce quite higher heat that may burn the plants to death. These grow lights must be placed a few metres above any type of hydroponic system. The grow lights can also be used carefully during winters to protect the plants from extreme cold. These grow lights are usually used for larger hydroponic plants (especially flowering plants) or for a larger area that is to be covered. These grow lights will run up the electric bill much more as compared to any fluorescent grow lights. Metal Halide (MH) lamps are popular because they have a low colour temperature. Generally MH lamps have a colour temperature of 6500

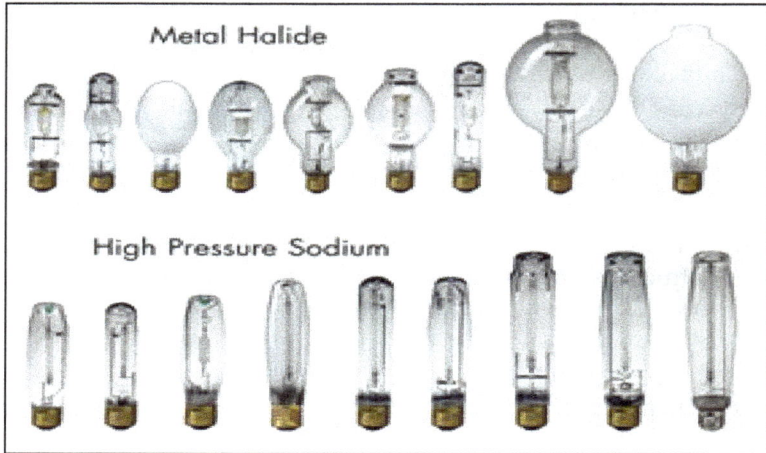

Kelvin. This means that they are a cooler running temperature and also emit a blue-white light that is a closer imitation of sunlight then HPS lamps. This closeness to natural sunlight is expressed as a CRI (colour rendition/rendering index). The CRI of most MH lamps is in the range of 60-70. This means that they are 60-70 per cent close to natural sunlight. While this CRI is quite high, MH lamps do not have a very high lumen output. As a result they do not last as long as HPS lamps. HPS lamps do not have as high a CRI as MH lamps but they make up for this by having a higher lumen output and a longer life. The CRI (colour rendering index) of HPS lamps is in the range of 25-28. So, at best a HPS lamp is 28 per cent close to natural sunlight. The colour temperature is around 2700K. This means that the colour emitted is red- orange. This is more like the colour of the sky at around sunset.

The sodium grow lights provide mostly red and yellow wavelengths and not much blue. Therefore, plants can become leggy under them. The metal halide contains very little red. For this reason these both can be combined in the hydroponic system. The life of sodium bulbs is up to 2 years and of metal halide is up to 1 year.

☆ **LED grow lights (light emitting diode):** Especially designed LED grow lights can give correct light spectrum wavelengths of reds and blues based on the nanometer scale of plant growth. They last from 10 to 15 times longer than HID bulbs and produce almost no heat at all and consumes very low power thus saves enormous amount of money on the electric bill as compared to any metal halide and high pressure sodium grow lights. They work great based on actual plant science and are 98 per cent efficient.

No matter what grow light system is being used for the hydroponic system, a **mylar** film or **foil** can be used to line the walls around the lights to intensify the

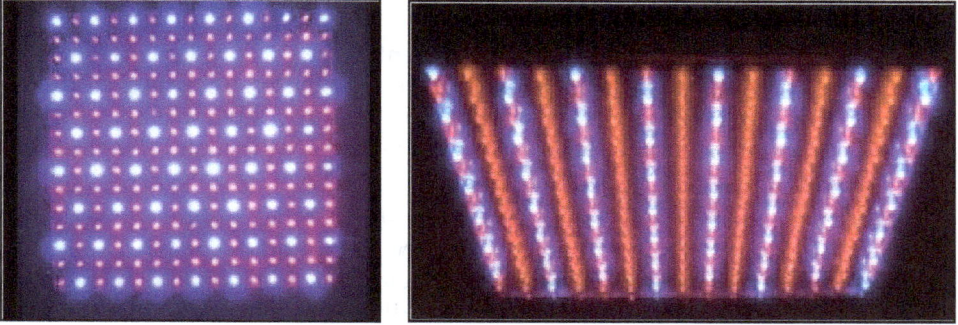

LED Grow Lights

light rays for better plant growth. It reflects light rays and bounces them back into the plant area thus giving more light where needed. Mylar film can be purchased from any good hydroponic store.

4.1.2 Ventilation and Circulation

Proper ventilation and air circulation is very important for a greenhouse or growroom to be successful because it assists in regulating temperature to the optimal level. It ensures movement of air and prevents build-up of plant pathogens such as *Botrytis cinerea*. Ventilation also ensures a supply of fresh air for photosynthesis and plant respiration, may enable important pollinators to access the greenhouse crop. Proper ventilation and air circulation is necessary for proper cooling, heating, CO_2 replenishment, and removal of undesirable gases, such as **ethylene**.

Greenhouses in the Westland Region (Netherland)

A well ventilated room will have higher yield as compared to a greenhouse or growroom with no ventilation. The only time this is not the case is if the room is controlled with a complete climate control system. In such a room, air-conditioners and CO_2 injection system are essential. CO_2 is then introduced not by airflow but through a designated injection system. It is important to note that CO_2 injection works best in a completely controlled climate.

Vents and recirculation fans (often controlled automatically) can be used for providing ventilation in a greenhouse.

4.1.2.1 Methods for Circulating Air

☆ **Vent-tube system:** It is not very efficient. It consists of a fan-jet connected to a perforated plastic tube that runs through the length of the greenhouse at ceiling height. The fan forces air through the tube, which moves the warm air in the roof space downward to displace the cooler air at the floor level.

☆ **Horizontal airflow system:** It is more efficient than the vent-tube system because it can move a larger amount of air around the plants. Large fans, hanging above the crop, are set up facing one direction in one section of the greenhouse, and in the opposite direction in the adjacent section of the greenhouse.

☆ **Vertical airflow system:** It is a very efficient but complicated system. This system uses fan-jets to move air along the roof, downward at the end walls, then along the floor through the crop, providing the best mixing of air and brings warm air down into the plants.

☆ **Exhaust fans:** Low-pressure propeller blade fans are placed on the end of the greenhouse opposite the air intake, which is often covered by evaporative cooling pads.

4.1.2.2 Carbon Filters

An indoor gardener may need to invest in some sort of air filtration system. The most common systems are carbon filters that offer the best air filtration system. Centrifugal fans are considered the best types of fans to use with carbon filters. Carbon filters are usually of solid heavy construction. High humidity can affect the longevity of carbon filters as well as the overall yield of the crop so it is important to have good airflow.

4.1.3 Temperature

The damage to plants are most often caused by cold weather and frosts and rarely damaged due to overheating. Maintaining proper temperature, especially in colder climates, is one of the major points of consideration in a greenhouse. Temperatures lower than the optimal temperature will alter the plant metabolic systems to slow growth. Night time temperatures should ideally be around 5°C below the daytime temperatures.

Some of the techniques used to heat the greenhouses are:

☆ Natural gas or electrical furnaces.

☆ Solar energy.

☆ Waste heat from livestock. For example, placing a chicken coop inside a greenhouse recovers the heat generated by the chickens, which would otherwise be wasted.

In small greenhouses, temperatures can be measured by using several thermometers placed throughout the greenhouse. These thermometers are adjusted

Temperature Requirements and Sensitivity to Frost of Various Vegetable Crops (From *Vegetable Production Guide for Eastern Ethiopia* (1969) by Haile Michael Kidane Mariam

Crop	Temp. (°C)		Stability to Frost	Temp. Range for Normal Growth and Development (°C)
	Minimal	*Optimal*		
Artichoke	–	–	Stable (perennial)	15.6–18.3
Asparagus	10	24	Stable (perennial)	15.6–18.3
Bean (common)	16	24	Sensitive	15.6–21.1
Beetroot	4	29	Partially stable	15.6–18.3
Broccoli	4	29	Stable	15.6–18.3
Brussells sprout	4	29	Stable	15.6–18.3
Cabbage	4	29	Stable	15.6–18.3
Carrot	4	29	Stable	15.6–18.3
Cauliflower	4	27	Partially stable	15.6–21.1
Celery	4	21	Partially stable	15.6–21.1
Chinese cabbage	4	29	Partially stable	15.6–21.1
Cucumber	16	24	Sensitive	18.3–23.8
Brinjal	16	29	Very sensitive	21.1–29.4
Endive	–	–	Partially stable	15.6–21.1
Garlic	2	27	Partially stable	12.8–23.8
Leek	2	27	Partially stable	12.8–23.9
Lettuce	2	24	Stable	15.6–18.3
Mustard	4	27	Stable	15.6–18.3
Onions	2	27	Partially stable	12.8–23.8
Parsley	4	27	Stable	15.6–18.3
Green Peas	4	24	Stable	15.6–18.3
Pepper	16	29	Very sensitive	21.1–29.4
Sweet pepper	16	29	Sensitive	21.1–23.9
Potato	–	–	Partially stable	15.6–18.3
Sweet potato	–	–	Very sensitive	21.1–29.4
Pumpkin	16	35	Very sensitive	18.3–23.9
Radish	4	29	Stable	15.6–18.3
Spinach	1	21	Stable	15.6–18.3
Squash	16	35	Very sensitive	18.3–23.9
Sweet corn	10	29	Partially stable	15.6–23.9
Swiss chard	4	29	Partially stable	15.6–21.1
Tomato	10	24	Sensitive	21.1–23.9
Turnip	–	–	Stable	15.6–18.3
Watermelon	16	35	Sensitive	21.1 –29.4

against each other. In large commercial operations, computer controlled systems are used to provide fully-integrated control of temperature, humidity, irrigation and fertilization, carbon dioxide, light and shade levels.

Temperatures above than the optimal temperature will affect the growth of plants and may even cause plant's death.

Some techniques used to cool the greenhouse are

☆ Fan Cooling.

☆ Evaporative Pad Cooling. It is used in combination with fans and is usually made of a cellulose material, aspen wood, or a multi-celled/honeycombed material.

☆ Fog or Mist Cooling. This type of cooling technique is used in low humidity environments only. Always clean water (free from soluble salt) should be used in order to prevent plugging of the mist nozzles.

For best results, the ventilation and circulation system must work together with the heating, cooling, and CO_2 systems. Thermostats can be used for controlling both cooling and heating devices in a grow-room or greenhouse.

4.1.4 Humidity

Plants need to transpire freely during photosynthesis and for this they need plenty of water, low to moderate humidity, and good air circulation. Humidity influences calcium uptake and hormonal distribution by controlling transpiration, ion pumping, and stomata opening and closing. High humidity coupled with low air movement will reduce cooling from transpiration and can lead to heat overload for the plant.

It is extremely important to monitor and control relative humidity around a hydroponic crop. High humidity increases the risk of moulds and mildew developing as well as encouraging an irresistible habitat for pests like spider mite. High humidity also reduces transpiration, which in turn reduces circulation of essential elements like calcium in the plants. Low humidity, or dry air, can decrease the likelihood of fungal diseases but it also increases transpiration in the plants which can in turn dry out the leaves and growing points excessively. This can produce lack of pollination in tomato crops because the growing point dies. This of course affects the overall harvest of the crop. Increased transpiration in the leaves also affects the fruit already on the plant. Fruit can die or flowers rot because less water is being channeled to the fruit/flower.

Plants see humidity in terms of vapour pressure deficit (VPD) which is the difference between the vapour pressure in the air and the vapour pressure inside the leaf. Low VPD (high humidity, greater than 90 per cent) and high VPD (low humidity, less than 50 per cent) are responsible for nutrient deficiency in plants. High humidity closes stomata and reduces gas exchange. Most greenhouse supply companies sell equipment to measure humidity. A sling psychrometer or humidistat can be used to measure greenhouse humidity.

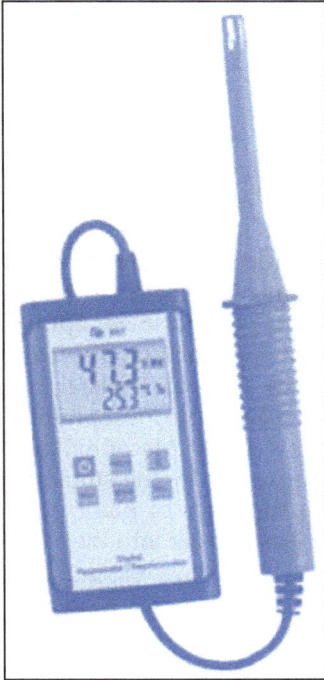

Sling Psychrometer   **Humidistat**

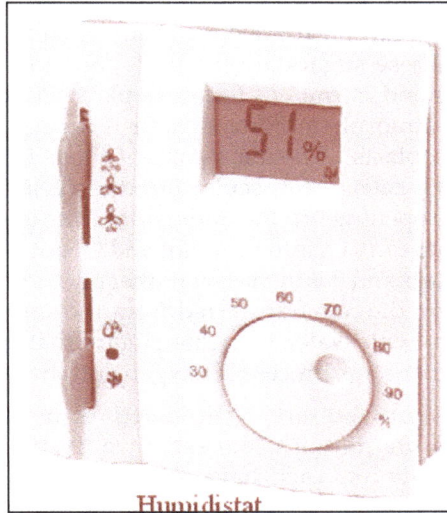

Misting and fogging systems can be used to increase humidity and decrease temperatures. However, if not used properly, these systems can greatly increase the incidence of mildews and plant diseases, and may corrode metal greenhouse structures.

Optimum Air Humidity for Vegetable Crops

Crop	Air Humidity (Per cent)	Crop	Air Humidity (Per cent)
Cabbage	50–75	Savoy cabbage	60–70
Brussels sprouts	60–70	Cauliflower	70–80
Turnip	60–75	Rutabaga	60–75
Tomato	45–60	Brinjal	50–60
Pepper	50–60	Cucumber	70–80
Onion	70–80	Welsh onion	70–80
Garlic	50–70	Carrot	60–70
Beet	60–80	Parsley	60–80
Celery	60–80	Parsnip	60–80

Source: Gobeev, A. B. and K. V.Guber (1980). Optimum Growing Conditions for Vegetable Crops[1]

Proper ventilation using 'inlet fans' and 'exhaust fans' will help in reducing humidity in the greenhouse. Inlet fans introduce less humid air into the greenhouse and the outlet fans (exhaust fans) remove the stale, humid air. Oscillating fans help to keep the air circulating. Air conditioning units also have a dehumidifying effect.

4.1.5 CO_2 Enrichment

Carbon dioxide is necessary for growth and plays a vital role in the process of photosynthesis. Photosynthesis describes the chemical process of the plants where the carbon from the carbon dioxide, water and light combine to make sugars. These sugars are the basis for carbohydrates, lignin, proteins, vitamins enzymes and hormones found in plants. In this process oxygen is released back into the atmosphere. Photosynthesis takes place in plant leaves and specifically in "chloroplasts". The main parts of a leaf are the upper and lower epidermis, the mesophyll cells, the vascular bundles (veins) and the stomata. The stomata are pore-like openings on the underside of the leaf. They allow air transfer, *i.e.* transfer of CO_2 and O_2 (CO_2 into the plant and O_2 out of the plant). The mesophyll cells have chloroplasts and it is in these cells that photosynthesis occurs. Mesophyll cells absorb the carbon dioxide transferred from the surrounding air by the stomata. Carbon combines with water to create sugars that form the basis of plant growth. The byproduct of this process is oxygen and this is released back into the atmosphere.

Chloroplasts are light reactive, meaning that they respond to light. CO_2, for plants with leaves can only be absorbed during daylight hours. Some plants like cacti have adapted to absorb CO_2 at night. If conditions are extremely hot and dry with minimal airflow, CO_2 absorption is lessened. CO_2 is one of the most critical elements in a plants development. It is directly related to the metabolism of the plant and as such an increase of CO_2 in the growing area will increase growth rates and ultimately yield. Using CO_2 in a greenhouse is not only viable but ultimately environmentally sustainable. If a greenhouse has been built in an industrialised area, CO_2 emitted by local factories can be harnessed into its greenhouse growing area. This kind of forward thinking and ingenuity provides one solution to excess CO_2 production in the world today and may be part of the solution for global warming. For the hobby grower or those not living near industrialised areas, CO_2 systems can be purchased. The most basic CO_2 system consists of attaching a regulator to a gas bottle which can then be set to a timer. The optimum CO_2 level in a greenhouse is 1500ppm. Usually the surrounding CO_2 levels are about 300ppm.

Plants can deplete the CO_2 in a closed greenhouse in a matter of hours, significantly reducing growth rates. Growers using CO_2 enrichment have claimed to a 20 to 30 per cent increase in tomato yields, and accelerating flowering and fruiting by as much as 10 days.

CO_2 can be injected in a greenhouse for better plant growth. CO_2 level of 1100–1500 ppm during the light phase will show a great increase in yields. It should be injected during daylight (light phase) where it is used by plants for photosynthesis. When CO_2 is being injected, the greenhouse should remain sealed with the inlet and outlet fans turned off. If the fans are not turned off, it will remove the CO_2 from the greenhouse before it is absorbed by the plants.

Bottled CO_2 with a regulator and solenoid valve can be used to inject CO_2 into a greenhouse. A specialized CO_2 controller can be used to monitor the CO_2 levels in the greenhouse. After a set time period, to allow for absorption of the CO_2, the fans in the greenhouse can again be switched on.

Natural gas or propane burners hooked up to sensors are specially designed CO_2 generators. Large commercial growers often use the fuel gases from a hot water boiler burning natural gas as a source of CO_2, or they will use bottled CO_2.

Note: *It is important that the CO_2 be free of contaminated gases, as tomatoes are extremely sensitive to many gases, especially ethylene. Plants enjoying elevated levels of CO_2 can be expected to increase fertilizer and water requirements.*

4.2 Growroom

A **grow room** (**growroom**) is a covered area or room of any size where wide range of plants is grown in soil or without soil (hydroponics system) under completely controlled environment. It is a full-room version of a grow box (growbox).

There are many reasons for utilizing a grow room for growing plants. Some use it to grow illicit cultivars while others may have no better alternative to indoor growing. Above all is its ability to have complete control over climatic variables such as temperature, humidity, light, CO_2, nutrition and the ability to shield out insect pests. When plants are grown using artificial light, sunlight or a combination of the two, the heat generated by high power lamps excessively increases the temperature of the grow room which may be not ideal for the growth of the plants. In such case a supplemental ventilation fan can be used.

Fully Automated Growroom System

Some of the advantages of grow room cultivation are:

☆ Off season crops can be grown.

☆ Exotic plants that are usually grown in other parts of the world in an entirely different climatic conditions, can be successfully grown in a grow room.

☆ Increased crop yields.

☆ Harvest times can be accurately scheduled with minimal losses.

☆ Most of the agrochemical usage can be avoided.

☆ Airflow into and out of a grow room can be filtered and sterilized making them ideal for use.

4.2.1 Lighting

Most grow rooms completely depend on artificial lighting. Different stages of plants require different levels of light. A lower light level for longer time periods (usually 18 hours of light and 6 hours of darkness) is required during the vegetative stage of a crop. A higher level of light for shorter time periods (usually 12 hours of light and 12 hours of darkness) is required during the flowering stage of a crop.

The most common varieties of lighting for indoor growing are:

☆ **High-intensity discharge lamps (HID):** HID lamps produce only a portion of the optimal wavelengths and wastes significant energy in the form of heat. These are most efficient in terms of Lumen (units) s of light output vs. power input.

☐ **Metal halide lamps** for vegetative stage growing

☐ **Sodium-vapor lamps** for flowering stage

☆ **Compact fluorescent lamps** and **Traditional fluorescent lamps:** Fluorescent light provides only a portion of the desired spectrum or incandescent. Compact fluorescent are slightly less efficient than HID lamps in terms of Lumen (unit) s of light output vs. power input. Traditional fluorescent lamps generally do not produce enough concentrated light to be a primary light source for most indoor growing operations. Fluorescent and compact fluorescent lamps are excellent light sources for young plants (seedlings). They are also excellent for supplementing the light generated by primary HID lamps.

☆ **Full spectrum indoor LED grow lights:** These are becoming more common in grow rooms due to their low energy requirements and very low heat output. LED grow lights are designed to optimize photosynthesis.

The lighting system in a grow room can be divided into 2 banks of lights which can be alternated. Only 1 bank of lights is used during the vegetative stage and both banks are used during the flowering stage to give more intensity. The adjustment of the height of the lights and the angles of reflector hoods can also increase or decrease light intensity.

Moving light which simulates the movement of the sun and provides light at a range of angles to the crop, can be a good idea for a grow room. In this system, a rail long enough to fully cover the growing area of the crop is installed above the crop and lights are suspended from a silent moving motor which rides on the rail. The motor moves along the length of the rail, pauses at the end and then slowly travels back in the other direction. This allows for more uniform growth of all the plants from top to bottom as they all receive the same amount of light with less shading of one plant over another.

4.2.2 Ventilation and Air-Circulation

Proper ventilation in a grow room helps in maintaining an optimal temperature for plant growth and is also required for adequate gas exchange which is very necessary for healthy growth of plants. Advanced grow room even include air conditioning to keep running temperatures down, as well as carbon dioxide to boost the plant's growth rate.

Inlet fans can be used to provide fresh air into the grow room and exhaust fans can be used to remove any old warm air. The balance of these fans is important. The exhaust fan should remove more air than what is coming into the grow room through the inlet fan. This is to create a negative pressure. This is important as a grow room with negative pressure will suck air into it when any doors or windows are opened thus preventing contagious or odours from escaping through opened doors and windows. A Fan Speed Controller can be used to balance inlet and extraction fan speeds in order to create a negative pressure. The exhaust fans should be switched off when the carbon dioxide is being injected as the fans will immediately remove the valuable CO_2 from the grow room.

4.2.3 Temperature

An ideal grow room should be able to maintain optimal day and night temperatures. It must be well insulated to limit any heating or cooling effects from the environment outside the grow room and should have different day and night temperature settings and a gradual blending of the two to simulate the natural outside environment. Night time temperatures (dark phase) should ideally be around 5°C below the daytime temperatures. Since the use of HID lighting will heat the air inside the grow room, hence air condition or an air cooled lighting system can be used to lower the grow room temperatures. The inlet and exhaust fans are also used to lower the grow room temperatures. (See also, 4.1.3 Temperature)

4.2.4 Humidity

Humidity influences calcium uptake and hormonal distribution by controlling transpiration, ion pumping, and stomata opening and closing. High humidity coupled with low air movement will reduce cooling from transpiration and can lead to heat overload for the plant. The level of humidity in a grow room affects the rate of plant transpiration, water uptake, transportation of nutrient, cell turgor (pressure and firmness) and overall growth of a crop.

Usually 55–75 per cent humidity depending on the type and age of plants, is maintained in a grow room. If filters are being used to filter the outgoing air, then humidity should be kept under 70 per cent to enhance their performance. Night time (dark phase) humidity levels should be lower than that of day time levels.

During periods of low humidity, misting and fogging systems can be used to increase humidity and decrease temperatures. However, if not used properly, these systems can greatly increase the incidence of mildews and plant diseases, and may corrode metal greenhouse structures.

A grow room which uses lights and CO_2 injection will make the crop grow vigorously with a high rate of transpiration. This will have the effect of raising the humidity. Proper ventilation using inlet fans and exhaust fans will help in reducing humidity in the greenhouse grow-room. Inlet fans introduce less humid air into the greenhouse grow-room and the outlet fans (exhaust fans) remove the stale, humid air. Oscillating fans help to keep the air circulating. Air conditioning units also have a dehumidifying effect.

4.2.5 CO_2 Enrichment

CO_2 enrichment in a grow room helps increase in crop yield. CO_2 level of 1100–1500 ppm during the light phase will show a great increase in yields.

CO_2 should be injected during daylight (light phase) where it is used by plants for photosynthesis. When CO_2 is being injected the grow room should remain sealed

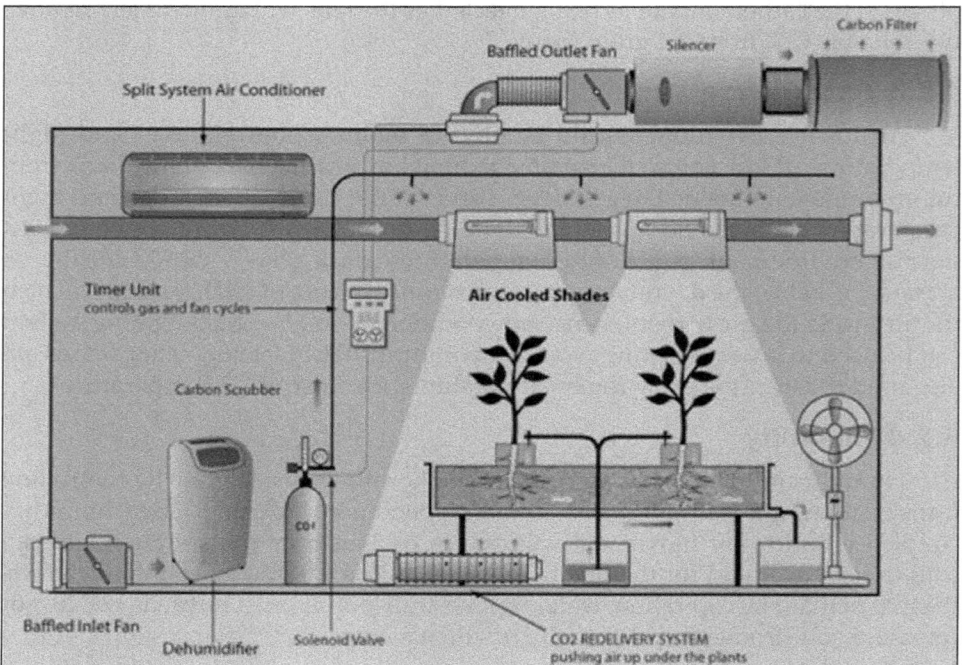

with the inlet and outlet fans turned off and the lights should be turned on. If the fans are not turned off it will remove the valuable CO_2 from the room before it is absorbed by the plants. At this time air conditioning units can be used to keep the room cool.

Bottled CO_2 with a regulator and solenoid valve is the best way to inject CO_2 into a growroom. A specialized CO_2 controller can be used to monitor the CO_2 levels in the grow room. After a set time period, to allow for absorption of the CO_2, the fans in the room can again be switched on.

4.3 Grow Box

Grow boxes, also known as grow cabinets, are partially or completely enclosed system used to grow plants indoors or in small areas year round. They are small version of growroom and come in different sizes and degrees of complexity. The grow box is sometimes also referred to as stealth cabinets especially when growers use it to grow illegal plants such as marijuana in the USA. The grow boxes can be hydroponic or soil-based. They help protect

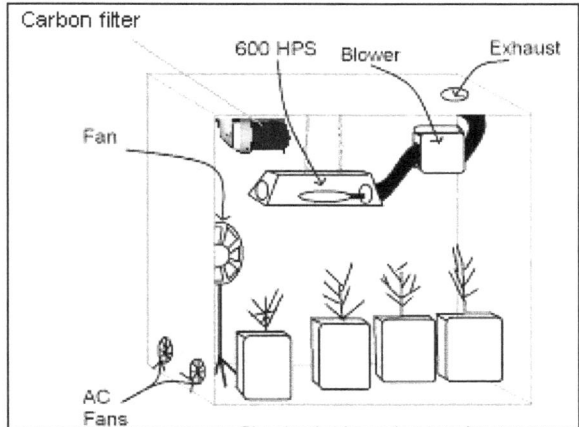

Simple Design of Grow Box

plants against pests or disease and are used for a number of reasons, including lack of available outdoor space or the desire to grow vegetables, herbs or flowers during cold weather months.

The most sophisticated grow box examples are totally enclosed, and contain a built-in grow light, intake and exhaust fan system for ventilation, hydroponics system that waters the plants with nutrient-rich solution, and an odour control filter. Some advanced grow box units even include air conditioning to keep running temperatures down, as well as CO_2 to boost the plant's growth rate. These advanced elements allow the gardener to maintain optimal temperature, light patterns, nutrition levels, and other conditions for the chosen plants.

4.3.1 Basic Equipments of Grow Cabinet

A grow cabinet may have different equipments required to improve plant growth and yields. These equipments, once all together, will allow its owner to control all conditions inside the cabinet to make them perfect for growth. Below is a table of different pieces of equipment on many commercially-built cabinets.

These units can come in variety of size. Depending on the size of the plants to be grown, growers can choose the right size grow box.

Equipments	Use/benefits	Types
Carbon filter	Prevents smells from escaping into areas around the grow box.	Cylinder filters in different sizes (heights and diameters).
Lighting systems	☆ Allow for control of seasons ☆ Increased energy uptake	CFL, HID, LED, *etc.*
Hydroponic systems	☆ Growing plants in nutrient solution ☆ Allows for increased nutrient uptake ☆ Better yields and growth rates over soil growing	DWC (deep water culture), aeroponics, drip systems, ebb and flow systems, floater systems, and wick systems
Ventilation	☆ Allows for circulation of air throughout cabinet system. ☆ Allows for controlling humidity and temperature conditions.	Includes:- Induct fan, rotating fan, light hood cooling fan
CO_2 systems	Allow for control of CO_2 levels to improve photosynthesis rates.	Includes: - Cans that slowly release it and are completely disposable. Tank systems that slowly release CO_2 from a tank at a set rate.
Water filtering system	Allow for filtering of nutrient solution to take out particles that are harmful for plants.	Many different types from inline filters to filtering systems placed in nutrient solution.
Temperature control	To control the root temperature as well as air temperature.	Exhaust fan system for ventilation. Water heater to increase the temperature. Some advanced grow box units even include air conditioning to keep running temperatures down.

4.4 Alarm Functions

An integrated detection and alarm system can be used in a professional greenhouse or grow room. It should be able to detect and alert the growers of failures of any equipment or any climatic variables which have exceeded the grower's pre-programmed set points for temperature, humidity, or lighting intensity.

Sources

1   Gobeev, A. B. and K. V.Guber (1980). Optimum Growing Conditions for Vegetable Crops.

⑤
Nutrient Solution for Hydroponics

Nutrient solution for a hydroponics system is prepared by dissolving plant nutrients, mostly in **inorganic** and **ionic** form, in the water. Although **organic** fertilizers cannot be used in conventional hydroponics due to their **phytotoxic** effects, some organic compounds such as iron chelates may be used. An essential element has a clear physiological role, and its absence prevents the complete plant life cycle. The main dissolved **cations** (positively charged ions) in the nutrient solution are **calcium, magnesium,** and **potassium.** The major nutrient **anions** in nutrient solution are **nitrate, sulphate,** and **dihydrogen phosphate.**

Organic fertilizers can be added in the **organic hydroponics** solution because the solution in **organic hydroponics** contains **microorganisms** which degrade organic fertilizers into inorganic nutrients.

If the grower has the knowledge of the nutrient concentration of each nutrient (macronutrient or micronutrient) for the plant species that is being grown, the nutrient solution for a hydroponics system can be prepared at home. The nutrient solution for hydroponics system is usually prepared by dissolving fertilizer salts in water. The selection of salts depends on various factors such as solubility in water, cost, availability, and ppm (one ppm is one part of one item in one million parts of another. For example, 180 ppm of calcium in a solution can be 180 microgramme per gramme or 180 milligramme per litre) requirement of all elements. Nutrients in the nutrient solution should be balanced. Apart from nutrients, water quality (salinity and concentration of potential harmful elements such as boron, sodium and chlorides) and pH should also be considered. The pH of the nutrient solution usually gets altered when plants absorb nutrients from it, especially at later stages of the plant's growth. Ammonium/nitrate is one of the major factors affecting the pH of the nutrient solution.

Note: One molecule of potassium nitrate (KNO_3) will yield one ion of potassium (K+) and one ion of nitrate, whereas one molecule of calcium nitrate (Ca(NO_3)2) will yield 1 ion of calcium (Ca++) and 2 ions of nitrate 2(NO_3-). Therefore, if a minimum number of cations is wanted while supplying sufficient nitrate (anions) then calcium nitrate should be used.

5.1 Essential Elements

Currently 17 elements are considered essential for most plants, these are carbon, hydrogen, oxygen, nitrogen, phosphorus, potassium, calcium, magnesium, sulphur, iron, copper, zinc, manganese, molybdenum, boron, chlorine and nickel (Salisbury and Ross, 1994).

Plants must obtain the following mineral nutrients from the growing media or nutrient solution:

☆ the primary macronutrients: nitrogen (N), phosphorus (P), potassium (K)

☆ the secondary macronutrients: calcium (Ca), sulphur (S), magnesium (Mg)

☆ the micronutrients/trace minerals: chlorine (Cl), manganese (Mn), iron (Fe), zinc (Zn), copper (Cu), boron (B) molybdenum (Mo), nickel (Ni)

The nutrient composition determines electrical conductivity and osmotic potential of the solution. Given below is the list of essential elements for most of the plants, their ionic forms absorbed by plants and their general ppm (mg per litre) requirements.

☆ **Carbon, hydrogen and oxygen**: These are absorbed from the air and water in presence of macronutrients and micronutrients and are involved in enzymatic process and oxidation-reduction process. The deficiency of these elements occur only in extreme conditions of water deficit and water logging due to which there is severely restricted uptake of hydrogen and oxygen by the plants. This results in wilting due to water deficit and yellowing of lower leaves of plants because of oxygen deficiency, which in turn also causes nitrogen deficiency and ultimately the death of plants.

a) **Carbon:** Carbon is primarily acquired from carbon dioxide in the atmosphere through stomatal pores in the leaves. It occurs in the cell walls, in chlorophyll, and in sugars manufactured by chlorophyll. It forms up to 50 per cent of the dry weight of a plant and is a constituent of all organic compounds that are found in plants

i) Ionic form absorbed by plants: CO_2

b) **Hydrogen:** It is normally obtained from water and helps in nutrient uptake from roots and is also essential for the formation of sugar and starch. It comprises 6 per cent of dry weight of the plant and is a part of all organic compounds of which carbon is a constituent. It is critical in the cation exchange in plant-soil/medium relationship and is a major constituent of plant structure when combined with oxygen to create water.

  i) Ionic form absorbed by plants: H_2O

c) **Oxygen:** It is essential for respiration and formation of sugar, starch, and cellulose. Oxygen is about 88 per cent of the composition of water and plays a critical role in plant's growth. Plants obtain the oxygen they need through the stomata on the leaves, through the roots via the water and through the process of photosynthesis. It is involved in anion exchange between the roots and surrounding medium.

  i) Ionic form absorbed by plants: O_2

✰ **Macronutrients:** Required in large quantities (g/kg dry matter) by the plants. Nitrogen, phosphorus and potassium are the primary nutrients for most of the plants.

a) **Nitrogen (N):** Nitrogen is essential for the formation of amino acids, co-enzymes and chlorophyll. It is necessary for the production of leaves and stem's growth and is also an essential ingredient in building plant cells. Nitrogen is converted to proteins that are necessary for the growth of new cells and is an essential element for the overall growth of a plant.

It is required for controlling oxygen level in plants. Oxygen from the nitrate (NO_3) form of nitrogen is used to break down carbohydrates produced by photosynthesis. It plays vital role in the formation of protein and is needed in highest concentrations at growing points such as young leaves, root tips and fruits and flowers. Nitrogen is a mobile element and moves easily to young buds, shoots and leaves, and slower to older leaves. Its **deficiency** first affects older leaves which may become pale yellow and may even die. New growth becomes weak and spindly. An abundance of nitrogen will cause soft, weak growth and will even delay flower and fruit production, if it is allowed to accumulate.

Gardeners recognize nitrogen as being available in two forms; ammonia-type nitrogen and nitrate-type nitrogen. Most plants prefer to obtain their nitrogen in the nitrate form. Urea and sulphate of ammonia are unsuitable for hydroponics as they need to undergo conversion from ammonia-type to nitrate-type by soil bacteria before plants can use them. In sunny climates, plants need more nitrogen than the same plants in a cooler climate. Plants such as lettuce, cabbage and silverbeet generally need more nitrogen than less leafy plants. The most popular nitrate-type fertilizers are potassium nitrate and calcium nitrate.

  i) Ionic form absorbed by plants: Nitrate (NO_3^-), ammonium (NH_4^+)

  ii) ppm: 100-250

        (1) Typical minimum concentration during growth phase: 160 ppm.

        (2) Typical minimum concentration during bloom phase: 130 ppm.

    iii) Fertilizer salts:

        (1) KNO_3 (potassium nitrate):

           (a) Molecular weight: 101.1

           (b) Supplies: $K+$, NO_3^-

        (2) $Ca(NO_3)_2$ (calcium nitrate):

           (a) Molecular weight: 164.1

           (b) Supplies: $Ca++$, $2(NO_3)^-$

        (3) NH_4NO_3 (ammonium nitrate)

           (a) Molecular weight: 80.1

           (b) Supplies: NH_4+, NO_3^-

b) Phosphorus (P): Phosphorus is important for the fruit producing vegetables such as peas and beans. It acts as a catalyst for energy within the plant and is an essential element for photosynthesis. It is useful for strong root growth and is necessary for seed, flower and fruit production. During seed germination and flowering, the requirement of phosphorus is very high. Phosphate (PO_4) is concentrated in seeds, fruits and merismetic tissue. It is mobile element so deficiency symptoms appear first in older leaves. Its deficiency will result in overall reduced plant growth and the leaves turn deep green and sometimes show brown or purple spots.

    i) Ionic form absorbed by plants: $H_2PO_4^-$, PO_4^{3-}, HPO_4^{2-}

    ii) ppm: 30-60

        (1) Typical minimum concentration during growth phase: 30 ppm

        (2) Typical minimum concentration during bloom phase: 60 ppm

    iii) Fertilizer Salts:

        (1) KH_2PO_4 (mono potassium phosphate)

           (a) Molecular weight: 136.1

           (b) Supplies: $K+$, $H_2PO_4^-$

c) Potassium (K): Potassium is required during all the growth stages of a plant, especially during fruit development. It stimulates root development and is essential for the good growth of plants. It functions as a catalytic agent. Potassium is used by plant cells during the assimilation of the energy produced by photosynthesis. Plants in cooler climates need more potassium than the same plants in hot, sunny climates. It helps plants to manufacture sugar and starch and is also responsible for hardness and strong root growth, water uptake, and activates enzymes that fight diseases. High potassium level is needed for protein synthesis. It increases chlorophyll in foliage and helps regulate stomata openings so that plants can make better use of light and air. Its deficiency results in taller plants that appear to be healthy, but older leaves become yellow between veins and then they turn dark yellow and die. Its deficiency also causes dropping of flowers and fruits. It is a mobile element so it can be found all over the plant but is concentrated in areas of high physiological activity.

i) Ionic form absorbed by plants: Potassium (K^+)

ii) ppm: 100-300

    (1) Typical minimum concentration during growth phase: 230 ppm

    (2) Typical minimum concentration during bloom phase: 300 ppm

iii) Fertilizer salts:

    (1) KNO_3 (potassium nitrate)

        (a) Molecular weight: 101.1 (potassium content = 38.6 per cent, nitrogen content= 13.85 per cent)

        (b) Supplies: K^+, NO_3^-

    (2) KH_2PO_4 (mono potassium phosphate)

        (a) Molecular weight: 136.1

        (b) Supplies: K^+, $H_2PO_4^-$

    (3) K_2SO_4 (potassium sulphate)

        (a) Molecular weight: 174.3 (potassium content= 44.87 per cent, sulphur content= 18.4 per cent)

        (b) Supplies: $2K^+$, SO_4^-

d) **Magnesium (Mg):** Magnesium is responsible for enzyme manufacturing, chlorophyll production and is also necessary for absorbing light energy. It is involved in the process of distributing phosphorus throughout plants. It helps the plants to utilize nutrients and to neutralize acids and toxic compounds produced by the plant. Its deficiency causes yellowness in older leaves from the centre to outward. The tips and edges of the leaves may become discoloured (turn lime in growing tips) and curl upward. It is a mobile element so its deficiency symptoms appear first in older leaves.

i) Ionic form absorbed by plants: Magnesium (Mg^{2+})

ii) ppm: 30-70

    (1) Typical minimum concentration during growth phase: 30 ppm

    (2) Typical minimum concentration during bloom phase: 30 ppm

iii) Fertilizer salts:

(1) $MgSO_4.7H_2O$ [Epsom's salts (magnesium sulphate)]

(a) Molecular weight: 246.5 (magnesium content= 9.86 per cent, sulphur content= 13.0 per cent)

(b) Supplies: Mg^{++}, SO_4^-

e) **Calcium (Ca):** Calcium is the main element required for cell wall formation. It encourages root growth and helps plants absorb potassium. It helps to reduce toxic effects of other mineral salts and aids in protein synthesis. It is a vital element and is the basis upon which the cell structure is built. It is an immobile element, therefore, its deficiency sign first appears in young growth resulting in deficient leaf tips, and edges and new growth will turn brown and may die. Its abundance in early life of plants causes stunted growth.

i) Ionic form absorbed by plants: Calcium (Ca^{2+})

ii) ppm: 80-140

(1) Typical minimum concentration during growth phase: 80 ppm

(2) Typical minimum concentration during bloom phase: 80 ppm

iii) Fertilizer Salts:

(1) $Ca(NO_3)_2$ (calcium nitrate):

(a) Molecular weight: 164.1

(b) Supplies: Ca^{++}, $2(NO_3)^-$

f) **Sulphur (S):** Sulphur plays an important role in root growth and chlorophyll supply. It is a building block for plant protein, amino acids and coenzyme A, vitamins thiamine and biotine. It assists in the production of plant energy and heightens the effectiveness of phosphorus. It helps in protein synthesis, water uptake, fruiting and seedling. It is present in large amounts in leaves which affect the flavour and odour in many plants. It is important for the health of root system and also affects nitrogen assimilation. It is an immobile element, so deficiency symptoms first appear in younger leaves. Its deficiency causes young leaves to become pale with slow growth and leaves tend to get brittle and stay narrower than normal.

i) Ionic form absorbed by plants: Sulphate (SO_4^{2-})

ii) ppm: 50-120

(1) Typical minimum concentration during growth phase: 60 ppm

(2) Typical minimum concentration during bloom phase: 60 ppm

iii) Fertilizer Salts:

(1) K_2SO_4 (potassium sulphate)

(a) Molecular formula: 174.3

(b) Supplies: $2K^+$, SO_4^-

(2) $MgSO_4.7H_2O$ (magnesium sulphate)

    (a) Molecular weight: 246.5

    (b) Supplies: Mg^{++}, SO_4^-

(3) $CuSO_4.5H_2O$ (copper sulphate)

    (a) Molecular weight: 249.7

    (b) Supplies: Cu^{++}, SO_4^-

(4) $MnSO_4.4H_2O$ (manganese sulphate)

    (a) Molecular weight: 223.1

    (b) Supplies: Mn^{++}, SO_4^-

(5) $ZnSO_4.7H_2O$ (zinc sulphate)

    (a) Molecular weight: 287.6

    (b) Supplies: Zn^{++}, SO_4^-

☆ **Micronutrients** (required in small quantities)

  a) **Iron (Fe):** Iron is required for formation of chlorophyll and is used in photosynthesis. It acts as an oxygen carrier and enzyme catalyst. It balances manganese and prevents manganese toxicity. It is an immobile element so deficiency symptoms first appear in younger leaves. Its deficiency turns leaves pale yellow or white but the veins remain green. It is a little difficult for plants to absorb iron and is moved slowly within the plant. Therefore, chelated (immediately available to the plant) iron in nutrient mixes should be used. Iron, in the inorganic form (*e.g.* as iron sulphate) when added to alkaline solutions (high in pH), precipitates out completely leaving no iron in the solution. Due to this reason iron for hydroponic nutrient solutions is best supplied as iron-EDTA (the iron salt of ethylene di-amine tetra acetic acid). Under alkaline conditions iron can combine with phosphates, carbonates and hydroxyl ions so it is important to maintain the correct pH for maximum iron availability.

    i) Ionic form absorbed by plants: Fe^{2+}, Fe^{3+}

    ii) ppm: 1.0-3.0

      (1) Typical minimum concentration during growth phase: 2.0 ppm

      (2) Typical minimum concentration during bloom phase: 2.0 ppm

    iii) Fertilizer Salts

      (1) FeEDTA (iron chelate)

        (a) Molecular weight: 382.1

        (b) Supplies: Fe^{++}

  b) **Chlorine (Cl):** Chlorine is an essential part of cytochromes and is required for photosynthesis. It functions as an enzyme activator in the process of releasing oxygen from water. It has a critical factor in drought resistance of plants because of its affect on tissue water content.

i) Ionic form absorbed by plants: Chloride (Cl^-)

ii) ppm: <75

c) **Manganese (Mn):** Manganese is a catalyst in the plant's growth process. It assists plant enzymes to reduce nitrates before producing proteins. It also plays a role in the formation of oxygen in photosynthesis. It is involved in carbohydrate metabolism and chlorophyll formation. It accelerates plant growth in conjunction with nitrogen. Its highest concentrations occur in the leaves. It is an immobile element so deficiency symptoms first appear in younger leaves. Its deficiency turns young leaves as mottled yellow or brown.

i) Ionic form absorbed by plants: Manganese (Mn^{2+})

ii) ppm: 0.5-1.0

    (1) Typical minimum concentration during growth phase: 0.5 ppm

    (2) Typical minimum concentration during bloom phase: 0.5 ppm

iii) Fertilizer Salts

    (1) $MnSO_4.4H_2O$ (manganese sulphate)

        (a) Molecular weight: 223.1

        (b) Supplies: Mn^{++}, SO_4^-

    (2) MnEDTA (manganese chelate)

        (a) Molecular weight: 381.2

        (b) Supplies: Mn^{++}

d) **Zinc (Zn):** Zinc is used in the formation of chlorophyll, respiration and nitrogen metabolism. It is an essential component in the energy transference process and should be present in little quantity for the growth of the plants. Utilization of zinc by the plant is directly related to the amount of light available to the plant. The more light available, the greater uptake of zinc resulting in higher metabolic activity. It is an enzyme activator and a component of indoleacetic acid (a plant growth hormone). It is a mobile element, so deficiency symptoms first appear in older leaves. Its deficiency results in stunting, yellowing and curling of small leaves. Its abundance can be toxic and may result in the plant death.

i) Ionic form absorbed by plants: Zinc (Zn^{2+})

ii) ppm: 0.2-0.6

    (1) Typical minimum concentration during growth phase: 0.2 ppm

       (2) Typical minimum concentration during bloom phase: 0.2 ppm

  iii) Fertilizer Salts

     (1) $ZnSO_4.7H_2O$ (zinc sulphate)

       (a) Molecular weight: 287.6

       (b) Supplies: Zn^{++}, SO_4^-

     (2) ZnEDTA (zinc chelate)

       (a) Molecular weight: 431.6

       (b) Supplies: Zn^{++}

e) Copper (Cu): Copper is needed in the production of chlorophyll. It is used to activate enzymes and is required for photosynthesis and respiration. It influences disease resistance of plants. It is an immobile element, so deficiency symptoms first appear in younger leaves. Its deficiency makes new growth wilt and causes irregular growth. Its abundance may cause plant death.

  i) Ionic form absorbed by plants: Copper (Cu^{2+})

  ii) ppm: 0.08-0.2

     (1) Typical minimum concentration during growth phase: 0.02 ppm

     (2) Typical minimum concentration during bloom phase: 0.02 ppm

  iii) Fertilizer Salts

     (1) $CuSO_4.5H_2O$ (copper sulphate)

       (a) Molecular weight: 249.7

       (b) Supplies: Cu^{++}, SO_4^-

f) Boron (B): Boron is necessary as a trace element to eliminate several undesirable features in vegetables, such as brittle stems, dying growing tips, and hollow stemmed cauliflowers. It is combined with calcium for the formation of the cell wall. It is an essential element for cell division and protein formation. It also plays an active role in pollination and seed production. It influences ratio in which anions and cations are taken in by the plant. It enhances the uptake of cations and limits the uptake of anions. It influences carbohydrate and nitrogen metabolism. It helps the plants to use calcium. It is an immobile element, so deficiency symptoms first appear in younger leaves.

  i) Ionic form absorbed by plants: BO_3^{2-}, $B_4O_7^{2-}$, $H_2BO_3^-$

  ii) ppm: 0.2-0.5

     (1) Typical minimum concentration during growth phase: 0.2 ppm

(2) Typical minimum concentration during bloom phase: 0.2 ppm

iii) Fertilizer Salts

(1) H_3BO_4 (boric acid)

(a) Molecular weight: 61.8

(b) Supplies: B^{++}

g) Molybdenum (Mo): Molybdenum helps in the formation of proteins, and in nitrogen metabolism and fixation. It converts nitrogen gas from the air into soluble nitrogen compounds by nitrogen-fixing micro-organisms. It is an immobile element, so deficiency symptoms first appear in younger leaves. Its deficiency causes leaves to turn pale and fringes to appear scorched. There may be irregular growth of leaves.

i) Ionic form absorbed by plants: Molybdate (MoO_4^{2-})

ii) ppm: 0.04-0.08

(1) Typical minimum concentration during growth phase: 0.05 ppm

(2) Typical minimum concentration during bloom phase: 0.05 ppm

iii) Fertilizer Salts

(1) $(NH_4)6Mo_7O_{24}.4H_2O$ (ammonium molybdate)

(a) Molecular weight: 1163.9

(b) Supplies: NH_4^+, Mo_6^+

h) Nickel (Ni): In higher plants, nickel is absorbed by plants in the form of Ni^{2+} ion. Nickel is essential for activation of urease, an enzyme involved with nitrogen metabolism that is required to process urea. Without nickel, toxic levels of urea accumulate, leading to the formation of necrotic lesions. In lower plants, nickel activates several enzymes involved in a variety of processes, and can substitute for zinc and iron as a cofactor in some enzymes.

Recent research tends to suggest that plants need silicon as a trace element as well. It is thought to have an influence on the strength of the cell wall. A suitable source of soluble silicon is sodium silicate, sold as "Waterglass" for preserving eggs. The usage quantity of "waterglass" would be only a few drops.

Certain plant species for their good growth may need other elements such as silica (Si), aluminum (Al), cobalt (Co), vanadium (V), sodium (Na), and selenium (Se). These are considered beneficial because some of them can stimulate the growth, or can compensate the toxic effects of other elements, or may replace essential nutrients in a less specific role (Trejo-Téllez *et al.*, 2007). The most basic nutrient solutions consider in its composition only nitrogen, phosphorus, potassium, calcium, magnesium and sulphur; and they are supplemented with micronutrients.

Note: *Typical minimum concentrations of elements above are given for high quality general purpose hydroponic formulations during the growth and bloom phase.*

There are many brands of fertilizers that are easily available with the garden stores. The product label for the fertilizer shows the list of ingredients (percentage of each ingredient) available in the fertilizer. The N-P-K (nitrogen, phosphate and potassium) is always listed first

Example: If in the list is specified **24-8-16,** it has 24 per cent nitrogen, 8 per cent phosphate and 16 per cent soluble potash as the macro-elements of this fertilizer. The minor elements may be listed as 0.02 per cent boron, 0.07 per cent copper, 0.15 per cent iron, 0.05 per cent manganese, 0.0005 per cent molybdenum and 0.06 per cent zinc.

Although ready-made nutrient formulations for various crops can be bought and mixed in freshwater to prepare nutrient solution, the growers of large commercial hydroponics systems prepare their own solutions through commonly used salts.

The commonly used salts and the required amounts to make 1000 litres of 1 ppm solution are given in the following table adapted from Jensen and Malter (1995). Multiplying the value for a salt by the number of ppm desired in the formula will yield the number of grammes to be used per 1000 litres.

5.1.1 Fertilizer Salts

Fertilizer Salts	Element Supplied	Grammes of Fertilizer Needed per 1000 Litres of Water to Provide 1 mg/l (ppm) of the Nutrient Specified
Boric acid [H_3BO_3]	B	5.64
Calcium nitrate [$Ca(NO_3)_2 \cdot 4H_2O$] (15.5-0-0)	N	6.45
	Ca	4.70
Cupric chloride [$CuCl_2 \cdot 2H_2O$]	Cu	2.68
Copper sulphate [$Cu(SO_4) \cdot 5H_2O$]	Cu	3.91
Chelated iron (9 per cent)	Fe	11.10
Ferrous sulphate [$FeSO_4$]	Fe	5.54
Magnesium sulphate [$MgSO_4 \cdot 7H_2O$] (epsom salts)	Mg	10.75
Manganese chloride [$MnCl_2 \cdot 4H_2O$]	Mn	3.60
Manganese sulphate [$MnSO_4 \cdot 4H_2O$]	Mn	4.05
Molybdenum trioxide [MoO_3]	Mo	1.50
Monopotassium phosphate [KH_2PO_4] (0-22.5-28)	K	3.53
	P	4.45
Potassium chloride [KCl] (0-0-49.8)	K	2.05

Contd...

Contd...

Fertilizer Salts	Element Supplied	Grammes of Fertilizer Needed per 1000 Litres of Water to Provide 1 mg/l (ppm) of the Nutrient Specified
Potassium nitrate [KNO₃] (13.75-0-36.9)	N	7.30
	K	2.70
Potassium sulphate [K₂SO₄] (0-0-43.3)	K	2.50
Zinc sulphate [ZnSO₄·7H₂O]	Zn	4.42

Source: Adapted from Jensen and Malter, 1995.

There are several formulations of nutrient solutions. Nevertheless, most of them are empirically based. The following table comprises some of them.

Nutrient	Hoagland and Arnon (1938)	Hewitt (1966)	Cooper (1979)	Steiner (1984)
N	210	168	200-236	168
P	31	41	60	31
K	234	156	300	273
Ca	160	160	170-185	180
Mg	34	36	50	48
S	64	48	68	336
Fe	2.5	2.8	12	2-4
Cu	0.02	0.064	0.1	0.02
Zn	0.05	0.065	0.1	0.11
Mn	0.5	0.54	2.0	0.62
B	0.5	0.54	0.3	0.44
Mo	0.01	0.04	0.2	Not considered

Concentration ranges of essential mineral elements according to various authors.

(Adapted from Cooper, 1988; Steiner, 1984; Windsor and Schwarz, 1990).

5.2 Silicon (Si) and Hydroponics

Almost all field grown plants are rich in silicon, but it is not present in most hydroponic solutions. In plants, silicon strengthens cell walls, improves plant strength, health, and productivity. Silicon, deposited in cell walls of plants, has been found to improve heat and drought tolerance and increases resistance to insect infestation and fungal infections. Silicon can help plants deal with toxic levels of manganese, iron, phosphorus and aluminium as well as zinc deficiency.

Although silicon is not considered to be an essential plant nutrient because most plant species can complete their life cycle without it, it is considered to be a beneficial element and some plant species can accumulate Si at concentrations higher than many essential macronutrients.

Silica is excluded from hydroponic nutrient formulas because it has a high pH and is unable to remain soluble (hold/remain stable) in concentrated nutrient formulas. Therefore, Si needs to be added to the nutrient tank as a separate element (additive).

We experimented by increasing the contents of Si into the nutrient solution of cucumber, rose, and courgette (zucchini) and found that these plants could benefit from enhanced Si concentration in the root environment, since total yield was increased and powdery mildew was suppressed.

5.2.1 Benefits of Si

☆ It protects the plants against insect and disease attack

  ❑ Increases disease resistance

  ❑ Increases resistance to pathogenic airborne fungi (eg. *Botrytis*)

  ❑ Increases resistance to waterborne pathogens

  ❑ Increases resistance to insects/pests

  ❑ Increases strength in cell structure

☆ It protects against toxicity of metals and stress

  ❑ Increases stress-tolerance

  ❑ Deals with toxic levels of manganese, iron, phosphorus and aluminium, as well as zinc deficiency

  ❑ Increases drought resistance

  ❑ Increases salt tolerance

☆ Increases yields

5.2.2 Silica and Fungi Suppression (*e.g. Botrytis*)

Si suppresses fungal pathogens such as *Botrytis*. The addition of Si could be beneficial to crops such as cucumbers grown in areas where the local water supply is low in this element, especially when grown in recirculating solution or in a medium low in Si, *e.g.* peat. Si treatment may reduce powdery mildew development by inducing host defense responses in plants.

It is believed that silicon deposition at sites of fungal pathogen penetration may be a common component of the host-defense response in a variety of plant families. Silicon is also deposited in the cell walls of roots where it acts as a barrier against invasion by parasites and pathogens. For instance, potassium silicate has been shown to act as a preventative against *Pythium ultimum*.

Soluble Si polymerizes quickly and disease development is suppressed only if Si is present in soluble form. To minimize disease development, Si must therefore be provided continuously in the nutrient feed in hydroponic systems. Therefore, a continuous source of soluble silicon is very important to combat pathogens. This can be from constant feeding in hydroponics or from retention in the growing medium with soils or soilless mixes.

5.2.3 Optimum ppm of Si in Solution

Liquid silicon as Si (not SiO_2 which is 46.743 per cent Si and 53.25 per cent O) is highly beneficial to plants in the range of 25-100 ppm in the nutrient solution. It is not included, at these levels, in nutrient concentrates. It needs to be added as a separate component by the grower. Since potassium silicate products are highly alkaloid, it is recommended to pre-dilute any Si product in about 5- 8 litres of water and pH should be adjusted to 5.8- 6.0 before adding to the nutrient tank/reservoir.

It is recommended that in coco substrate, Si should be used at the lower end of the scale *i.e.* between 25 - 35ppm, and in inert media and water-based systems it can be at higher rates of between 50–75ppm.

5.3 Harmful Elements (Sodium and Chloride)

Chloride deficiency is rare because it is required by plants in minute quantities and most water sources contain chloride concentration well above than required by the plants. The excess of chloride may prove to be toxic for plants and will typically cause chlorosis (leaf yellowing) which results in poor growth or yield. Toxic levels can occur due to the use of potassium chloride and/or ammonium chloride. For re-circulating hydroponics systems, **sodium** can be very harmful, since it builds up with time in the hydroponic solution. The concentration of sodium and chloride for most hydroponically grown plants is less than 75 ppm.

Toxic levels of **ammonium** can occur due to the use of excessive ammonium nitrate in the formulation. Ammonium concentration should represent no more than ~20 per cent of the required total nitrogen concentration. An excess can cause damage to roots and stem bases, particularly in younger plants, which results in poor growth and yield. It is generally accepted that plants will not uptake nitrogen in the ammonium (NH_4^+) form. It must first be oxidized into nitrate (NO_3^-).

5.4 Nutrient Consumption by Plants

Plants consume nutrients in varying amounts depending on their needs. While carbon is primarily acquired from carbon dioxide in the atmosphere through stomatal pores in the leaves, virtually all other nutrients come in through the roots. In hydroponics, especially in medium culture hydroponics, there may be rapid changes in the status of water and nutrients. Therefore, the changes such as EC, pH and nutrients level should be regularly monitored and corrected using simple devices and methods, so that the water and nutrients are efficiently used by the plant roots. Although monitoring pH and EC will give an indication of changes in the nutrient solution, it cannot indicate the changes in preferential uptake of particular ions. Therefore, in a large commercial closed hydroponics system, analysis of nutrient solution is necessary. In case analysis is not possible, the nutrient solution should be completely changed every one to two weeks.

To know the changes occurring in the medium (especially after fertigation), the fertigation input line and the drainage points should be regularly monitored.

Rough estimates of nutrient consumption can be obtained by checking the changes in the nitrate (NO_3-) content of the nutrient solution. Compared to the drip-line point, it will go up and down, reflecting changes in plant consumption. The nitrite (NO_2-) concentration can also tell about the status of the growing medium. Due to over-irrigation, water may accumulate in the medium resulting in water logging and decreased availability of oxygen in the medium. Thus resulting in the chemical transformation of ammonium to nitrate causing nitrite (nitrite anion is toxic to plant roots) to appear and accumulate in the medium.

5.4.1 Nutrient Lockout

Some most common causes of nutrient lockout that are often overlooked are:

☆ Unbalanced pH: Nutrient lockout may occur if the pH of the nutrient solution is not within the desirable range for the plants being grown. Growers should know exactly what the pH should be for the plants they are growing. Growers should always invest in a good pH meter and should check it often. It is ok if small fluctuations occur, but if the pH drops below 5.2 or rises above 6.4, it is advised to correct it manually. If the grower is providing everything to the garden and it is still not growing as much as it should, then it is most likely a pH problem. The grower must remember that most pH up and down are composed of nutrient salts like phosphate.

☆ Improper temperature: Temperatures can also affect the plants nutrient absorption. The slightest change in temperature can alter the plants growth. It is recommended to keep a thermometer in the reservoir and should be periodically checked. The nutrient solution must stay within the desirable range at the hottest and coolest times of the day. The uptake of the nutrients will decrease dramatically if the solution gets below 15 C.

☆ Nutrient strength can be a commonly overlooked factor of nutrient uptake. If the nutrient solution is too weak, the plants won't get everything they need. If the nutrient solution is too strong, it can cause some nutrients to block the absorption of others, such as iron. Since tap water already has elements in it, it is important to measure first and then add the nutrients as needed. It is important to use a ppm/EC meter to check the strength of the nutrients regularly.

The temperature, pH and ppm of the nutrient solution should be regularly monitored, and if required, should be adjusted slowly, to ensure the plants are getting everything they need to thrive.

5.5 Factors that Affect Nutrient Solution

Various factors affect uptake of nutrients by the roots and root function. Given below are some points to be taken care of for preparing and maintaining nutrient solution.

5.5.1 Water Quality

If poor quality water is being used to prepare nutrient solution, it can cause toxicity problems, diseases, pH problems and the blockage of drippers and plumbing. Water quality is of utmost importance in any hydroponic system. The majority of a plant's weight is made up of water and it is the life blood of a plant as it carries the dissolved nutrients which the plant feeds from. Plants which are grown in clean water grow faster, healthier and with less problems. Water that is suitable for drinking by humans is not necessarily ideal for growing plants. This is because many plants have a lower tolerance for certain elements (*i.e.* sodium) than what humans would do.

The water quality also depends on the dissolved solids in the water. There may be such components in the water that could affect the availability of nutrients to the crops. For good soft water, the right combinations of nutrients are added but for hard water, or water contaminated with sodium, sulphide, or any number of heavy metals, the water has to be filtered using "reverse osmosis". Therefore, the analysis of water for nutrient solution is to be done in the lab. Common problems with water are total dissolved salts (salinity), pH, alkalinity, carbon dioxide, hardness, precipitates due to hardness, salts, iron, corrosiveness, contamination/pollution, cloudiness/turbidity, colour, odour, slimes and algae. Poor quality water can lead to a number of plant growth problems such as stunted growth, mineral toxicity, elemental deficiency symptoms, build up of unwanted elements in plant tissue, bacterial contamination, *etc.*

It is common to find high levels of salts in well water or municipal water supplies. Calcium and magnesium carbonates are among the most common ingredients in tap water and in well water. In fact, water "hardness" is defined as a measure of the water's content of calcium and magnesium carbonates, or sulphates. Since calcium and magnesium are important plant nutrients and water with reasonable levels of these elements is good for hydroponic cultivation, but, if their levels are too high, it becomes a problem. Usually, for most hydroponics system, the calcium content should be less than 200 ppm and the magnesium content should be less than 75 ppm. An excess can cause other important elements in the nutrient solution to "lock-out" and become unavailable. For example, excess calcium can bond with phosphorous to make calcium phosphate, which is not very soluble and therefore not available to the crop. The key is to start with pure water and add the right combination of nutrients.

The salts (ions) dissolved in water can be found by having the water chemically analysed. If the town water is being used, the water authority will normally be able to provide the information and if bore, dam or stream water, *etc.* is to be used, a sample is to be sent for analysis.

Water for nutrient solution may contain variety of harmful and beneficial microorganisms. Water can be sterilized before using in the nutrient solution to remove harmful microorganism (*e.g. Pythium*). UV radition, heat treatment, membrane filtration, ozone, monochloramine, chlorine and hydrogen peroxide are commonly/widely used methods of sterilization in hydroponics. Water sterilization or the nutrient solution sterilization will kill both beneficial and pathogenic bacteria. It is important to note that hydrogen peroxide act only as root disease preventative and once root disease is present in the crop they are ineffective and other means for controlling the disease should be sorted out.

Note: The selection of fertilizers and their concentration in the hydroponic nutrient solution depends greatly on the quality of the raw water as the raw water for hydroponic nutrient solution may contain minerals. Therefore, it is necessary to test the raw water prior to deciding on a fertilizer formula. Minerals such as calcium, magnesium, sulphur, and trace elements such as boron, manganese, iron and zinc may be present in the source water. These elements must be factored in when adjusting the hydroponic nutrient solution. Additionally, raw water might contain high concentrations of unwanted minerals, such as sodium, sulphide, chloride or fluoride, which is unsuitable for hydroponics.

5.5.1.1 Common Ions in Water Supplies that Influence Plant Nutrition

The most common ions in water supplies which influence plant nutrition are:

☆ **Sodium (Na^+) and Chloride (Cl^-):** These are the constituents of common salt and normally occur together and are not taken up to any degree by most plants, especially sodium. Therefore, they tend to accumulate if present in significant amounts. Chloride is actually a trace element (micronutrient), but is usually available in far greater concentration than required for nutrition. This is why it is virtually never included in formulations.

☆ **Iron (Fe ...):** Although iron is a micronutrient, in this form it rapidly oxidises and precipitates as rust, which makes it unavailable as a nutrient. In practice this can give problems, especially by blocking drippers, so the precipitation is best speeded up by aerating the water, followed by settling or filtration.

☆ **Calcium (Ca^{++}) and Magnesium (Mg^{++}):** These are constituents in hard water. They are usable in the nutrient solution as major nutrients (macronutrients). Their presence should be allowed for in calculating formulations.

☆ **Bicarbonate (HCO_3):** It is also a constituent in hard water. It is not a nutrient but is alkaline and will raise the pH. It will need to be neutralised by acid, typically phosphoric or nitric. The amount of equivalent phosphorus or nitrogen added should be allowed in calculating formulations.

☆ **Boron (B):** Boron is the micronutrient with the narrowest range. If present in the water it can be omitted from the formulation. It can become a problem if its concentration is over 1ppm, or lower for sensitive crops. This only occurs in a few water supplies.

5.5.1.2 High Salt Levels in Water Supplies

The level of salts which can be tolerated in the water depends on their composition and which crop is to be grown. Even 50 ppm of sodium can be toxic to plants such as lettuce, straw berry and rose. In contrast tomato could cope with over 200 ppm. Because of the accumulation that will develop, water supplies with increasing dissolved solids will make management more critical. Beyond a certain limit, depending upon the crop, recirculating systems will become unmanageable. Non-recirculating systems can continue to be used but with increased percentage run-off. Eventually, water may become so bad that it will be unusable in any system. In this case the only possible solution is to remove the bulk of the salts from the water, typically with reverse osmosis.

5.5.1.3 Water Treatment

If managed correctly.reverse osmosis is capable of reducing the dissolved salts to very low levels. The major and most expensive component of a reversed osmosis (RO) machine is the membrane. The efficiency and life of the membrane is very dependent upon the water quality being pumped through it. Often, chemical pre-treatment, such as removing iron, is necessary for the equipment to work effectively, and to give reasonable membrane life.

An alternative is to collect rainwater. This may be used directly if sufficient is available or mixed with the poorer water supply to make it more acceptable.

Suspended matter should be removed. Good final filtration with a sand filter for example, can avoid many of the dripper blockages experienced by some growers. The water should be free of diseases (pathogens). Town water is usually safe but other sources need to be sterilised, most commonly by chlorinating. Water also needs to be free of plant poisons (phytotoxins).

5.5.2 Temperature and Oxygen

Temperature and rootzone oxygen concentrations affect the rate of root respiration. As the concentration of oxygen decreases in the rootzone, the roots will experience hypoxia. Hypoxia occurs when the rate of respiration is affected because oxygen has become a limiting factor. Root respiration affects a wide variety of functions in root tissue that may affect crop production.

Maintaining proper temperature of nutrient solution will enhance plant's growth and quality. A high quality **aquarium heater** can be used to warm nutrient solution in the winter and **chillers** can be used to cool nutrient solution in the summer (if high nutrient temperature becomes a problem). Although for small hydroponics systems aquarium heater and aquarium shade cloth cover can be used to prevent sudden temperatures shifts from occurring, greenhouse will do the same for larger systems. Also a larger volume of water of 1,000 litres and more will offer stability and moderate sudden temperature swings in the system. If the nutrient solution is too cold, seeds won't germinate, cuttings will not root and plants will grow slowly, or stop growing and die. If it's too hot, then again the seeds won't germinate, cuttings won't root and plants will die from oxygen deficiency or from temperature stress. Most plants prefer a root zone temperature range of between 62°F (17° C) and 80°F degrees (27° C), cooler for winter crops, warmer for tropical crops. When adding water to the reservoir, it should be allowed to come to the same temperature as the water in the reservoir. Plants do not like rapid temperature changes, especially in the root zone.

The nutrient solution's temperature for most plants can be maintained between 62°F and 80 F (17-27 C).

☆ The temperature above 22 C may cause the dissolved oxygen (DO) dipping too low. In such case use of proper aeration is necessary.

☆ At temperature below 16 C, plants tend to slow their metabolism.

☆ Some growers use this to their advantage and reduce nutrient temperature towards the end of the flowering stage to aid in ripening.

The consumption of oxygen (O_2) will increase with the increase in the temperature. Therefore, if the aeration in the root is not adequate, there will be an increase in the relative concentration of carbon dioxide (CO_2) in the root environment. The concentration of oxygen in the nutrient solution also depends on crop demand, being higher when the photosynthetic activity increases. Much decreased dissolved oxygen may inhibit root growth and shows a brown colouration, which can be considered as the first symptom of lack of the oxygen. The influence of temperature for uptake of water and nutrients varies with plant types. For example, in rose plants, if the nutrient solution is made generally cold, the uptake of NO_3^- gets increased and the production of roots are thin and white, but the uptake of water gets decreased. Each plant species has a minimum, optimum, and maximum temperature for growth, which requires the implementation of heating or cooling systems for balancing the nutrient solution temperature.

5.5.3 Electrical Conductivity (EC)

EC can be maintained by mixing special nutrient blends for different kinds of plants and for each stage of the crop's life-cycle. The nutrient solution should be checked regularly for its electrical conductivity (EC). It is the dissolved salts in most water that allows it to conduct electricity and is usually measured in S/cm which just means that the material has a certain conductance in S (Siemens) per centimeter. Purified water will show no, or very low, salt content (conductivity) when tested with a dissolved solids meter, also called an electrical conductivity (EC) meter. An EC meter or ppm (parts per million) meter can be used to check the EC of nutrient solution. An EC meter measures an electrical current in the solution and reads the conductivity produced from the motion of the mineral ions. Low conductivity means low nutrient concentration, often resulting in nutritional deficiencies and slow plant growth. High conductivity means more food for your plants. A very high conductivity can burn or kill the plants.

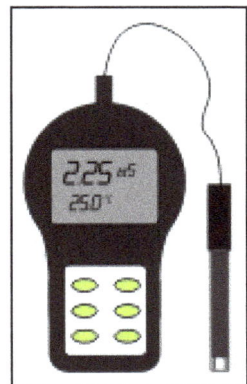

EC Meter

EC should be measured regularly only at a constant pH. An EC measured at pH 5 and an EC measured at a pH 7 will be completely different because the ions which determine pH have a very large effect on the EC value. Another important fact is that the conductimeter should be calibrated using a solution of known conductivity. If it is not, comparison between measurements can be meaningless.

The EC should be measured when the solution is prepared and then at regular intervals (once, twice, thrice a day depending on the type and stage of plant). If the EC of the nutrient solution becomes too high, water can be added to lower it to the original value but if EC becomes too low (70 per cent of original value), nutrients should not be added, as the nutrient solution has been substantially changed in composition by the plant and it needs to be disposed off and a fresh one needs to be prepared. This is because the EC reading does not give the exact ionic composition of the solution and it is not known which nutrient the plant took up. To know the exact ionic composition of the solution, some fancy atomic emission analysis can be done but the easiest thing would be to dispose off the nutrient solution and start with a fresh solution.

Note: The EC reading doesn't provide information regarding the exact mineral content of the nutrient solution. In closed hydroponics systems, the hydroponic nutrient solution is re-circulated and elements which are not absorbed in high quantities by plants (such as sodium, chloride, fluoride, etc.) or ions released by the plant, build up in the hydroponic nutrient solution. In such case there is a need for more information about the contents of the nutrient solution that EC cannot provide.

Analysis of Nutrient Solution

Plants absorb nutrients in varying amounts depending on their needs, and monitoring EC alone cannot indicate changes in preferential uptake of particular ions. Therefore, in a commercial system, especially in closed system, analysis of nutrient solution for the concentration of each nutrient is absolutely necessary. If it is not feasible to analyse the nutrient solution, then the nutrient solution should be completely changed every one or two weeks.

Analysis of Plant Tissue

Plant tissue analysis is also necessary in a commercial hydroponics system as it will help the growers in finding out the problems that plants may be having in absorbing nutrients from the nutrient solution. This problem may be due to fluctuating pH levels, high cation exchange capacity of the media, high humidity, or diseases and nematodes problems.

Symptoms due to excessive EC: If the total salt content in water is excessive, it should not be allowed to touch foliage. The following symptoms generally occur when the total salt content (EC) in water is excessive:

    ☆ Wilting of leaves and stems, followed by reduced growth.

☆ Leaves become scorched around the edges and may wither or die.

☆ Leaves may drop off and shoots die back.

Many of these symptoms also indicate other problems, such as a lack of water, disease, malnutrition and excessive light or heat.

5.5.4 The pH (Potential Hydrogen)

The pH is a parameter that measures the acidity or alkalinity of a solution. The pH is usually measured on a scale of 1-14 and represents the concentration of hydrogen ions in solution. This value indicates the relationship between the concentration of free ions H+ and OH- present in a solution. Generally, it is used to determine whether a hydroponic solution is acidic or basic (alkaline). A solution is acidic if it

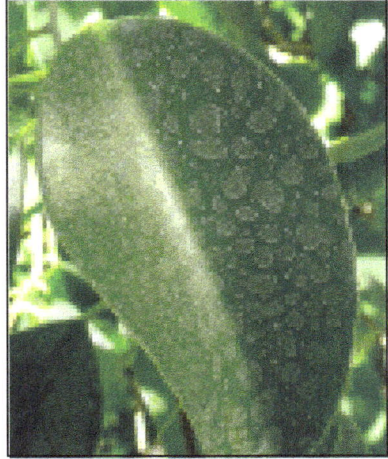

Leaf Burn Due to Excessive EC Nutrient Solution Touching the Foliage.

has more positive hydrogen ions and is alkaline if it has more negative hydroxyl ions.

TDS

TDS stands for 'total dissolved solids' and is measured in ppm or parts per million. One ppm is one part by weight of the mineral in one million parts of solution. In a laboratory, TDS is found by allowing the liquid to evaporate and weighing the left over particles. In hydroponics, TDS is estimated using a conversion from the measure of electrical conductivity (EC). The growers either use TDS or EC as the standard of measurement. TDS is a conversion from EC, and different manufacturers use different conversion rates. Different TDS meters may show a discrepancy of as much as 600 ppm when reading the same solution. When calibrating TDS meters, the correct calibrating solution from the correct manufacturer should be used.

Conversion of readings from TDS to EC and vice versa

To get the approximate TDS value, simply multiply the EC reading (in mirco-siemens/cm) by 1000 and divide by 2. To get an EC value, multiply the ppm reading by 2 and divide by 1000.

For example,

if EC is 1 : 1 x 1000/2 = 500ppm

and if ppm is 500 : 500 x 2/1000 = 1EC

CF

CF stands for 'conductivity factor' and relates to EC. It is simply a measure of EC multiplied by 10.

Changing the pH of a nutrient solution affects its composition, bioavailability and the distribution of elements among their various chemical and physical forms like: free ions, soluble complexes, chelates, ion pairs, solid and gaseous phases and different oxidation. An important feature of the nutrient solutions is that they must contain the ions in solution and in chemical forms that can be absorbed by plants, so in hydroponic systems the plant productivity is closely related to the nutrient uptake and the pH regulation. Each nutrient shows different responses to changes in pH of the nutrient solution.

A pH reading refers to the acidity or alkalinity of any liquid. A **1(one)** on the scale represents a low ion concentration or acid. Pure water has a balance of hydrogen and hydroxyl ions. Pure water is considered neutral at a pH of **7 (seven)**. If the pH of solution is below 7, it is considered as acid and if the pH of solution is above 7, it is considered as alkaline. A **14** (fourteen) on the scale represents the highest concentration of ions (basic/alkaline).

A well-buffered nutrient should be used that promotes **pH** stability in the nutrient solution.

☆ The usual pH range should be >=5.5 and <=6.5.

☆ Perfect pH range is 5.8-6.3.

☆ The pH can be allowed to float between 5.5 and 7.0.

☆ For vegetative growth, the optimal pH range is 6.0-6.3.

☆ For flowering and fruiting, the optimal pH range is 5.7-5.9.

The pH levels preferred by plants vary and the pH of the nutrient solution should be within reasonable range. Micronutrients are more available at lower pH, but when pH levels drop below 5.5, there is a risk of micronutrients toxicity, as well as impaired availability of calcium and magnesium. If pH is too high, iron, manganese, boron, copper, zinc and phosphorus may become unavailable even if these elements are sufficient in the nutrient solution. A pH that is too low will reduce the availability of potassium, sulphur, calcium, magnesium and phosphorus. The plants are unable to absorb iron at higher pH and results in an iron deficiency (weak yellowish leaves). Special chelates are available at the fertilizer shops that assure iron availability even at higher pH ranges, but higher pH can damage plants in other ways.

The best way to maintain reasonable pH is to mix fresh nutrient with the water and let it stand for a while to stabilize, then the pH should be tested and adjusted. The city water supplies often require to add a bit of pH down (usually phosphoric acid) to lower the pH to the range for most plants, between 5.8 and 6.3. As the plants grow, the pH should be occasionally tested and adjusted (if required). The pH can be allowed to float between 5.5 and 7.0 without adjustment, as frequently adding chemicals into the system to maintain a perfect pH of 5.8 to 6.3 can do a lot of damage. It is common for pH to drift up for a while, then down, and up again. This change is an indication that the plants are properly absorbing nutrients. The pH should be adjusted only if it wanders too far. A pH below 5.5 or above 7.0 is not good but a sudden drift in pH can be due to malfunctioning of pH meter. Also, all pH measuring methods are temperature dependent. The instructions that come with meter and test kit must be properly followed.

Note: Ammonium/nitrate is one of the major factors affecting the pH of the nutrient solution.

The pH can be changed in a hydroponic solution by adding a tiny bit of the pH up or pH down adjusters. The growers should always check the ppm of the hydroponic solution first because bringing up the ppm to the correct level may also correct or partially correct any pH problem in the hydroponic solution. For acidifying the hydroponic nutrient solution, sulphuric acid, phosphoric acid and nitric acid can be used. Two aspirin tablets or a teaspoon of distilled white vinegar can be added per gallon of water to lower the pH level from 8 to about 6. There are several products that can be used to quickly and easily adjust the pH up or down as needed.

There are several methods to check the pH of the nutrient solution, such as using paper test strips, liquid test kits, digital meters, and pH meters.

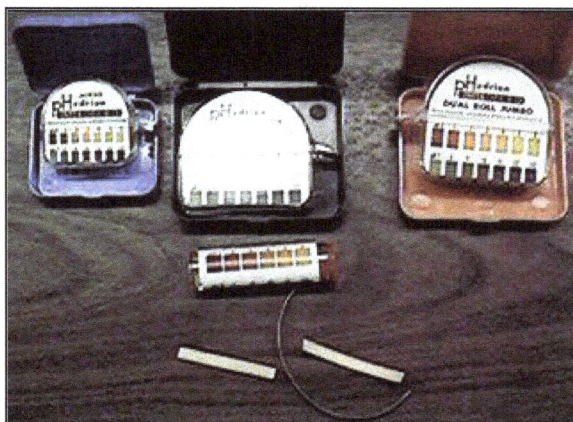

pH Test Strips

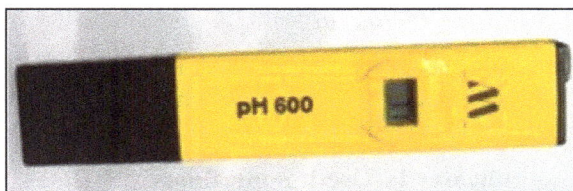

pH Meter

5.5.4.1 Test pH using pH Test Strips

The pH test strips contain a dye and when dipped into the solution its colour gets changed. The strip should be dipped into the solution for a very small duration (about a second). The strip can then be compared to a colour chart which determines the pH level of the solution being tested. To get the accurate readings, the test strips should be read quickly because the strips continue to change colour.

The pH reading alone cannot tell that how much acid or alkali is needed to adjust the pH of nutrient solution. For example, in two different samples of water of equal pH, one may require 2 times more acid than the other to achieve pH adjustment. This situation can occur even with samples taken at different times from the same source. It is caused by seasonal changes and inconsistency in water treatment at the supplier's treatment plant. The reason for this is the concentrations of bicarbonate and carbon dioxide present in the water.

5.5.4.2 Media Quality and pH

The poor quality growing medium is another cause of unstable pH. Industrial grade rockwool and gravel may result in a very high pH levels that can cause the nutrient pH to rise constantly to dangerous levels.

A new growing medium can be tested by simply putting some of the medium (rockwool, gravel, sand) into a clean small container (a cup or a jar) and then immersing it in distilled or "deionized" (chemically pure) water. After allowing it to sit for some time, the pH of the water should be tested and recorded. This test should be performed occasionally for about a week until it has stabilized. The pH may rise to 8.0, 9.0 or as high as 10.0 and may lead to death of plants. The pH of media is one of the main reasons that **water culture (solution culture)** hydroponic methods are gaining popularity over **media-based** hydroponics. If lesser medium is used in hydroponics, the pH instability and salt accumulation would also be less. Also, the **solution culture** systems require less water and nutrient than media-based methods, due to higher efficiency and reduced evaporation.

5.5.4.3 Bicarbonate–HCO$_3^-$

Bicarbonate (HCO$_3^-$) may be present in the undergroundwater (bore water). It is alkaline and therefore raises pH. High alkalinity make-up water may result in an increase in the pH of nutrient solutions. Therefore, frequent pH check is required whenever high alkalinity make-up water is added in the nutrient solution. Alkalinity is removed by lowering the pH with acid.

5.5.4.4 High pH Buffering Capacity

Bicarbonate is weakly alkaline and is unable to raise pH much above 8.3. Therefore, when it is present at a high concentration it provides a high pH buffering capacity, *i.e.* it resists pH change when acid is added.

5.5.4.5 When Groundwater is Used, sometimes pH Rises Soon after it is Lowered

When acid is added to a bicarbonate laden nutrient, the by-products are carbon dioxide (CO$_2$) and water (H$_2$O). The presence of free (*i.e.* uncombined) CO$_2$ tends to lower the pH because it reacts with water to form carbonic acid (H$_2$CO$_3$). This is the reason why pH sometimes rises soon after it is lowered, or soon after water is placed in the nutrient reservoir.

The concentrations of CO$_2$ above about 0.5mg/l in water are unstable when such waters are exposed to the atmosphere at sea level pressures. In such condition, excess CO$_2$ (above 0.5mg/l) will slowly escape from the water into the atmosphere.

This loss of acidity causes a corresponding rise in pH. This is the reason why pH sometimes rises from its minimum value soon after it is lowered.

It is advisable to use aeration before using water having high alkalinity or high CO_2. This will remove CO_2 from water.

It is also advisable to reduce the nutrient solution's pH to about 5.0 when it is first made, or after top-up water is added.

5.5.5 Change-Interval of Nutrient Solution

Timely change of the nutrient solution prevents salt accumulation or deficiencies and eliminates the wastes that plants discard into the nutrient solution. Also, as plants transpire, moisture and nutrient levels drop in the nutrient solution and the EC or strength of the nutrient can rise to dangerous levels. The change interval of nutrient solution depends on the species, the number and size of the plants that are being grown, stage of plants, water quality, the capacity of the reservoir, quality of nutrients being used, environmental conditions (such as temperature and humidity), and on the type of hydroponic system being used. It all requires a bit of monitoring and record keeping. A hydroponics system should be started with a fresh nutrient solution and the records such as date, pH, and EC or ppm of the solution should be timely noted down. The solution level will usually drop as the system runs. The EC/ppm level should be recorded and then freshwater should be introduced in the reservoir and again should be tested for nutrient concentration. If the nutrient strength has dropped significantly, a bit of nutrient should be added accordingly. The amount of water added to top-up the reservoir should be recorded carefully. Testing the hydroponic nutrient solution frequently helps in deciding on the timing for replacing the nutrient solution or diluting it with freshwater.

The same procedure should be followed each time when the reservoir is topped-up with the water. When the total amount of water added equals the capacity of the reservoir, the nutrient solution of the reservoir should be drained and replaced with fresh nutrient solution.

5.5.5.1 Cleaning of Nutrient Solution Reservoir

It is advisable to clean the reservoir every one to two weeks or when changing the nutrient solution. The grower can follow the following steps to clean the reservoir.

☆ Remove all plant matter and debris that can be grabbed from the water

☆ Drain the reservoir

☆ Wipe out as much residue as possible with a paper towel

☆ Refill the reservoir and add fresh nutrient solution

It is recommended to sterilize the reservoir in between crops to prevent disease and pests. The grower can use a 10 per cent hydrogen peroxide or 10 per cent bleach solution for sterilization. If the reservoir is too large to wash in a sink, the grower may need to sterilize it where it stands, by filling it up with a bleach or h_2O_2 solution and allowing it to sit for several hours.

The beauty of h_2O_2 is that it will break down on its own and do not need rinsing. If a bleach solution is being used, it should be rinsed properly until there is no residue left.

Note: Wear gloves when handling concentrated peroxide or bleach, otherwise it may burn hands.

5.5.6 The Salt Accumulation in the Medium

A salt build-up (due to transpiration and evaporation) can be seen in the medium because of improper management practices. Although some salts are absorbed by the plant, there is a sharp increase in the concentration and a build-up of some undesirable salts. This salt build-up may interfere with plant growth because there is no space to buffer this salt build-up. Since the higher concentration of salt deposits are harmful due to exosmosis, therefore, these should immediately be washed away to lower their concentration to the required strength. This problem of salt build-up can be avoided by ensuring sufficient drainage (by supplying extra water) at every irrigation cycle. The irrigation water should pass through whole volume of medium to ensure proper dilution and leaching away of the salts in excess at the drainage point. A 30-50 per cent increase in water volume during daily irrigation cycles will solve the problem. Regular chemical treatment may prove to be effective in preventing excess salt build-ups.

Accumulation of salts is common in recirculating systems where the nutrient solution is being re-circulated. Hence, the nutrient solution will need to be frequently completely replaced. Build-up may occur when multiple volumes of salty top-up water are added to the reservoir to compensate for transpirational and evaporative losses (when water evaporates, it is essential to understand that all salts are left behind). Thus, the salt content, and its contribution to the total EC, rises proportionately.

5.5.7 Hardness

Hardness of water may be due to the combined concentrations of calcium and magnesium, and when hard water is being used in re-circulating systems, the calcium level may become excessive after several additions of top-up water. This can result in a white precipitate (calcium sulphate) forming in the system. This can cause the blockage of plumbing components such as pipes and drippers. It also reduces the amount of available sulphate.

A water sample containing more than 150 mg/litre of bicarbonates (HCO_3) could be considered as hard. It is characterised by high levels of mineral salts and usually has a very high pH. The hydroponic grower will usually add phosphoric acid (H_3PO_4) to lower the pH. As it takes a significant amount of phosphoric acid to lower the pH, the levels of phosphates in the solution will increase. Over time, the phosphate will accumulate and the high levels

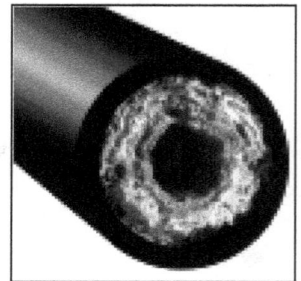

Plugging of Pipe Due to Ca Precipitation which Cements and Chokes the Internal Body.

will affect the uptake of other nutrients, such as zinc for instance. Major nutrient imbalances may occur if the water in the system is hard.

5.5.7.1 Identifying Hard Water

☆ It is harder to get soap to produce lather in hard water.

☆ Hard water also often leaves a buildup of lime on the pots and pans.

☆ The surest way to identify the hard water is to get a water quality analysis from the local water company, or from the local hydroponic shop.

5.5.7.2 Dealing with Hard Water

There can be two choices to deal with hard water.

1. Hydroponics growers should use specially formulated hydroponics nutrients for hard water. These formulas should be more acidic and should also be especially formulated to take account of the minerals, such as calcium, that are usually present in hard water. By lowering these elements in the nutrient solution it is possible to ensure that the final solution is as close as possible to ideal levels of the major elements.

2. It would be better if the grower uses reverse osmosis unit (RO). RO systems filter the water before it enters the reservoir, and removes about 98 per cent of inorganic salts.

5.5.8 Iron in Water

Iron is an essential trace element for plant growth in its natural chemical form. However, if exposed to the atmosphere, it will not form a stable solution in water at pH values above 3.

Iron is common in bore or groundwater. Water containing iron above ~0.3mg/l can be clear and colourless when first drawn to the surface but on exposure to air, the iron changes to iron oxide which makes the water appear brown and cloudy. When this water is kept still for some time, the iron oxide often settles to form a brown precipitate. This is a common cause of blocked drippers and brown staining on surfaces that it comes into contact with.

To avoid blockage, iron should be removed from make-up water prior to use. To remove iron oxidation by aeration, followed by 48 hour gravity settling can be tried. If aeration is unsuccessful, add a small amount of chlorine, followed by 5 minutes of gentle agitation then gravity settling.

5.5.9 Corrosion

Corrosion of metals will increase with total salt content and acidity (CO_2). In natural groundwater, CO_2 is usually the source of corrosion. With corrosive waters, pumps with stainless steel, ceramic or plastic parts should be used and reservoirs and plumbing must also be of plastic.

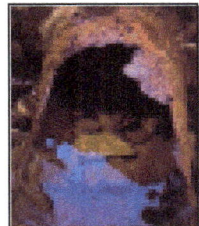

Corrosion in Iron Pipe

5.5.10 Pathogens

Pathogens in the nutrient solution from contaminated water supply or from sick plants may spread the disease to the rest of the crops. It is also common regional and seasonal problem. The best way to avoid such problems is working clean, planting only healthy disease-free plants, and closely monitoring the crop. On encountering disease problem, sick plants should be immediately removed from the system, the nutrient solution in the reservoir should be drained, reservoir should be properly washed and fresh nutrient solution should be added to the reservoir. If possible flush the system by running freshwater without nutrient for a day. Then drain and refill with fresh nutrient. Flushing between every three or four nutrient changes can help in maintaining cleanliness in the root zone and in the hydroponic system. Periodic flushing is especially helpful for gravel systems to remove salt accumulation in the medium.

The improper growing conditions such as improper light and air ventilation may invite pests or diseases in the system. Therefore these conditions should also be taken care of properly.

5.5.11 Beneficial Micro-Organisms

Beneficial fungi and bacteria can be added in the nutrient solution to enhance the growth of roots and to keep the water clean and plants healthy. Enzymes also work to break down dead roots. Using a combination of these three additives will result in a robust expanded root zone that will absorb nutrients faster and more efficiently.

5.6 Nutrient Additives

Nutrient additives are used to supply those elements that are not compatible with concentrated nutrient formulations. For example, a meaningful amount of silica is not stable in concentrated nutrients and therefore must be stored in a separate container. These additives provide benefits such as flowering enhancement, disease prevention, root initiation and increased fruit weight.

Additives are also required for supplying those substances which have a dose rate that varies throughout the growth cycle at a different rate than the base nutrient. This is often the case with plant growth regulators (PGR's) or hormone additives.

5.6.1 Common Additives

5.6.1.1 Phosphate Potassium (PK) Additives

PK additives can be used during the flowering or fruiting phase of a plant. Many flowering additives are available; however PK additives are the most common type. Although bloom nutrients contain phosphorus and potassium, heavy fruiting plants such as tomatoes and peppers have an especially high demand for these elements. Hence, PK additives are used in conjunction with bloom nutrients to help ensure no deficiencies occur. These additives should generally be used from the onset of budding until flowers or fruits are fully formed. Growers must select a pH neutral PK additive as an alkaline PK additive will cause a significant increase in nutrient solution pH. These can quickly destabilize the nutrients if pH is not adjusted immediately after dosing.

5.6.1.2 Silica Additives

Stable silica solutions are highly alkaline, therefore it cannot be included in concentrated nutrient formulations at meaningful levels. It must be used throughout both vegetative and flowering phases. Electron microscopy and X-ray analysis both confirm that existing silica within the plant is not mobile and cannot benefit new growth. To benefit new growth, silica must be present at all times in the nutrient solution. Various research projects have shown that the presence of silica in plant tissue produces several beneficial side effects:

- ☆ Silica produces healthier and stronger plant growth which enables superior leaf orientation and therefore greater rates of photosynthesis and growth. When silica is taken up by the roots, it is deposited in the cell walls of the plant as a solid, rigid 'quartz-like' matrix.
- ☆ Increases plant tolerance to heat stress or wilting.
- ☆ Increases the weight and shelf-life of fruit due to the physical accumulation of silica in plant cells.
- ☆ Increases resistance to fungal diseases, particularly mildews and botrytis. It resists fungal ingress by accumulating around the points of fungal attack.
- ☆ Improves the healing rate and neatness of pruning wounds. This property is especially beneficial in commercial cropping of plants such as tomato and cucumber. Regular pruning of these species threatens the plant's survival due to the risk of disease penetration through the site of the pruning wound.
- ☆ Increases a plant's tolerance to harmful chemicals such as sodium and chloride.

Growers should note that many commercial silica solutions have a poor shelf-life. This is evident when the concentrated solution turns cloudy.

5.6.1.3 Humic and Fulvic Acids

Humic acid and fulvic acid additives are known to increase the cation exchange capacity of media and act as a **'chelator'** to aid nutrient uptake. They also help enhance root development, improve the water holding capacity of soils and promote the growth of beneficial bacteria. Humic acid is derived from leonardite, a coal like material formed via the decomposition of plant matter. This is often the cause of colour in surface waters. Fulvic acid is the fraction of humic acid that is soluble in water under all pH conditions. Fulvic acid is regarded for providing superior efficacy than humic acid.

5.6.1.4 Nutrient Disinfectants (Sterilizing Agents)

Plants in hydroponics systems do not have a protective root cover as in plants that are grown in soil. The nutrient solution is continuously exposed to the atmosphere, thus allowing the free access of fungi such as *Pythium, Phytophthora* and *Fusarium*. These diseases spread rapidly when excess water is present and as a preventative strategy, the nutrient solution and medium needs to be regularly disinfected.

There are a variety of disinfectants available for hydroponic systems and the most popular are oxidizing agents which do not enter the plant. Therefore, disinfection treatment can be continued throughout the fruiting period without causing potential harm to the consumer. Examples of oxidizing agents include chlorine dioxide, sodium hypochlorite, monochloramine, hydrogen peroxide, UV and ozone. However, their relative suitability for hydroponics is largely determined by their half-life duration and oxidation potential.

☆ **Half-life duration:** A long half-life means that when the disinfectant is in the nutrient solution it remains chemically active for a longer period of time. This ensures that entire system (root zone, plumbing, etc) is treated each time when the nutrient pump is turned on. Without a sufficiently long half-life the disinfecting chemical becomes inactive before it can reach the root zone. This is a deficiency with ozone and UV. They only disinfect nutrient at the site of application. Further, they can also destroy chelated trace elements (*e.g.* iron EDTA).

  ❑ Monochloramine has the longest half-life of all the oxidizing agents. Chlorine dioxide and sodium hypochlorite have relatively short half-life.

☆ **Oxidation potential:** This is the measure of an oxidizing agent's power or ability to kill fungi and other organic organisms such as algae and slimes. Sufficient oxidation potential also helps eradicate food sources that are responsible for attracting fungus gnat and other root zone pests. Although monochloramine, chlorine dioxide and sodium hypochlorite have sufficient oxidation potential, only monochloramine is gentle enough for hydroponics. Chlorine dioxide and sodium hypochlorite can cause harm to roots, organic media and organic growth promoters.

5.6.1.5 PGR Additives (Plant Hormones)

PGRs are compounds that affect root growth, flowering, stem elongation or shortening. They can be either synthetic compounds that mimic the role of natural hormones, or they can be hormones extracted from plant tissues. They are applied as either a foliar spray or a liquid drench to the root zone.

There are 5 groups of PGR's:

1. **Auxins**: These produce several growth responses in plants including the formation of adventitious roots, promotion of apical dominance, growth, flower formation and fruit-set. Common examples of auxins include NAA (naphthyl-acetic-acid) and IBA (indole-butyric-acid). These are used to stimulate root growth in cuttings.

2. **Gibberellins**: These are used to stimulate cell division and elongation. They break seed dormancy and induce flowering. Gibberellic acid, for example, can be used to influence the timing of flowering, flower gender and flower size.

3. **Cytokinins**: These can be useful for stimulating cell division, inducing shoots and delaying aging and death (senescence). Benzyl-amino-purine (BA6) is a common cytokinin used for inducing side branching. This is useful for the development of short, bushy plants.

4. **Ethylene**: Used for promoting ripening, inducing leaf drop (abscission) and senescence. Ethylene is commonly used to ripen fruit in preparation for sale.

5. **Abscisic acid (ABA)**: This is a plant growth inhibitor. It causes abscission of leaves, fruits and flowers and causes stomata to close. During periods of drought when stomata are closed, high concentrations of ABA are found in leaf guard cells.

5.6.1.6 Seaweed (Kelp) Additives

Seaweed additives can provide micro-nutrients, cytokinins, auxins, gibberellins, polysaccharides, amino acids, and proteins. These all are very beneficial to plants.

5.6.1.7 Calcium, Magnesium, Iron Additives

Combination of calcium (Ca), magnesium (Mg) and iron (Fe) additives are used to help overcome or prevent the deficiency of these elements. The deficiency of these 3 elements can cause common symptoms such as blossom-end-rot (death of the end part of fruit), upward leaf curl, withered flower/fruit set, interveinal chlorosis and the death of terminal buds and root tips.

Excessive pH can cause both calcium and iron to become insoluble. Excessive humidity restricts the distribution (translocation) of nutrients throughout the plant and is a common cause of calcium deficiency.

This additive should be avoided with excessively hard water because it can cause the build-up of white scale in pipes and eventually causes blockage, and may cause reduction in the amount of available sulphate which is an essential plant nutrient.

5.6.2 Negative Side-Effects of Additives

Although additives can provide profitable yields, they can cause negative side-effects such as.:

☆ **Chemical compatibility:** Some additives are unstable when mixed with nutrient solutions, or other additives. It is necessary for the grower to follow a nutrient manufacturer's recommended feed schedule.

☆ **pH change:** Additives will cause some change on the nutrient solution pH. For example, alkaline conditions are needed to stabilize concentrated solutions of silica and many PK flowering additives. These additives can cause a significant pH change when they are added to nutrient solutions. Therefore, the nutrient's pH should be immediately checked and corrected after adding additives.

☆ **Algae and slimes:** Carbohydrate (or sugary) additives often induce the growth of algae and slimes in nutrient solutions. This can attract pests and cause blockage of plumbing.

5.7 Hydrogen Peroxide (H_2O_2)

Hydrogen Peroxide is made up of oxygen and water. When it breaks down, a single atom of oxygen is released along with a single water molecule. It leaves no residue or waste behind. The single atom of oxygen is what makes it so useful. Hydrogen Peroxide has two main uses in hydroponics:

☆ Disease fighting: It can be added into nutrient solution to help fight root rot. At higher concentrations it can be used to sterilize growing medium, gardening tools and plastic hydroponic trays and pots.

☆ Aeration: It can be added regularly to nutrient solution to increase the amount of dissolved oxygen in solution.

5.8 Nutrition in Closed Systems and Open System

Although the root zone solutions are almost the same for both closed system and open system feeds, the feed solutions are usually different for both the systems due to the rate of nutrient-uptake.

5.8.1 Nutrition in Closed Systems

A healthy plant will take up the nutrient balance it requires provided the solution around the roots contains nothing at a deficient or toxic level. In a totally closed system the only nutrient usage is what the plant takes up. If the nutrient solution in the root zone is to remain in balance, then what is fed in must have exactly the same balance as the plant uptake. If not, the solution within the system gets out of balance and continues to get further out of balance. If this happens then solution must be discarded from the system.

For better growth rates there needs to be maximum nutrient uptake. However, uptake is more difficult and slow for some nutrients than for others. Increasing the strength of root zone solution for those nutrients will increase the driving force and boost their rate of uptake.

5.8.2 Nutrition in Open Systems

With the open system's feed, there is not only plant uptake, but run-off as well. The feed is therefore part way between supplying the uptake and providing the required root zone solution strength. By the time it runs off, it has built up to the required balance in the medium.

If an adequate run-off is maintained then an imbalance goes out with it. For this reason the feed composition is nowhere near as vital as with the closed system.

Therefore, nutrient management is far easier in an open system. Due to this reason, most commercial hydroponics systems are designed to be open systems.

5.8.3 Fundamental Management Techniques for All Systems

☆ Sample the feed regularly (preferable daily), especially, the solution around the plant roots (or the run-off in open systems).

☆ Analyse all samples for pH and, especially, EC.

☆ Occasionally get a full chemical analysis done. If needed this could be tied to a leaf analysis.

☆ Record the results in a diary. Include other information such as weather conditions, crop performance and symptoms, pests, diseases.

☆ Manage by watching for trends and adjusting gently to correct them. Try to avoid taking severe action.

☆ Calcium, magnesium and sulphur should be far higher, up to double the strength of the feed because they have a slow uptake. In comparison, nitrogen should be about the same strength but potassium should be low because its uptake is faster.

☆ The intended balance in the root zone may be built up gradually or it may be done by having a special starter solution. Starter solutions will obviously be higher than normal in calcium, magnesium and sulphur.

5.9 Organic Nutrients in Hydroponics System

Many brands of organic hydroponics nutrients are available with the hydroponics stores. Managing these nutrients is quite difficult and requires a good skill and care as compared to regular hydroponic nutrients. These organic nutrients may cause problems, if not managed correctly.

Regular hydroponic nutrients contain mineral elements (such as nitrogen, calcium, etc.) that are immediately available to the plant. Organic nutrients (such as bone meal) must be broken down before the plant can absorb the minerals. Organic nutrients work great in soil because of the micro bacteria which increase the rate of this process. In hydroponics there are no beneficial micro-organisms and so the growth can be much slower as nutrients are not as quickly available to the plants. The plants are also at greater risk for nutrient deficiencies and need to be watched a bit more closely. Further organic nutrients may clog drip emitters, tubing and pumps. It will be easier to grow organically using an ebb and flow system (with no drippers) as opposed to a drip system. Many growers use a mineral based nutrient and supplement it with organic additives for the best of both worlds. A separate mineralization tank can be added to develop a colony of beneficial micro-organisms. In its simplest form it contains some type of porous media, water and something to provide aeration (usually air stone). The organic hydroponic nutrients, manure or fish waste may then be added either manually or can be configured so that they may be directed automatically to the solid mineralization device for mineralization and breakdown and then the mineralized nutrients may easily be added into the nutrient solution reservoir either manually or automatically. See Aquaponics (Chapter-17).

5.10 Conclusion

In short, when fresh nutrient solution is prepared, it should be measured carefully. At later stages too, notes on observations (of EC drift, pH drift, total water usage, temperature range, *etc.*) and comments on crop health and progress should be kept. Diseases should be carefully looked for and sick plants must be destroyed immediately.

Note: *Plants can tolerate a lot of stress and still produce well.*

⑥

Substrates or Growing Medium

Hydroponic growing media are used not only to provide support for the plants but are also used to distribute appropriate oxygen and carbon dioxide, *vis-a-vis* providing the place from where the plants will absorb nutrients. A good hydroponic medium must be eco-friendly, pH stable, must have a good nutrient and moisture holding capabilities and must drain well. Different media are appropriate for different growing techniques. The growing media should be used correctly in conjunction with the other aspects of the hydroponic growing system.

Some of the hydroponic growing media are explained below:

6.1 Hydroton (Expanded Clay Aggregate)

Hydrotons or grow rocks or expanded clay aggregates are the baked clay pellets that have been heated at temperatures of 1200°C or more and have very high capillary action. These are mostly used in those hydroponic systems where all nutrients are controlled in water solution. The clay pellets are inert, porous, light in weight, pH neutral and do not contain any nutrient value. The expanded clay aggregates are ecologically sustainable and re-usable after they are cleaned, sterilized (using white vinegar, chlorine bleach or hydrogen peroxide) and rinsed completely. Features of hydroton are:

☆ It can be used in almost all hydroponic systems

☆ Reusable

Expanded Clay Aggregate

☆ Eco-friendly

☆ Does not compact

☆ Porous

☆ Provides lots of oxygen to the root zone

☆ Light in weight

☆ Almost free from nutrients and pH neutral

☆ Needs to be watered about 4 to 8 times a day or every 2 to 4 hours

☆ If mixed with coco peat it stays moist much longer

Disadvantages

☆ Drain and dry very fast, roots may dry out

☆ Harder to find

☆ Roots sometimes get entangled in the pellets but can be easily cleaned

6.2 Coco-Fibre (Cocopeat or coir) and Coco Chips

6.2.1 Coco Fibre

Coco-fibre is made from the husk of the coconut and is a natural **organic** grow and flowering medium being used in most of the hydroponic systems. It has millions of capillary micro sponges that allow it to quickly absorb and slowly release the nutrients as and when required by the plants. It also provides lots of oxygen and aeration to the root zone of plants. It can be used either alone or can be mixed with other growing media. It should be washed properly (using at least hot water) before use, to remove dust or remains and to stabilize **pH**.

Coco-peat has come to the fore in the rose growing industry as an ideal medium as it has good drainage properties but at the same time retains good moisture content. It is also very stable and is unaffected by fertilizers applied to it, it's consistency is

Rose Plants in Cocopeat

very fine which is perfect for the development of fine root hairs which are responsible for the absorption of nutrients and water required by the rose plants. Coco-peat does not affect the pH and EC. At the end of its life expectancy, it is bio-degradable and does not present a hazard to the environment.

Coco fibre does tend to colour the water, but that diminishes over time. Most of the colours can be leached out by soaking it in warm/hot water a few times before use.

Some of its features are:

☆ Holds nutrients and moisture for a long period

☆ Provides lots of oxygen and aeration to the roots

☆ The coco peat promotes rapid development of roots

☆ It allows for exchange of gases like carbon dioxide

☆ It allows the nutrient solution to drain quickly

☆ It has a natural pH of 6.5 to 7.0

☆ It has some anti-fungal properties

☆ It acts as a great hormone rich medium for the plants

☆ It can hold up to 8 times its dry weight in water

☆ It is eco-friendly

☆ It needs to be watered about 3 to 5 times a day or every 3 to 5 hours

☆ It can be used in almost all hydroponic systems, either alone or can be mixed with other growing media to increase their water holding capabilities

☆ It is compactable, *i.e.* can be purchased compressed and expanded at home thus saves money on shipping

Disadvantage: Built up salts in this medium can often be a problem.

6.2.2 Coco Chips

The coco fibre particle size is about the same as potting soil, while the coco chips particle size is more like small wood chips. The larger size of the coco chips allows for bigger air pockets between particles, thus allowing even better aeration for the roots.

Coco Chips

6.3 Growstones

Growstones are made from the wastes of glass and have both more air and water holding space as compared with perlite, peat and parboiled rice hulls. They are light weight, unevenly shaped, porous, and reusable and have good wicking ability. They can wick water up to 10 cm above the water line. The waste glass bottles are crushed into a powder, melted and then some calcium carbonate is added in it. After the mixture is cooled down, it is broken into small pieces to form what looks like lava rocks. It is extremely lightweight, sustainable, and a great medium in almost any application.

Growstones

Its major drawback is that it is hard to clean, as roots will stick to it after harvest. Also, it is difficult to move the plants to another grow area as the plant roots get gripped to the growstones resulting in root damage.

6.4 Parboiled Rice Hulls

Rice hulls are agricultural byproducts (shells that surround rice) and decay slowly over time. These are porous and allow drainage and have less water holding capability than most of other media such as growstones. Parboiled rice hulls are prepared by steaming and drying the rice hulls after the rice has been milled from them. This kills any spores, bacteria, and microorganisms, leaving a sterile and clean product. Phosphorus and potassium are all leached out from rice hulls to some extent. Its pH value is quite stable (slightly acidic) and ranges from 5.7 to 6.5. Since rice hulls do not readily absorb water, therefore, they should be matured for some time or mixed with kuntan, perlite or other substrates having a good water retaining capacity.

Rice Hulls

Fresh and/or composted rice hulls tend to have high manganese content. But problems with Mn toxicity can be avoided as long as the pH is above 5. This is below normal range for hydroponics anyway.

Rice hulls are used as fresh, aged, composted and parboiled, or carbonized. Fresh rice hulls are typically avoided as a hydroponic growing medium because of the high probability of contaminants such as rice, fungal spores, bacteria, decaying bugs, and weed seeds.

6.5 Kuntan Culture

Kuntan used as medium is the carbonized rice hulls and is very light in weight having a great water retention capacity. It is usually used in drip method of hydroponics.

Disadvantage: It has high pH values and much water soluble potassium.

6.6 Rockwool (Mineral Wool)

Rockwool is an inert substrate made from molten rock, basalt or slag that is spun into bundles of single filament fibres, and bonded into a medium capable of capillary action. The rockwool is formed into blocks, sheets, cubes, slabs, or flocking. It is a sterile, porous and non-degradable medium which sucks up water easily. Also, it is protected from most common microbiological degradation. The nature of rockwool may differ depending on the raw materials. It should be pH-balanced

Rockwool Cube

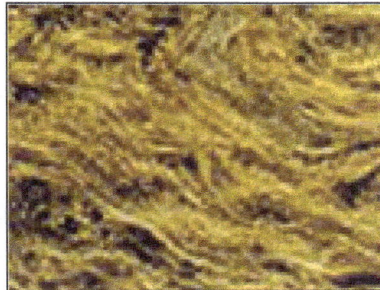

Rockwool Sheets

before use, by soaking it in pH balanced water. It is the most widely used growing medium in hydroponic systems. Some of its features are:

☆ It is well suitable for seed starting

☆ It can hold water and nutrients for a very long period

☆ It has a very good oxygen levels, *i.e.* 18 per cent

☆ It needs to be watered about 1 to 5 times a day depending on what is being grown

☆ It is available in various sizes

Disadvantages

☆ Not environment friendly

☆ It has a high and unstable pH

☆ Its cation exchange capacity (CEC) is extremely low

☆ The fibres and dust in rockwool are bad for lungs, and irritate the skin

☆ Difficult to dispose off as thin fibres of melted rock will last essentially forever

6.7 Oasis Cubes

Oasis cubes are inert medium similar to rockwool cubes but they don't become waterlogged as easily as rockwool cubes. It is an inexpensive medium and can be used for the germination and further seedling growth. But oasis cubes are more like the rigid green or white floral foam used by forests to hold the stems in their flower displays. Oasis cubes are an open cell material which means that the cells can absorb water and air. The open cells wick moisture throughout the material, and the roots can easily grow and expand through the open cell structure.

Oasis Cubes   Lettuce in Oasis Cubes

Advantages

☆ Inexpensive

☆ No presoaking

Disadvantages

☆ Not organic

☆ Useful for germination only, not as a full growing medium

6.8 Perlite

Perlite is a volcanic rock that has been naturally fused (fusion of granite, obsidian, pumice and basalt) at high temperature. It is a soil-free growing medium that has helped to add aeration to soil mixes for years. Perlite has a neutral pH, excellent wicking action, and is very porous. It is very light weight and floats over water. Therefore, it is usually mixed with other media such as coco coir or vermiculite to increase their oxygen retention level and is rarely used alone. To use it in hydroponics systems such as 'top drip' or in NFT, right pots should be used. It is usually not suitable for hydroponics systems such as 'Ebb and Flow' systems. It can be used loose or in plastic sleeves immersed in water. It is capable of holding more air but less water. It cannot be used in most of the net pots as it will fall through the holes. Needs to be watered about 3 to 5 times a day or every 3 to 5 hours depending on what is being grown. It is environment friendly medium.

Perlite

6.9 Vermiculite

Vermiculture is very much similar to perlite and is made using volcanic rock that is superheated into lightweight pebbles. It can float over the nutrient solution. It holds more water than perlite and has a natural wicking property. It has a relatively high cation-exchange capacity, meaning it can hold nutrients for later use. It cannot be used in most of the net pots as it will fall through the holes. Vermiculite can be mixed with perlite to provide more oxygen to the root zone and to enhance moisture holding properties. Like perlite, it is also usually used in combination with other types of media to create a highly customized media for specific hydroponic applications.

Vermiculture

6.10 Pumice

Pumice is similar to perlite and is a lightweight, mined volcanic rock that is crushed and used in some hydroponics systems. It has high oxygen retention level.

Pumice

6.11 Sand

Sand is an easily available and extremely cheap (almost free) but heavy growing medium that must be sterilized before every use. It is quite heavy and does not hold water very well and may become waterlogged. Sand is like rock but smaller in size.

Because the particle size is smaller than regular rock, moisture doesn't drain out as fast. Sand is also commonly mixed with vermiculite, perlite, and or coco coir. All these help retaining moisture as well as help aerate the mix for the roots.

Sand

Advantages

☆ Cheap

☆ Easy to find

Disadvantages

☆ Heavy

☆ Low water retention

☆ Small size may affect certain hydroponic systems

6.12 Gravel

Gravel is the same medium that is used in aquariums and is quite heavy, relatively cheap, drains well, easy to clean, and will not become waterlogged. It should be cleaned properly before use. The plant roots may dry out if the system does not provide continuous water.

Gravel

Advantages

- ☆ Inexpensive
- ☆ Readily available
- ☆ Drains well
- ☆ Easy to clean
- ☆ Reusable
- ☆ It will not become waterlogged

Disadvantages

- ☆ Heavy
- ☆ Plant roots may dry out
- ☆ Not suitable for certain hydroponic systems
- ☆ pH unstable

6.13 Wood Fibre

It is produced from steam friction of wood and is able to keep its structure for a very long time. It is great and efficient pure organic medium for hydroponics. It is biodegradable and may not be sterile and may attract pests.

Wood Fibre

6.14 Brick Shards

Brick shards are bricks that have been crushed into small pieces and have similar properties to gravel but may alter the pH. It is inexpensive and needs extra cleaning before reuse to get rid of brick dust.

Advantages

- ☆ Inexpensive
- ☆ Easy to clean
- ☆ Drains well

Brick Shards

Disadvantages

☆ May affect pH

☆ Requires more thorough cleaning

☆ Heavy

☆ Plant roots may dry out

6.15 Sheep Wool

The sheep wool has a greater air capacity which may decrease with use and its water capacity may increase with use.

Sheep Wool

6.16 Polystyrene Packing Peanuts

Polystyrene packing peanuts are inexpensive, very light in weight, have excellent drainage and are mostly used in closed-tube systems. Biodegradable peanuts should not be used in any sort of hydroponic system as they will decompose into sludge. Typically used in NFT systems. There is possibility that the plants may absorb styrene from these media and may pose a contamination risk.

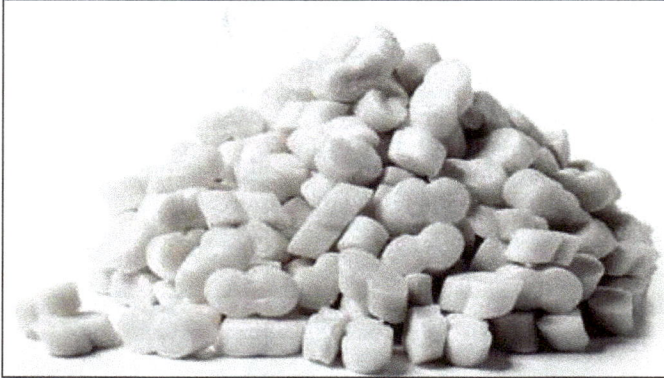

Polystyrene Packing Peanuts

6.17 Sawdust Culture

Sawdust is kept in U-shaped or V-shaped grooves or packed in polyethylene bags and crops are cultivated in these bags. The characteristics of sawdust may differ more or less depending on the types of tree. This medium is widely used where large-scale lumber industries exist.

Sawdust

Disadvantages: Some sawdust may contain elements not suitable to the growth of crops.

6.18 Water Absorbing Crystals (Water-Absorbing Polymers)

Although water absorbing polymer crystals are being used in many industries (such as in kid's diapers and in sport's industries), these are not very popular as hydroponics media. The crystals expand to many times their size as they soak up water. One pound of the crystals can hold as much as 50 gallons of water. The crystals come in many sizes, everything from a powder, to marble and even golf ball size. Depending on the size of the crystals they can take more than an hour or two to fully absorb. Once they dry out, they can be stored and reused again over and over.

Most large nurseries carry them as soil amendments. These crystals can be mixed into the soil to help retain moisture in the soil. Florists use them in vases to keep flowers fresh, and the coloured ones make for a nice decorated display.

Water Absorbing Polymer

The water absorbing polymer crystals are not a common hydroponic growing medium, it is growing in popularity and is quite inexpensive, and reusable. The larger size crystals are better suited for use in hydroponics. The larger size helps retain some of the air pockets between the crystals. Also by mixing some river rock or other similar growing media with the crystals will help increase the air pockets between the crystals.

6.19 Poly (Polyurethane) Foam Insulation

Poly foam is not commonly used in hydroponics, but can be used as an alternative to using rockwool or oasis cubes as starter cubes with great results. Poly foam is cheap and easy to find. Any hobby store or place that sells fabrics should carry it. It is most commonly used as furniture foam, and is also referred to as "foam batting." It comes in sheets or rolls of different sizes and thickness.

Polyurethane

6.20 Composted and Aged Pine Bark

Pine bark was considered a waste product, but has found uses as a ground mulch, as well as substrate for hydroponically grown crops. Pine bark is considered better than other types of tree bark because it resists decomposition better, and has less organic acids that can leach into the nutrient solution than others. Bark is generally used as fresh, composted, or aged. Pine bark can be found at places that sell ground mulch.

Pine Bark

6.21 Pine Shavings

Pine shavings are the inexpensive hydroponic growing media used by many commercial growers, usually in large scale hydroponic drip irrigation systems. It should not be confused with saw dust. Saw dust will become compact and water-logged easily. Before using it, the grower must make sure that the pine

Pine Shavings

shavings were made from kiln dried wood, and does not contain any chemical fungicides.

Good source to find pine shavings are pet supply stores. It is used for things like hamster and rabbit bedding. It should be made sure to read the package to be sure it doesn't have any chemical additives like fungicides or odour inhibitors. It is also used as bedding in horse stalls. The larger the air pockets between the shavings, the better aeration to the roots. Pine shavings are a wood product, so they absorb water easily, and thus can easily become water-logged.

6.22 River Rock

River rock can be purchased from home improvement stores or from pet supply's stores (with the fish and aquariums). River rock is fairly inexpensive, and comes in many different sizes. River rock is rounded with smooth edges from tumbling down the river. Though manufactured river rock is rounded using large mechanical tumblers, it has the same end result with smooth edges.

River Rock

River rocks must be cleaned and sanitized before they are used. It will be better to wash them properly and then should be soaked overnight in bleach water. Then these rocks can be rinsed and used. Though using rock as a growing media is inexpensive and easy, it will get heavy quickly. River rocks are not porous, therefore they do not hold and retain moisture in the root zone of hydroponic systems. Rock is uneven so it has a lot of air pockets between the rocks so that the roots can get plenty of oxygen, but water easily drains down to the bottom. The growers need to adjust their watering schedules so the roots don't dry out between watering. Coco chips or other growing media that hold moisture can be mixed with river rocks. Because of the good drainage property of rock, it aids in the drainage of other hydroponic growing media that might otherwise become saturated from sitting in water

6.23 Floral Foam

Floral foam can be used as a growing medium in hydroponics as well, and is similar to the oasis cubes, though the cell size is larger in the floral foam. The main disadvantages of using floral foam are; it can crumble easily and can leave particles in the water; it easily gets water-logged. Floral foam absorbs water easily.

Note: Application of a biostimulator consisting of humic acid, lactic acid and Bacillus subtilis may improve yield in all substrates.

Floral Foam

⑦
Nutritional Disorders

Nutritional disorders include deficiencies of nutrients, toxicities or excesses of nutrients and other elements, and excessively high or low acidity. Plant growth can be optimized by diagnosing nutrient deficiencies on crop-by-crop basis. Nutrient deficiency diagnosis is very subjective and requires careful observation. Nutritional disorders may involve not only nutrient levels, but also other factors such as temperature, humidity, day length and disease. It can be very complex and may produce a syndrome which does not resemble any single disorder.

The nutrient deficiency in plants can be categorized as **acute deficiency** and **chronic deficiency.** Acute deficiency occurs when a nutrient is suddenly no longer available to a plant and chronic deficiency occurs when there is insufficient supply of a nutrient that is unable to meet the growth demands of the plant. The hydroponics studies favour the development of acute deficiencies because the nutrients are fully available and when a nutrient is depleted, the plant suddenly faces an acute deficiency.

Symptoms such as tip burns, chlorosis or necrosis, which are characteristic to some nutrient deficiencies, can also be associated with other stresses, such as plant disease, chemical spray and/or salinity stress. The location on the plant where deficiency symptoms are expressed can help in diagnosing the nutrient disorder.

Before going into more detail, one should have a look on external leaf structure and some common leaf abnormalities resulting from nutrient deficiencies.

Nutrient deficiencies are usually uniform throughout the crop whereas the effect of plant diseases and chemical spray is not uniform across the affected plants. Knowing the spray history can prove to be helpful in distinguishing between

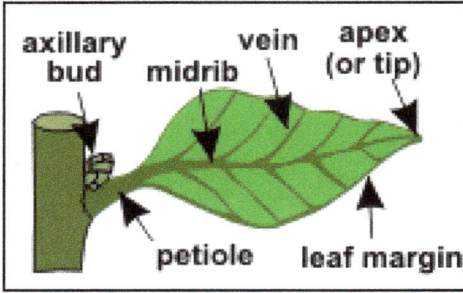

External Leaf Structure.

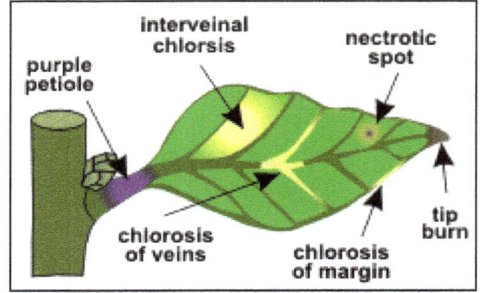

Abnormalities Resulting from Nutrient Deficiencies.

nutrient deficiencies symptoms and chemical spray burns. Also, if symptoms occur soon after a chemical was applied, they might be related to the chemical that has been sprayed. Salinity damages, such as tip burns might be misinterpreted as nutrient deficiencies. For example marginal leaf burns might be interpreted as potassium deficiency.

7.1 Indentifying Plant Deficiencies

Correct and timely diagnosis of nutritional deficiencies will help the gardener to respond and fix the issue before the plants experience major stress. Signs and symptoms of deficiencies vary depending on the growth stage, plant species and

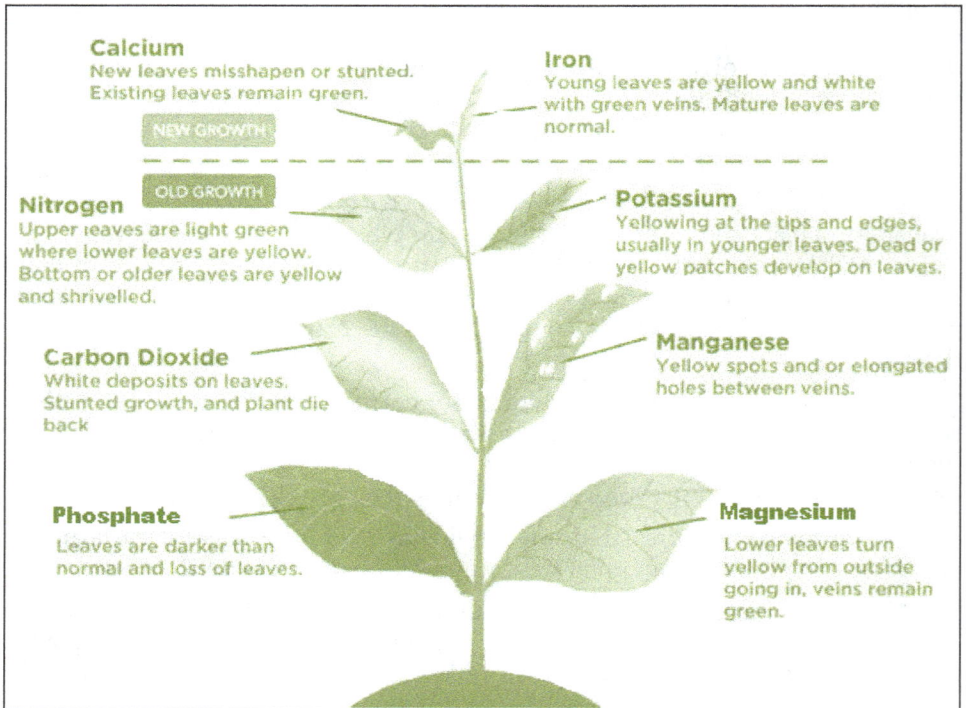

Calcium
New leaves misshapen or stunted. Existing leaves remain green.

NEW GROWTH

Iron
Young leaves are yellow and white with green veins. Mature leaves are normal.

OLD GROWTH

Nitrogen
Upper leaves are light green where lower leaves are yellow. Bottom or older leaves are yellow and shrivelled.

Potassium
Yellowing at the tips and edges, usually in younger leaves. Dead or yellow patches develop on leaves.

Carbon Dioxide
White deposits on leaves. Stunted growth, and plant die back

Manganese
Yellow spots and or elongated holes between veins.

Phosphate
Leaves are darker than normal and loss of leaves.

Magnesium
Lower leaves turn yellow from outside going in, veins remain green.

other factors such as climate and growing conditions. Gardeners need to become familiar with the wide range of deficiencies and understand how they can affect each crop.

Gardeners can use a plant journal to record the conditions of their garden by making notes and taking photographs of any strange or unusual appearances or behaviour of plants. Pictures can be very helpful in identifying the deficiencies.

Pictures of deficiency symptoms shown in text books may often appear different than in the live plants. Even most experienced professionals often find it difficult to diagnose the deficiencies because several deficiencies can occur at the same time and some elements produce similar symptoms. Also, sometimes, symptoms can be confused with those caused by pests, diseases, genetic abnormalities, and under/ over watering.

The most important is early detection, when the deficiency symptoms are often unique and are easier to distinguish and also other causes (such as EC, pH, humidity, *etc*) can be checked in the early stages.

7.2 Nutrient Mobility and Deficiency Symptoms

Major point in identifying the symptoms of nutrient deficiencies is the effect of nutrient mobility on deficiency symptoms. Mobile nutrients such as nitrogen, phosphorus and potassium can be moved from older leaves to younger ones. Therefore the deficiency symptoms of mobile nutrients will first develop on the older mature leaves and then to actively growing parts (younger leaves). The **nitrogen, phosphorus** and **potassium** are **very mobile** nutrients whereas **magnesium, sulphur, iron** and **manganese** are **moderately mobile** nutrients. Mobility of sulphur varies with species. Deficiency symptoms of **immobile** nutrients such as **calcium** and **boron** will first appear in the newer growth because these nutrients cannot be moved from the older leaves to the new growth.

The main kinds of nutrient's deficiency symptoms are: chlorosis (yellowing), purpling, local necrosis (tissue death), stunting, poor quality (under-developed, cracked or few) buds, flowers or fruit, poor root development, distorted leaves, *i.e.* cupped or twisted.

7.2.1 Overall Chlorosis (Yellowing)

☆ Chlorosis occurs due to deficiencies of elements required for photosynthesis or chlorophyll production. In chlorosis, leaf or whole plant turns light green or yellow. It can also be more localized. For example, yellowing of the veins themselves or between the veins (interveinal chlorosis).

☆ May indicate deficiencies of nitrogen or sulphur, especially when it occurs mostly on older leaves.

☆ Chlorosis due to nitrogen deficiency is a familiar sight in soil-grown plants.

☆ Chlorosis may also precede death of the older leaves during the onset of damage from excessive levels of sodium, potassium, chloride or phosphate.

☆ Chlorosis of part of the mature leaves is sometimes a sign of magnesium deficiency.

☆ Non-nutritional causes of generalized chlorosis include old age, lack of root aeration, or lack of light.

7.2.2 Interveinal Chlorosis (Yellowing)

☆ Yellowing between leaf veins; with the veins forming a green pattern.

☆ It is a symptom of deficiencies of iron, manganese or sometimes zinc or copper.

☆ These deficiencies (especially iron) may arise from alkalinity (excessively high pH) in the nutrient solution, which makes these metals less soluble.

☆ Non-nutritional causes of interveinal chlorosis include virus diseases.

Some ornamental plants are naturally yellow between the veins.

7.2.3 Purpling

☆ Purple or red discolouration usually occurs on stems or along leaf petioles, veins or margins due to abnormal levels of anthocyanin that accumulates when plants are stressed. These symptoms can also be caused by physical stresses such as cold, drought and diseases.

☆ Causes an overall dark green colour with a purple, red, or blue tint, and is the common sign of phosphate deficiency.

☆ Some plant species and varieties respond to phosphate deficiency by yellowing instead of purpling.

☆ Purpling is natural to some healthy ornamentals.

☆ It also results from drought and aluminum toxicity. These two disorders are common in soils but perhaps not in hydroponics.

7.2.4 Necrosis (Tissue Death, Papery Appearance, Dry Patches, Scorching)

☆ Leaf burn or death of patches or spots on leaves.

☆ Generally happens in the later stages of deficiency where the affected plant part becomes stressed to the point that it becomes brown and dies.

☆ It can be due to deficiency of potassium, magnesium and of the less common molybdenum deficiency.

☆ Non-nutritional causes include frost, severe drought (or low humidity), and sun scorch.

☆ The shape, position, and colour of the dead areas vary depending on the disorder and the plant, and these features can help sharpen the diagnosis.

7.2.5 Stunting

☆ Plant is shorter than normal.

☆ Overall stunting of the whole plant is a sign of all nutrient disorders, and the hardest to detect unless normal plants of the same kind and age are nearby for comparison.

☆ Some cases of disorders produce acute stunting of growing points or young leaves, resulting in characteristic growth patterns.

☆ Zinc deficiency produces little leaf in many species, especially woody ones.

☆ The younger leaves are distinctly smaller than normal.

☆ Zinc deficiency may also produce **rosetting** where the stem fails to elongate behind the growing tip, so that the terminal leaves become tightly bunched.

☆ Deficiencies of boron, calcium, potassium, and sometimes zinc can kill growing points, and especially in woody plants, this can lead to dieback of the shoot.

7.2.6 Root Stunting

☆ It is characteristic of calcium deficiency, acidity, aluminum toxicity, and copper toxicity.

☆ Some species may also show such symptoms when are deficient in boron.

☆ The shortened roots become thickened, the laterals become stubby, peg like, and the whole system often discolours, brown or gray.

7.2.7 Dieback

☆ Leaves or growing points die rapidly and dry out.

☆ It may be due to boron or calcium deficiencies.

7.2.8 Mottling

☆ Irregular spotted surface.

☆ Blotchy pattern of indistinct light and dark areas often associated with virus diseases.

7.2.9 Checkered Appearance (Reticulation)

☆ Patterns of small veins of leaves remain green while interveinal tissue becomes yellow.

☆ May be due to manganese deficiency.

7.2.10 Spindle Formation

☆ Growth of stem and leaf petioles are very thin and succulent.

☆ May be due to nitrogen deficiency.

7.2.11 Nutrients and their Deficiency Symptoms

The following is a general guideline to follow in recognizing the response to nutrient deficiencies and toxicities.

Nutrients	Deficiency Symptoms	Toxicity
Nitrogen (N)	☆ Its deficiency first effects **older leaves** which may become pale green or yellow (chlorotic) due to lack of chlorophyll and may even die. ☆ Yellowing (chlorosis) may proceed from light green to yellow from older leaves to new growth. ☆ Leaf growth may become stunted and leaf tips of older leaves burn and may even fall. ☆ New growth becomes weak and spindly and branching is reduced. ☆ Overall restricted growth of tops, roots and shoots (especially lateral shoots). ☆ In some plants, stems, petioles and lower leaf surfaces may turn purple. ☆ There is no interveinal striping or yellow patches. Yellowing is uniform over the entire leaf including the veins. ☆ Leaves approach a yellowish white colouring under extreme deficiency. ☆ In some instances, an interveinal necrosis replaces the chlorosis.	☆ An abundance of nitrogen will cause soft, weak growth and even delay flower and fruit production if it is allowed to accumulate. ☆ Plants often become dark green and in the early stages abundant with foliage. ☆ If excess is severe, leaves will dry and begin to fall off. ☆ Root system will remain under developed or deteriorate after time. ☆ Fruit and flower set will be inhibited or deformed. ☆ Higher susceptibility to disease and pests.

Plants/leaves Showing Nitrogen Deficiency

Contd...

Contd...

Nutrients	Deficiency Symptoms	Toxicity
Phosphorus (P)	☆ This deficiency symptom appears first in **older leaves**. ☆ Phosphorus deficiency symptoms are not very distinct and thus difficult to identify. ☆ The plants are dwarfed or stunted and develop very slowly in comparison to other plants growing under similar environmental conditions but without phosphorus deficiency. ☆ Phosphorus deficient plants are often mistaken for unstressed but much younger plants. ☆ Leaves turn dark green or grey with blotches. ☆ Its deficiency will result in overall reduced plant growth and the leaves turn dull dark green to bluish green and sometime show brown or purple spots.	☆ Its excess is rare and usually buffered by pH limitations. ☆ Excess phosphorus can interfere with the availability of other elements such as iron, manganese, copper and zinc.

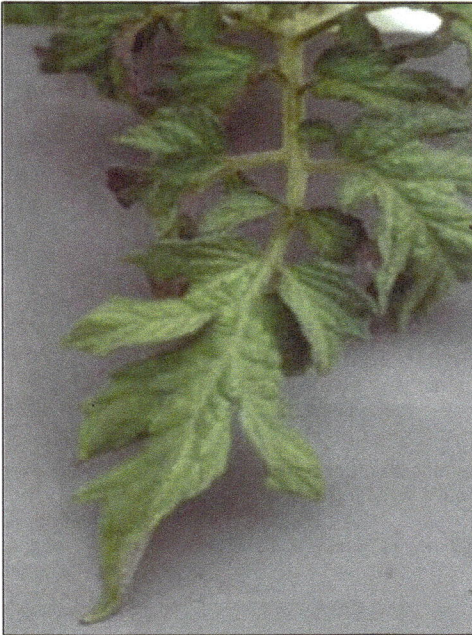

Leaves Showing Phosphorus Deficiency

☆ Leaves will also tend to be smaller and dark green, and may or may not include necrotic patches.
☆ Bud size will be smaller and underdeveloped, as will the root mass.
☆ May show purpling of petioles and the veins on underside of younger leaves.
☆ Stems also may turn purple.

Contd...

Contd...

Nutrients	Deficiency Symptoms	Toxicity

Leaves showing phosphorus deficiency

☆ Long term deficiency results in poor root formation.
☆ Plant maturity is often delayed.
☆ Some species such as tomato, lettuce, corn and the brassicas develop a distinct purpling of the stem, petiole and the undersides of the leaves.
☆ Under severe deficiency conditions there is also a tendency for leaves to develop a blue-gray luster.
☆ In older leaves under very severe deficiency conditions a brown netted veining of the leaves may develop.

Nutrients	Deficiency Symptoms	Toxicity
Potassium (K)	☆ Its deficiency first effects **older leaves.** ☆ Its deficiency results in taller plants that appear to be healthy. ☆ Small dead spots may appear at tips and between veins.	☆ Its excess is rare. ☆ Excess potassium can aggravate the uptake of magnesium, manganese, zinc and iron.

Leaves Showing Potassium Deficiency

Contd...

Contd...

Nutrients	Deficiency Symptoms	Toxicity
	☆ Margins cup downward with brown spots. ☆ **Older leaves** become yellow between veins and then they turn dark yellow, brown and they curl up and may even die. ☆ Older leaves are initially chlorotic but soon develop dark necrotic spots (dead tissue) and as the deficiency becomes more severe, these necrotic spots progress inward and also upward towards younger leaves. ☆ Stem and branches may become weak and easily broken. ☆ Its deficiency also causes dropping of flowers and fruits, and poorly developed roots.	
Magnesium	☆ The expression of symptoms is greatly dependent on the intensity to which leaves are exposed to light. Deficient plants that are exposed to high light intensities will show more symptoms. ☆ Its deficiency first effects **older leaves.** ☆ It deficiency causes advanced interveinal chlorosis in leaves, with necrosis developing in the highly chlorotic tissue. ☆ In its advanced form, magnesium deficiency may resemble potassium deficiency. ☆ The symptoms generally start with mottled chlorotic areas developing in the interveinal tissue of **older leaves** and as the deficiency becomes severe, inter-veinal chlorotic mottling will proceed towards the younger leaves. ☆ The interveinal laminae tissue tends to expand proportionately more than the other leaf tissues, producing a raised puckered surface, with the top of the puckers progressively going from chlorotic to necrotic tissue. ☆ In some plants such as the *Brassica* (*i.e.*, the mustard family, which includes vegetables such as broccoli, brussel sprouts, cabbage, cauliflower, collards, kale, kohlrabi, mustard, rape, rutabaga and turnip), tints of orange, yellow, and purple may also develop. ☆ Its deficiency causes yellowness in older leaves from the centre to outward but leaf veins stay green.	☆ Its toxicity is rare and not generally exhibited visibly.

**Symptoms of Magnesium Deficiency Showing
Yellowing in Older Leaves.**

Contd...

Contd...

Nutrients	Deficiency Symptoms	Toxicity

☆ Leaves develop brown spots.

**Symptoms of Magnesium Deficiency Showing
Brown Spots in Leaves.**

☆ The tips and edges of the leaves may discolour
(turn lime in growing tips) and curl upward

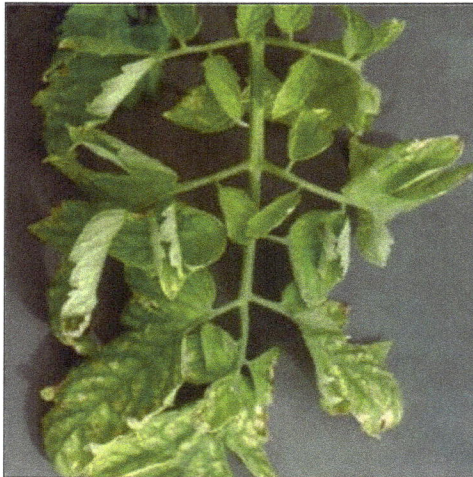

**Discolouration and Upward Curling Due
to Magnesium Deficiency.**

☆ Buds do not develop.
☆ In some crops, the interveinal yellow patches are
followed by necrotic spots or patches and marginal
scorching of the leaves.

Calcium

☆ Its deficiency sign first appears in **young growth**
resulting in deficient leaf tips (without brittle tissue)
and edges.
☆ Common symptoms of calcium deficiency include
blossom-end rot of tomato (burning of the end part
of tomato fruits), tip burn of lettuce, blackheart of
celery and death of the growing regions in many
plants. All these symptoms show soft dead necrotic
tissue at rapidly growing areas, which is generally

☆ Difficult to distinguish
visually.
☆ May precipitate with
sulphur in solution and
cause clouding or residue
in tank.
☆ Its abundance in early life
of plants causes stunted
growth.

Contd...

Contd...

Nutrients	Deficiency Symptoms	Toxicity
	related to poor translocation of calcium to the tissue rather than a low external supply of calcium.	

☆ New leaves stop developing and stay small and dry.

☆ Young leaves become small and distorted or chlorotic with irregular margins, spotting or necrotic areas.

☆ New growth will turn brown to black and may die.

☆ The scorched and die-off portion of tissue is very slow to dry so that it does not crumble easily.

☆ Bud development is inhibited.

☆ Long term deficiency stunts growth and causes poor or under-developed root formation.

☆ Fruit may be stunted or deformed.

☆ Young leaves are affected first.

☆ **Boron** deficiency also causes scorching of new leaf tips and die-back of growing points, but calcium deficiency does not promote the growth of lateral shoots and short internodes as in boron deficiency.

Calcium Deficiency Showing Blossom-End Rot of Tomato

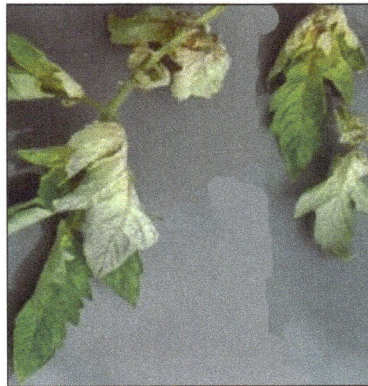

Leaves Showing Calcium Deficiency

Contd...

Contd...

Nutrients	Deficiency Symptoms	Toxicity
Sulphur	☆ Its deficiency signs first appear in **younger leaves.** ☆ Symptoms may vary between plant species. For example, in corn, sulphur deficiency shows up as interveinal chlorosis; in wheat, the whole plant becomes pale while the younger leaves are more chlorotic; in potatoes, spotting of leaves might occur ☆ Its deficiency causes **young leaves** to become pale yellow with slow growth. ☆ Veins become chlorotic (turns yellow). ☆ The initial symptoms are the yellowing of the entire leaf including veins usually starting with the younger leaves. ☆ Yellowing starts from the top, and progresses down, including the veins of the leaves. Treat the same as Mg deficiency.	☆ Leaf size will be reduced and overall growth will be stunted. Leaves yellowing or scorched at edges. **Leaf Showing Sulphur Toxicity**

Leaves Showing Sulphur Deficiency

☆ Leaf tips may yellow and curl downward.
☆ Leaves tend to get brittle and stay narrower than normal.
☆ Resembles nitrogen deficiency in that older leaves become yellowish green and the stems thin, hard and woody.

Sulphur Deficiency in Corn and Soybean

Contd...

Contd...

Nutrients	Deficiency Symptoms	Toxicity
	☆ Some plants show colourful orange and red tints rather than yellowing on the underside of the leaves and the petioles have a more pinkish tone and is much less vivid than that found in nitrogen deficiency. ☆ The stems, although hard and woody, increase in length but not in diameter.	

Micronutrients

Nutrients	Deficiency Symptoms	Toxicity
Iron	☆ Its deficiency signs first appear in **younger leaves.** ☆ Its deficiency will mask magnesium deficiency. ☆ The newer growth will exhibit the interveinal chlorosis typically differentiating it from a magnesium deficiency. ☆ Interveinal chlorosis evolves into an overall chlorosis, and ends as a totally bleached leaf that often develops necrotic spots. ☆ Its deficiency turns leaves pale yellow or white but the veins remain green. ☆ Leaves dry out and turn crinkly. ☆ In severe cases, the new leaves become completely lacking in chlorophyll but with little or no necrotic spots. ☆ The chlorotic mottling on immature leaves may start first near the bases of the leaflets so that in effect the middle of the leaf appears to have a yellow streak.	☆ Its excess accumulation is rare but could cause bronzing or tiny brown spots on leaf surface.

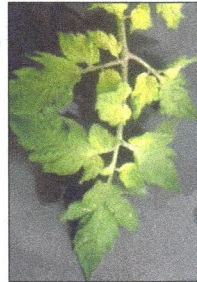

Leaves Showing Iron Deficiency

Nutrients	Deficiency Symptoms	Toxicity
Chlorine	☆ Chlorosis and wilting of the **young leaves**. The chlorosis occurs on smooth flat depressions in the interveinal areas of the leaf blade. ☆ In more advanced cases, there often appears a characteristic bronzing on the upper side of the mature leaves. ☆ Wilted chlorotic leaves become bronze in colour.	☆ Plants are generally tolerant of chloride, but some species such as avocados, stone fruits, and grapevines are sensitive to chlorine and can show toxicity even at low chloride concentration in the soil. ☆ Burning of leaf tip or margins. ☆ Bronzing, yellowing and leaf splitting.

Contd...

Contd...

Nutrients	Deficiency Symptoms	Toxicity

☆ Reduced leaf size and lower growth rate.

Leaves showing chlorine deficiency

☆ Roots become stunted and thickened near tips.

Manganese

☆ Its deficiency signs first appear in **younger leaves.**
☆ Its deficiency exhibits a general chlorosis, followed by yellowing patches and necrotic patches between the veins of the larger leaves.
☆ Shows netted veins of the mature leaves especially when they are viewed through transmitted light.
☆ Turns young leaves mottled yellow or brown, and growth virtually stops.
☆ Interveinal chlorosis on younger or older leaves followed by leaf shedding.
☆ In many plants it is indistinguishable from that of iron.
☆ On fruiting plants, the blossom buds often do not fully develop and turn yellow or abort.
☆ As the deficiency becomes more severe, the new growth becomes completely yellow.
☆ As the stress increases, the leaves take on a gray metallic sheen and develop dark spots and necrotic areas along the veins.
☆ A purplish lustre may also develop on the upper surface of the leaves.

☆ Chlorosis or blotchy leaf tissue due to insufficient chlorophyll synthesis.
☆ Growth rate will slow and vigour will decline.

Leaves Showing Manganese Deficiency

Contd...

Contd...

Nutrients	Deficiency Symptoms	Toxicity

☆ Grains such as wheat and barley are extremely susceptible to manganese deficiency. They develop a light chlorosis along with gray specks which elongate and coalesce, and eventually the entire leaf withers and dies.

Zinc

☆ In some plants, its deficiency first affects **older leaves** and in others it affects **immature leaves.**

☆ Its deficiency results in stunting, yellowing and curling (downward) of small leaves.

☆ Small young leaves with twisted leaflets.

☆ Growth rate of plant slows or gets stopped completely.

☆ Chlorosis may accompany reduction of leaf size and a shortening between internodes.

☆ Leaf margins are often distorted or wrinkled.

☆ Excess zinc interferes with iron causing chlorosis from iron deficiency.

☆ Its abundance can be toxic and may result in the plant death

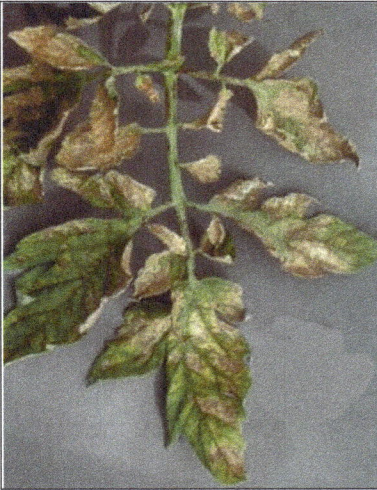

Leaves Showing Zinc Deficiency

☆ **In some plants, the interveinal chlorotic mottling first appears on the older leaves and in others; it appears on the immature leaves.**

☆ It eventually affects the growing points of all plants.

☆ The interveinal chlorotic mottling may be the same as that for iron and manganese except for the development of exceptionally small leaves.

☆ When zinc deficiency onset is sudden such as the zinc left out of the nutrient solution, the chlorosis can appear identical to that of iron and manganese without the little leaf.

Contd...

Contd...

Nutrients	Deficiency Symptoms	Toxicity

☆ As the deficiency progresses, these symptoms develop into an intense interveinal necrosis but the main veins remain green, as in the symptoms of recovering iron deficiency.

☆ In many plants, especially trees, the leaves become very small and the internodes shorten, producing a rosette like appearance.

Leaves Showing Zinc Deficiency

Copper

☆ Its deficiency signs first appear in **younger leaves.**

☆ Copper deficiency may be expressed as overall light chlorosis along with the permanent loss of turgor in the young leaves.

☆ Its deficiency makes new growth wilt and causes irregular growth.

☆ Growing tips die off first along with the crispy leaves.

☆ The new shoots will die from the tips and margins first, often going brown or even white before they die.

☆ Young leaves often become dark green and twisted. They may die off or just exhibit necrotic spots.

☆ Recently matured leaves show netted, green veining with areas bleaching to a whitish gray.

☆ Reduced growth followed by symptoms of iron chlorosis, stunting, reduced branching, abnormal darkening and thickening of roots.

☆ This element is essential but extremely toxic in excess.

☆ Its abundance may cause plant death.

Leaves Showing Copper Deficiency

Contd...

Contd...

Nutrients	Deficiency Symptoms	Toxicity
	☆ Leaves on top half of plant may show unusual puckering with veinal chlorosis. ☆ Absence of a knot on leaf where petiole joins the main stem of plant beginning about 10 or more leaves below growing point. ☆ Trees under chronic copper deficiency develop a rosette form of growth. Leaves are small and chlorotic with spotty necrosis.	
Boron	☆ Its deficiency sign first appears in **young growth.** ☆ Its deficiency may result in limited budding, bud break, distorted shoot growth, short internodes, increased branching, flower buds falling and inhibition of fruit and seeds development. ☆ Growing tips (leaves and petioles) may become light green to yellow. ☆ Rosetting of terminal growth because of shortening of internodes. ☆ Terminal buds may die and new growth may form at lower leaf axils. ☆ Its deficiency may show slight chlorosis to brown to black scorching of new leaf tips and die-back of the growing points similar to calcium deficiency. ☆ Leaves show various symptoms which include drying, thickening, distorting, wilting, and chlorotic or necrotic spotting. ☆ Also the brown and black die-back tissue is very slow to dry so that it can be crumbled easily. ☆ Both the pith and epidermis of stems may be affected as exhibited by hollow stems to roughened and cracked stems. ☆ Root tips often become swollen and discoloured. ☆ Internal tissues may rot and become host to fungal diseases.	☆ Yellowing (chlorotic) of leaf tip followed by necrosis of the leaves beginning at tips or margins and progressing inward. Later leaves may fall and in severe cases may cause plant death. ☆ Some plants are especially sensitive to boron accumulation. **Leaves Showing Boron Toxicity**

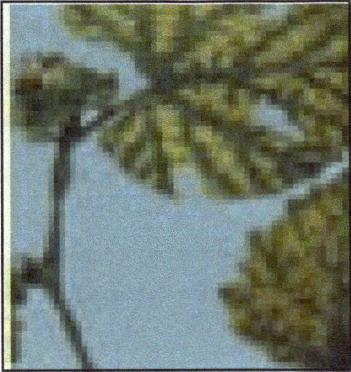

Boron Deficiency Symptoms in Pepper Leaves

Contd...

Contd...

Nutrients	Deficiency Symptoms	Toxicity
Molybdenum	☆ Its deficiency signs first appear in older **leaves.** ☆ Mottling of older leaves with veins remaining light green. ☆ Leaf margins become necrotic and may curl upward. ☆ Exhibit yellowing necrotic leaves from the tips inward, with necrotic lesions present from the tips inward. ☆ A very distinctive feature is that it occurs in the middle of the plant. It also may spread to the rest of the plant if not stopped, eventually killing the plant. ☆ Its deficiency causes leaves to turn pale and fringes to appear scorched. ☆ There may be irregular growth of leaves. ☆ Often interveinal chlorosis which occurs first on older leaves, then progressing to the entire plant. ☆ Younger leaves twist and may die. ☆ Older leaves show interveinal chlorotic blotches, become cupped and thickened. ☆ Chlorosis continues upward to younger leaves as deficiency progresses. ☆ In the case of cauliflower, the lamina of the new leaves fail to develop, resulting in a characteristic whiptail appearance. ☆ In many plants there is an upward cupping of the leaves and mottled spots developing into large interveinal chlorotic areas under severe deficiency.	☆ Used by the plant in very small quantities. ☆ Excess may cause discolouration of leaves depending on plant species. ☆ The leaves turn a very brilliant orange.

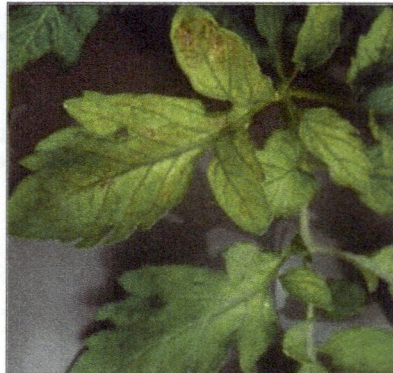

Leaves Showing Molybdenum Deficiency

Note:

☆ Silicon or sodium deficiencies might occasionally develop in some plant species.

☆ It is difficult to induce chlorine and molybdenum deficiencies..

7.3 Effect of Plant Nutrition on Plant Diseases

The resistance of plants to diseases is generally related to genetics. However, the ability of the plant to express its genetic resistance to a particular disease is affected by mineral nutrition. Some nutrients have a greater impact on plant diseases than

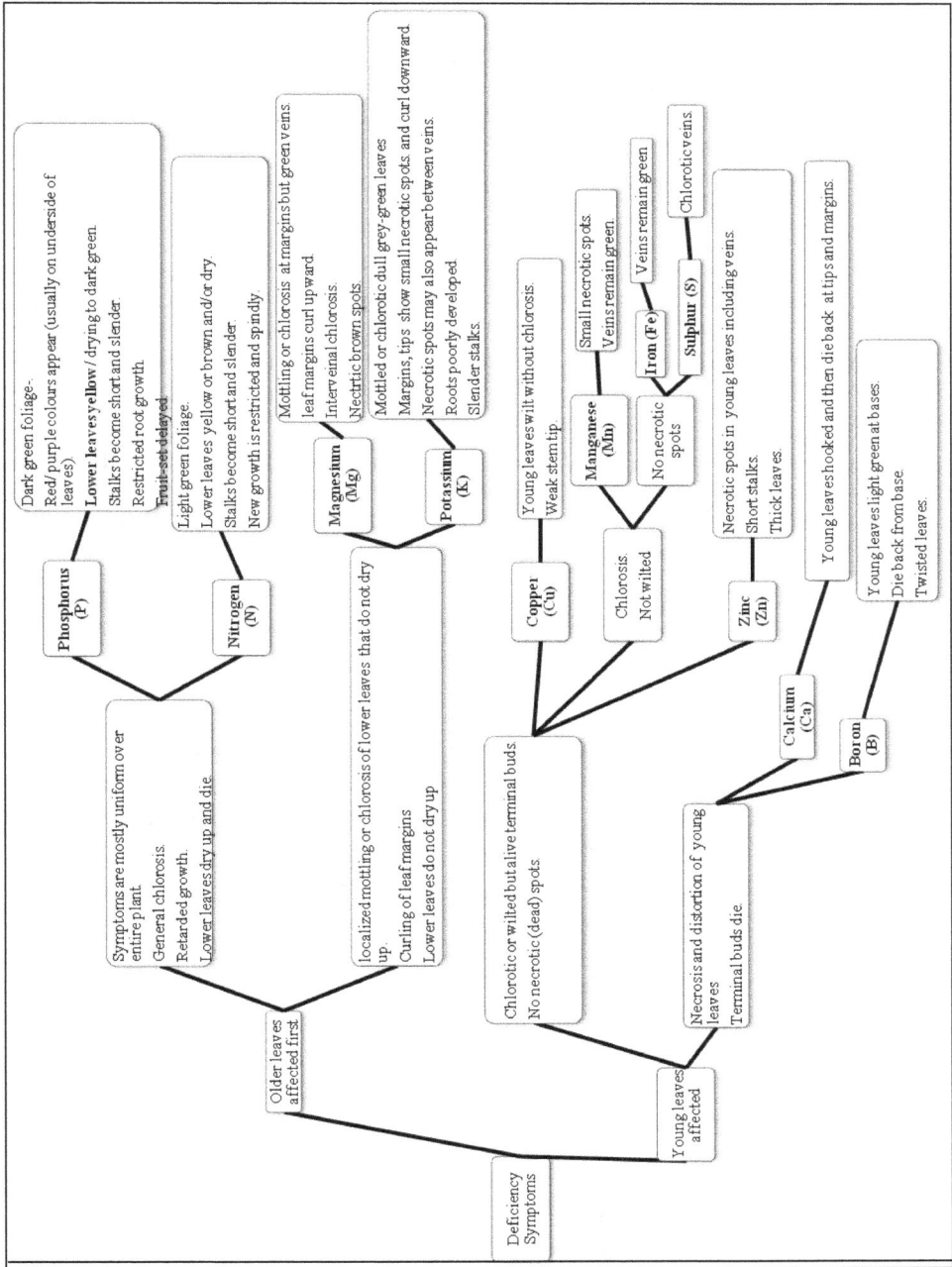

Deficiency Symptoms

Older leaves affected first

- Symptoms are mostly uniform over entire plant
  - General chlorosis.
  - Retarded growth.
  - Lower leaves dry up and die.

Phosphorus (P)
- Dark green foliage-
- Red/purple colours appear (usually on underside of leaves).
- **Lower leaves yellow**/ drying to dark green
- Stalks become short and slender
- Restricted root growth
- **Fruit-set delayed**

Nitrogen (N)
- Light green foliage.
- Lower leaves yellow or brown and/or dry
- Stalks become short and slender.
- New growth is restricted and spindly.

- localized mottling or chlorosis of lower leaves that do not dry up.
  - Curling of leaf margins
  - Lower leaves do not dry up

Magnesium (Mg)
- Mottling or chlorosis at margins but green veins.
- leaf margins curl upward
- Interveinal chlorosis.
- Necrotic brown spots

Potassium (K)
- Mottled or chlorotic dull grey-green leaves
- Margins, tips show small necrotic spots and curl downward
- Necrotic spots may also appear between veins.
- Roots poorly developed
- Slender stalks.

Young leaves affected

- Chlorotic or wilted but alive terminal buds.
  - No necrotic (dead) spots

Copper (Cu)
- Young leaves wilt without chlorosis.
- Weak stem tip

Manganese (Mn)
- Chlorosis.
- Not wilted
- Small necrotic spots
- Veins remain green

Iron (Fe)
- No necrotic spots
- Veins remain green

Sulphur (S)
- Chlorotic veins

Zinc (Zn)
- Necrotic spots in young leaves including veins.
- Short stalks.
- Thick leaves

- Necrosis and distortion of young leaves
  - Terminal buds die

Calcium (Ca)
- Young leaves hooked and then die back at tips and margins.

Boron (B)
- Young leaves light green at bases.
- Die back from base
- Twisted leaves.

others. A particular nutrient may decrease the incidence of one disease, but increase the incidence of others.

Adequate nutrition makes plants more healthy and resistant to diseases by:

☆ Forming a mechanical barrier such as thickness of cell walls. Strong cell walls act as a physical resistance to fungal diseases and prevent infections. Calcium plays a major role in the ability of the plant to develop stronger cell walls and tissues.

☆ Synthesis of natural defence compounds such as antioxidants, phytoalexins and flavonoids.

When some nutrients are below a certain level, the released compounds contain higher amounts of sugars and amino acids that stimulate the establishment of the fungus. Bacteria march into the plant tissue through wounds, through sucking insects and through the stomata. Then they spread within the intercellular spaces. The bacteria release enzymes that dissolve the plant tissue. **Calcium** is known in its ability to inhibit such enzymes.

Another mechanism in which bacteria spread within the plant is in the xylem (the vessels that transport water in the plant). The bacteria forms slime within the vessels and block them. As a result, stems and leaves wilt and die. Certain plant nutrients suppress the ability of bacteria to form this slime.

Viruses are transmitted to plants by sucking insects and fungi. Although silicon is not a plant nutrient but inhibits the feeding ability of some sucking pests like aphids. As a result, virus infection is reduced.

7.4 Foliar Feeding

Foliar feeding is a technique of feeding plants by applying liquid fertilizer through their foliage. It involves spraying water-dissolved fertilizers directly on the leaves. These can consist of a combination of elements or just one element which the plants may be deficient in (*e.g.* calcium).

In hydroponics the balanced nutrient solution fed to the plants through the root zone is usually the only nutrition that is required. Foliar feeding can be used to supplement an existing nutrient solution feed programme in the following situations.

☆ The crop may be showing signs of deficiencies. In this situation a foliar feed containing specific elements to target the deficiency can then be applied. This provides a more immediate (but temporary fix) reversal of the deficiencies while the nutrient solution formula is adjusted.

☆ The crop may be suffering from a root disease problem which may be preventing the full balance of nutrients being taken up. A foliar feed containing specific elements to target the deficiency can then be applied. This provides a more immediate reversal of the deficiencies while the grower works to bring the root disease under control.

☆ The climate may be very dry (rapid transpiration) or very humid (low transpiration) resulting in a poor uptake of the nutrients by roots.

☆ Conditions such as high or low nutrient solution's pH, temperature stress, root disease, presence of pests that affect nutrient uptake, nutrient imbalances in the nutrient solution, *etc*.

☆ During key stages it can improve yield and quality.

7.4.1 Limitations of Foliar Feeding

☆ Nutrients applied in foliar application cannot meet the entire nutrient requirements of the crop.

☆ Applying high concentrations of nutrients by foliar application might result in leaf burn, as water evaporates and salts remain on the leaves.

7.4.2 Improving the Effectiveness of Foliar Feeding

Various factors affect the effectiveness of foliar feeding:

☆ **pH of the foliar spray solution**–Generally, acidic pH improves the penetration of nutrients through leaf surfaces. In addition, pH affects foliar absorption of nutrients in three other ways:

  ☐ pH affects the charge of the cuticle (a waxy layer covering the leaves) and therefore its selectivity to ions occurs.

  ☐ The ionic form of nutrients is pH dependent, and therefore pH can affect the penetration rate.

  ☐ pH might affect the phytotoxicity of the sprayed compounds.

☆ **Use of surfactants (adjuvant)**–Surfactants contributes to a more uniform coverage of the foliage. They increase the retention of the spray solution by reducing the surface tension of the droplets.

Without Surfactant   **With Surfactant**

☆ **Time of the day** - The best time to foliar feed is early morning or late evening, when the stomata are open. Foliar feeding is not recommended when temperature exceeds 80 F (27 C).

☆ **Droplet size**–Smaller droplets cover a larger area and increase efficiency of foliar applications. However, when droplets are too small (less than 100 microns), a drift might occur.

☆ **Spray volume** - Spray volume must be sufficient to fully cover the plant canopy, but not too high (otherwise it will not run off the leaves).

7.5 Recommended Treatments when there is a Nutrient Deficiency or Toxicity

☆ First check the nutrient being used contains an adequate mix of all macro and trace elements. Any well known nutrient manufacturer will have a well formulated nutrient.

☆ A deficiency in any nutrient element will occur for a number of reasons, the most common are

☐ insufficient nutrient strength (*i.e.* not strong enough)

☐ incorrect pH

☐ temperature in the growing area as well as the water can affect nutrient uptake. As a gardener it is important to provide the best growing environment for your plants, any deviation from this can produce deficiency or toxicity symptoms.

To treat for toxic levels of any nutrient element:

☆ Flush with pH adjusted freshwater and then adjust nutrient to correct EC/CF.

☆ Check that nutrient strength (EC/CF) and pH are at the right level. Maintain optimum EC and pH for the grown variety and growth stage of plants.

☆ Check and maintain water temperature.

☆ Check and adjust (if required) glasshouse/room temperature (day and night). It may vary depending on type of crop.

☆ Check watering system. Make sure that there is neither too much nor too little water being fed to plants. If the medium is kept too wet it can inhibit root development and contribute to overall ill health and potential root rot of the plant. The medium in any hydroponic system needs to be kept moist but not saturated, so growers should allow ample time between feeds for the medium to begin to dry out before it is fed again.

⑧

Plant Enemies (Insect Pests and Diseases)

Although hydroponically grown plants are quite safe from soil borne pests and diseases, these cannot be completely eliminated because of some faults that may occur even in a fully controlled environment of hydroponics. Apart from common plant pests, several diseases such as mildew, moulds and viruses may get developed in the hydroponics system. Most diseases are caused due to poor pH level; stale air in the plant environment or insufficient aeration in the vicinity as well as around the roots in the nutrient solution; high humid conditions, stagnant water due to over-watering or land being marshy or improper drainage and improper lighting, *vis-à-vis* algal growth; excess of nutrients causing toxicity or their deficiencies; and infection through diseased plants or contaminated tools, soils and water.

In re-circulated hydroponics systems the repeated use of nutrient solution may result in root diseases and nutrient element insufficiencies. Therefore, the nutrient solution will require reconstitution, filtering, sterilization, and may even require to be replaced within the growing period.

In the fully controlled **aeroponics** environment, there is rarely any disease transmission because of reduced plant-to-plant contact, due to each spray pulse being sterile, and also as any plant observed with any disease symptom in the aeroponics culture is removed immediately to avoid further infection.

8.1 Prevention and Control of Pests and Diseases

The care and maintenance should be routinely performed to prevent pests and disease. On seeing any sign of disease, quick actions should be taken to cure it.

Pests and Diseases can be controlled as explained below:

☆ Using plants that have been bred for good resistance to many diseases:

- ❏ Today, technological advancements in agriculture allow growers to select a cross-bred species that is best suited for a particular region's pathological profile. Breeding practices have been perfected over centuries, but with the advent of genetic manipulation even finer control of a crop's immunity traits is possible.

☆ Use of pathogen-free seeds.

☆ Appropriate plant density.

☆ Adequate nutrition.

☆ Proper and clean growing environment such as:

- ❏ Proper Light.
- ❏ Proper air circulation.
- ❏ Control of moisture and temperature.
- ❏ Proper drainage of water.
- ❏ Only disease free planting material should be used.
- ❏ Removing dead matters and debris from the system.

☆ Proper maintenance of nutrient solution

- ❏ Only clean disease free water should be used. Water samples can be tested for the presence of pathogens by plant pathology laboratories. The water should be treated if pathogens are found, but a failure to find pathogens in any one sample cannot be taken as evidence that the water supply is disease free.
- ❏ Maintaining optimum temperature, pH and EC.
- ❏ Reconstitution, filtering and sterilization of nutrient solution, especially in re-circulated systems.
- ❏ Use of beneficial fungus and bacteria: These can be used to enhance the growth of roots and to keep them clean and healthy. Enzymes also work to break down dead roots.
- ❏ Incorporating Fungicide in nutrient solutions: Treatments of recirculating systems with chemical fungicides is sometimes recommended, more commonly for ornamental crops than food crops.

☆ Crop rotation: Usually in soil gardens.

- ❏ Crop rotation may prevent a parasitic population from becoming well-established. For example, an organism affecting leaves would be starved when the leafy crop is replaced by a tuberous type, *etc.*

☆ Beneficial life forms present a popular and successful way to keep pests away from the hydroponics garden. These life forms combat certain types of fungi and bacteria, crowd out or eat spider mites and other pests, and do it all without making any damage to the plants.

☆ And in severe cases, by the use of pesticide.

❏ The products marketed to fight plant disease include harmful chemicals which will get into the nutrient solution and into the plant cell. Therefore, first mild pesticides should be used.

☆ Integrated:

❏ The use of two or more of these methods in combination offers a higher chance of effectiveness.

The best practice to get rid of plant diseases is prevention. However, if the plants get infected by one or other diseases, always simple and safe control methods should be tried before using pesticides. For example, the following steps can be followed:

a) Physical control methods:

1. Sometimes a strong blast of water from a hose may kill and wash away the pests.

2. Pests like caterpillars can be hand-picked and dropped into soapy water.

3. Potato beetles can be removed by shaking off the plants onto a sheet.

4. Aphids, cabbage worms and white flies are attracted to yellow colour so they can be trapped by covering a piece of yellow poster board with sticky glue and hanging it among the infested plants.

b) Beneficial life forms: These life forms combat certain types of fungi and bacteria, crowd out or eat spider mites and other pests, and do it all without making any damage to the plants. Pests and diseases in hydroponic gardens can be controlled to a great extent by recruiting beneficial bugs that often eat the harmful bugs and helps to reduce or avoid the use of pesticide in the hydroponic garden.

c) Homemade organic pesticides: If physical controls don't work, the next thing to try is a homemade organic pesticide. Some examples of homemade pesticides are:

1. A homemade garlic spray can be applied to the upper and lower surface of infected leaves and stems. This can be repeated every few days until the problem is eradicated. To prepare the garlic spray:

i) Crush approximately 15-20 garlic cloves and a small amount of water in a blender. Squeeze the mixture using a clean cloth and put the liquid into a clean spray bottle.

2. Mix water, baking soda, lemon juice and small amount of dish detergent into a spray bottle and apply or mist onto the infected foliage or stem. The nutrient solution reservoir should be covered properly before applying the spray.

3. Mix 1 tablespoon of baking soda and ½ teaspoon of liquid soap in 1 gallon of water.

4. Put neem leaves and fruits in water. Boil it for fifteen minutes. Remove the debris and pour the neem water into the spray bottle. Apply or mist onto the infected plant.

5. Neem oil is another good choice for pest control, attacking more than four hundred pest types often found in both traditional and hydroponic gardens. Spray oil onto foliage to chase pests away. Insects absorb the natural oil and are unable to reproduce, which results in a steady decrease in population. Commercial pesticides containing neem oil can be used for serious infestations.

d) In some cases where the above control methods fail, the use of pesticides may be necessary. First organic pesticides should be tried.

1. Horticultural oils, insecticidal soaps and botanical insecticides are all organic pesticides. Botanicals are non-toxic soaps and include plant compounds combined into efficient formulas for easy use. Most are safe to use on food crops because they break down quickly and have no long-lasting effect on the environment. However these are not only toxic to the pests and diseases but can also be toxic to insects, birds and soil microorganisms that are part of a healthy garden.

e) Harsh pesticides should only be used if all the above methods are unable to combat the pests.

Note: The pests and diseases to be controlled should be identified carefully and only then anything should be applied. Also, the instructions on product's labels should be strictly followed.

8.2 Mineral Nutrition and Resistance of Plants to Diseases

The resistance of plants to diseases is generally related to genetics. However, the ability of the plant to express its genetic resistance to a particular disease is affected by mineral nutrition (adequate nutrition makes plants more healthy and resistant to diseases) and surrounding environment. Some nutrients have a greater impact on plant diseases than others. A particular nutrient may decrease the incidence of one disease, but increase the incidence of others.

Adequate nutrition makes plants more healthy and resistant to diseases by:

☆ Forming a mechanical barrier such as thickness of cell walls. Strong cell walls act as a physical resistance to fungal diseases and prevent infections. Calcium plays a major role in the ability of the plant to develop stronger cell walls and tissues.

☆ Synthesis of natural defence compounds such as antioxidants, phytoalexins and flavonoids.

When some nutrients are below a certain level, the released compounds contain higher amounts of sugars and amino acids that stimulate the establishment of the fungus. Bacteria enter into the plant tissue through wounds, through sucking insects and through the stomata. Then they spread within the intercellular spaces. The bacteria release enzymes that dissolve the plant tissue. **Calcium** is known in its ability to inhibit such enzymes.

Another mechanism in which bacteria spread within the plant is in the xylem (the vessels that transport water in the plant). The bacteria forms slime within the vessels and block them. As a result, stems and leaves wilt and die. Certain plant nutrients suppress the ability of bacteria to form this slime.

Viruses are transmitted to plants by sucking insects and fungi. Although silicon is not a plant nutrient, it inhibits the feeding ability of some sucking pests like aphids. As a result, virus infection is reduced.

8.3 Plant Enemies

Plant pathogens include fungi, oomycetes, bacteria, viruses, nematodes, protozoa and many harmful insects.

8.3.1 Fungal Diseases

The main causes of fungal diseases are high humidity and poor air circulation. Fungal diseases may be controlled through the use of fungicides and other agriculture practices, however new races of fungi often evolve that are resistant to various fungicides. Fungal problems are best dealt with prevention approaches such as:

☆ By maintaining good sanitation in and around the greenhouse hydroponics.

- The greenhouse floor should be kept clean by removing any debris.
- If possible, greenhouse floor can be covered using a white plastic groundcover. This will accomplish several things:
  - ✦ Reduce pest and pathogens which reside in soil.
  - ✦ Reflect additional light to the plants.
  - ✦ Decrease humidity by preventing soil moisture from evaporating up from the warm greenhouse floor.
- The plant debris should be composted far from the greenhouse air intakes.
- Weeds should not be allowed to grow in or near the greenhouse.
- Any affected plants should be burned.

☆ All growing media should be sterilized before use.

Some common fungal diseases are explained below:

Anthracnose

It is also known as 'black dot root rot' and it appears as small, sunken spots on ripe fruit. Leaves become dark and dry. In some plants such as cucumbers, pale green ulcers appear near the top of the fruit, they then become covered with mould and turn pink, then black and are covered with white mould. As the disease progresses the fruits become yellow and die. Anthracnose usually develops in high humidity and temperature conditions and may develop due to over-watering. To prevent it, the substrate for seedlings must be sterilized and room must be regularly ventilated. It can be controlled by:

☆ Removing or cutting off damaged foliage and fruits.

☆ Spraying with a mild fungicide.

Gray Mould (*Botrytis*)

It is a sooty grayish-white and furry growth on the green fruits (*e.g.* tomato) and stems. It is the most common fungal disease which is usually aggravated in cool and humid conditions with stale air. The infection usually enters the plant through damaged or dead parts. If not detected at early stage, it may prove to be fatal. If it gets detected at an initial stage, it can be controlled to a great extent without using pesticides but by following the common precautions such as:

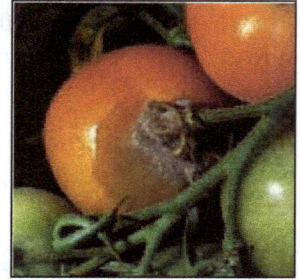

Gray Mould (*Botrytis*)

☆ By cleaning it using a soft-dry cloth (at initial stage) and then washing the plant gently.

☆ By decreasing the humidity and increasing air circulation and temperature.

☆ Removing and destroying the affected plants.

☆ Also, the dead matter and debris should be completely removed from the system.

☆ If required, first mild and natural pesticides should be used and harsh pesticides should be used only if mild pesticides are not effective.

Powdery Mildew (*Erysiphe*)

It first appears as small white or grayish spots or patches on the tops of the leaves, and then progressing to a fine, pale gray-white powder, all over. Growth of infected plants become slow; leaves turn yellow and the plant dies. It thrives in cool temperature, dim light and high humidity. Its treatment includes:

☆ Decrease of humidity

☆ Increase in air circulation and ventilation

☆ Increase in light intensity

☆ Finally using of pesticides

Powdery Mildew

Buckeye Rot

Appears as large, dark brown soft spots on fruits. Also can cause damping-off of seedlings, crown and root rot. Moderate moisture and temperatures favour this disease. Fungicides can help to control buckeye rot.

Damping-off (*Pythium, Rhizoetonia, Phytophthora, etc.*)

It is caused by soil-borne fungi under too-moist planting media, and usually affects the newly planted seedlings. It affects the base of the stem and causes toppling over and death of the plants. Since there is no cure therefore situations causing such conditions should be prevented:

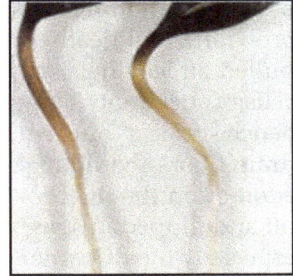

Damping Off

☆ Only sterile, fast draining media should be used for seedlings.

☆ The damaged sections must be removed.

Root Rot

The root rot is common in both indoor and outdoor plants, more serious being in indoor plants with poor drainage. A common belief that hydroponics growing method would exclude soil borne pathogens is not correct. Microorganisms similar to those found in soils, rapidly establish themselves in hydroponic systems. Certain species of water mould genera of *Phytophthora* and *Pythium* from parasitic Oomycetes and *Fusarium* cause many of the root rot problems, destroying the plant from the bottom. Among the species of *Phytophthora*, the most aggressive one is *P. cinnamomi*. *Phytophthora* species are generally host-specific whereas, *Pythium* species can infect a large range of hosts. Most species of *Pythium* are plant parasites, but *Pythium insidiosum* is an important pathogen of animals. The three most commonly encountered species of *Pythium* are *P. aphanidermatum*, *P. irregulare* and *P. ultimum*. *P. aphanidermatum* and *P. irregulare* may occur in 'ebb and flow' systems because they form a swimming spore stage that can move in water. This is likely to occur only if irrigation times are long (30 minutes or longer) or if pots sit in puddles of water because the bench or floor does not drain completely. If *Pythium* infects a cutting bed or if contaminated water is used in propagation, large losses occur. *Pythium ultimum* is primarily associated with soil and sand. Almost all plants are susceptible to *Pythium* root rot. Root tips are very important in taking up nutrients; hence these are attacked and killed first. *Pythium* can also rot the base of cuttings. Usually a plant with root rot does not survive but through vegetative propagation, it can be regenerated. The airborne spores from root rot causing agents and which are also carried by insects and other organisms in the system may contaminate other plants therefore it is suggested to remove and destroy such plants immediately. Root rot that occurs in hydroponics is usually due to poor aeration in nutrient solution resulting in root zone oxygen starvation. Oxygen level available to the roots greatly affects the nutrient uptake of the plant. Root rot is a condition when the roots of the plant rot or decay, mainly due to soggy root environment, overwatering and poor drainage. Excess water makes it very difficult for the roots to get the air that they need resulting in decay. Aeration in nutrient solutions can be accomplished using air pumps, air stones, air diffusers and by adjusting the frequency and length of watering cycles. Overly warm nutrient or waterlogged media greatly affect the oxygen content in nutrient solution. As the water warms up it loses its capacity to hold oxygen. The nutrient temperature should be below 25 °C. If a plant is affected with root rot, these become stunted, pale and die, by mid-day the infected plants may show wilting but recovering in the night, on uprooting root tips exhibit browning and death, outer root tissues also show browning which can easily be pulled off leaving a strand of vascular tissues exposed, and under microscope the cells of such roots show round and thick-walled spores. To prevent this disease all bench surfaces, nutrient solution reservoirs, benches, all equipments and flood and drain floors should regularly be cleaned and disinfected, proper temperature and aeration in the nutrient should be maintained, the damaged roots should be cut off and dipped in some effective fungicide, only clean water free from pathogens, especially the RO water or UV-treated, chlorinated, ozonised or heated and cooled ones should be used in nutrient solution to avoid contamination, no contaminated debris should be allowed to fall in the nutrient solution, nutrient solution should be reconstituted and/or replaced within the growing period and after every crop, filtered and sterilized, and propagation and growing media should also be properly sterilized. Addition of hydrogen peroxide to nutrient solution will destroy single-celled organisms by liberating oxygen. Some biological control agents (micro-organisms) that may prove to be effective are *Candida oleophila* (yeast), *Bacillus subtilis* and *Streptomyces griseoviridis* (bacteria), and *Gliocladium catenulatum*, *Trichoderma harziamum* and *T. virens* (all fungi). *Trichoderma* may enhance biomass production promoting root development.

Pythium Rot

It is same damping off, but it attacks at a later stage of the plant. Appears as a sunken brown spots on fruit (such as tomatoes), outlined with concentric rings. It can be controlled by:

☆ Using low concentrations of surfactants that prove to be effective in killing the mobile zoospores in re-circulating (closed) hydroponic systems thus preventing spread of the disease.

☆ However, surfactants are not effective in curing diseased plants.

Wilts (*Verticullum* and *Fusarium*)

Wilts initially appear as small spots on the leaves of infected plants (*e.g.* tomato, peppers or eggplants). The lower leaves start to curl up, dry out and wilt. Sometimes, portions of the plant may wilt suddenly. *Fusarium* wilt causes an overall yellowing of the plant. Sometimes, it starts with older leaves. Browning of vascular tissue in stem indicates the fungus is clogging those tissues, causing wilting. Roots turn brown, and tap root may rot away. These wilts can be prevented:

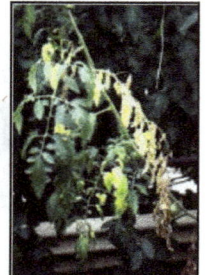

Wilts

☆ By choosing wilt resistant varieties.

☆ By destroying all infected vines.

☆ By using only fresh, clean medium for each planting.

☆ By removing the infected plant from the system.

Early Blight

Irregular dark brownish spots and patches can be seen on lower leaves or stems of infected plants. These spots may enlarge and create a pattern of bull's-eye. Its symptoms on green fruits may not appear at initial stages but as the infection increases, symptoms become visible. Fungicides may not be very effective. It is best treated with harsh chemicals. Its preventive measures are:

☆ Using resistant varieties.

☆ Decreasing humidity and increasing air circulation.

☆ Removal of older leaves.

Leaf Mould

It spreads rapidly, especially in greenhouses. Its symptoms include gray or brown spots on lower side of the leaf and pale areas on upper surface. It can be controlled by:

Leaf Mould

☆ Decreasing humidity and increasing the ventilation.

☆ Fungicides are also effective.

Late Blight

Its effect can be seen as irregular greenish-black water-soaked spots on older leaves, leading to leaf drop. Petioles bend downward. Under humid conditions, a white fuzzy growth may give plants a frosted look. Fruit lesions are brown, firm, and irregular. It is most common when day temperatures are high and night temperatures are low. It can be controlled with fungicides.

Rust

This fungus is often found on the lower surface of leaf as a powdery uneven red patch that may be slightly raised. It thrives at high levels of humidity. It is usually cured by using harsh chemicals in small amounts.

Club Root

This disease transforms previously healthy roots into balls or clumps of club-shaped tubers, resulting in stunted growth. Fungicides can be used for its treatment.

Crown and Stem Rot

Crown and stems of the plants becomes pulpy and subsequently rot.

☆ Remove rotted sections of the plant and spray with fungicide

8.3.2 Bacteria

Bacterial diseases are much common in sub-tropical and tropical regions of the world. Most bacteria that are associated with plants actually do no harm to the plant itself. However, a small number of known species are able to cause diseases. Bacterial and viral diseases spread at a very high speed, especially in a closed hydroponics system, where nutrient solution is re-circulated.

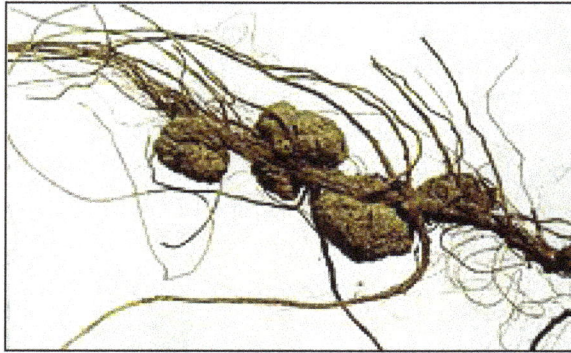

Crown Gall Disease Caused by *Agrobacterium*

Some bacterial diseases that may infect hydroponics plants are:

☆ **Bacterial canker**: It is mainly caused by infected seeds and thrives well in moderate temperatures and high humidity. This disease first appears on older leaves, causing unilateral wilting on one side of the leaf which later dies. Often, the petioles remain on the plant, which helps distinguish bacterial canker from other diseases. Small raised white spots with browning at centres may appear on fruits. Once the plants are infected by bacterial canker, it is difficult to cure. The spread of the disease can be controlled to a great extent by removing the infected plant and sterilizing all equipment and media.

☆ **Bacterial spot**: Initially, small, dark, water-soaked spots appear on leaves which progresses to dried, cracked lesions surrounded by yellow colouration. Spreads rapidly *via* wounds, such as pruning or sucking insects. This disease flourishes in warm, moist temperatures. The spread of this disease can be controlled at initial stage through copper sprays. Also, all the dead matters and debris should be destroyed.

☆ **Bacterial wilt**: This disease is highly contagious and can cause severe damage to the garden as there is no known effective control for it. Symptoms first appear as wilting of older leaves, followed by whole plant. It can be quickly diagnosed by placing a freshly cut infected stem in a glass of water. A white, milky stream of bacterial ooze flowing from the cut confirms the presence of this disease.

Bacterial Wilt

8.3.3 Viruses

There are many types of plant viruses causing great loss to crop yield. It is not economically viable to effectively control them, the exception being when they infect perennial species, such as fruit trees. Plant viruses are mostly transmitted from plant to plant by a vector or through tools. This is often by an insect (for example, aphids), but some fungi, nematodes, and protozoa have been shown to be viral vectors. In many cases, the insects and viruses are specific for virus transmission such as the beet leafhopper that transmits the curly top virus causing disease in several crop plants.

Some precautionary measures that should be taken to prevent bacterial and viral infections in a hydroponics system are:

☆ Water being used for nutrient solution should be clean and free from pathogens.

☆ Although UV and ozone water purifiers can be used in hydroponic systems, they can be very expensive and have adverse effects on the minerals in the solution.

☆ Removal of any weeds that may be susceptible to diseases.

☆ The clothes and shoes of persons entering the greenhouse grow-room should be free from soil.

☆ If possible, shallow trays of bleach solution should be kept at the entrance of the greenhouse so that the workers can clean off their shoes before entering the growing areas.

☆ Hands and tools must be cleaned regularly to prevent spread of disease within a greenhouse.

☆ Tobacco or tobacco products should be strictly kept away from greenhouse because these products may contain various tobacco viruses, such as tobacco mosaic virus.

☆ Smoking or chewing tobacco in the greenhouse must be strictly prohibited.

☆ After handling tobacco or its products, workers must wash their hands properly.

☆ Aphids can transmit viruses, so control of insect-pests will reduce the risk of viral problems.

Some viral diseases that may infect hydroponics plants are:

Tobacco Mosaic Virus (ToMV)

This virus remains alive for years on tobacco leaves and can badly affect many hydroponics crops such as cucumbers, peppers and tomatoes. It is very hard to eradicate this virus. Its symptoms are seen as disfigurement of the leaves and stunted growth. This virus is usually transmitted by sucking insects or hands and tools of workers in the greenhouse. The best practices to prevent this virus are:

☆ Use resistant cultivars.

ToMV

☆ Keep the harmful insects away from the greenhouse.

☆ Tobacco or tobacco products should be strictly kept away from greenhouse because these products may contain tobacco mosaic virus.

☆ Smoking or chewing tobacco in the greenhouse must be strictly prohibited.

Tomato Yellow Leaf Curl Virus (TYLCV)

It is transmitted by the tobacco whitefly, *Bemisia tabaci,* and is a serious problem in the Middle East. It causes interveinal chlorosis and upward curling of leaf edges. Fruit set is greatly reduced. Particularly destructive if young plants are attacked, as very little fruit will be produced. Its preventive method is a complete control of whitefly.

Cucumber Mosaic Virus

Dark green blotches appear on short deformed fruits. The green skin becomes covered with white, yellow and dark-green spots. The symptoms become more evident as the temperature rises inside the greenhouse. To prevent it, the soil or compost must be sterilized.

Other mosaic viruses, such as **common mosaic, aucuba mosaic**, and **cucumber mosaic** can all infect tomatoes and various other crops. Symptoms include mottled areas of the leaves, yellowing, and stunted growth.

8.3.4 Algae

It is a type of plant with no stems or leaves and can be usually seen as greenish vegetative growth in water or on damp surfaces. It covers the top of the plant media, and may infest the nutrient reservoir. Algae growth is caused by stagnant water, excessive moisture, and light. Although algae do no harm to the plants, it will take up the plant nutrients and alter the EC. Some measures that may help in preventing algal growth in hydroponics greenhouses are:

Algae

☆ The nutrient solution reservoir should be opaque. This will prevent algae growth in nutrient solution.

☆ If using fine, porous medium, like coco coir or perlite mixes, top each pot with 2.5–5.0 cm of gravels to provide a dry barrier. This not only holds down the lighter media, it helps prevent algae and fungus gnats.

☆ If algae get developed, it should be scraped off and overly-damp conditions that caused it should be corrected.

☆ System should be sterilized between plantings with 10 per cent bleach solution.

8.3.5 Protozoa

Protozoa are very small organisms which often live inside larger animals or plants. There are a few examples of plant diseases caused by protozoa. They are transmitted as zoospores that are very durable, and may be able to survive in a resting state in the soil for many years. They have also been shown to transmit plant viruses. When zoospores come into contact with a root hair they produce a plasmodium and invade the roots. It can be prevented by maintaining good sanitation in and around the hydroponics system.

8.3.6 Nematodes

Nematodes are microscopic, multi-cellular and worm-like organisms, most of them living freely in the soil. Some nematodes are even beneficial parasites that make no harm to plants or earthworms, instead they can be introduced in the garden to control most of the soil-dwelling insects, such as moths, fleas, loopers, weevils and borers. These enter the body of its prey and destroy the bad bug. Some of the species of nematodes can parasitize plant roots. They are a problem in tropical

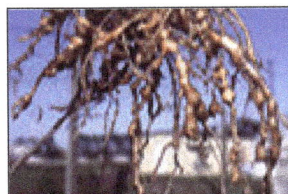

Root-knot Nematode Galls

and subtropical regions of the world, where they may infect crops. Nematodes have also been shown to transmit plant viruses. It can be prevented by maintaining good sanitation in and around the hydroponics system, especially when a soil plant is being transplanted in hydroponics system.

8.3.7 Mites and Insects

Although hydroponics gardens are quite safe from pests (insects), many harmful insects may find their way into the growing space. These insects feed off plant foliage causing leaves to turn pale and then drop. Some insects target and destroy roots (*e.g.* fungus gnat) and fruits. Insects also spread disease from one plant to another *via* sap transfer. Therefore, plants should be regularly checked for signs of pests and diseases. Bright coloured (bright yellow) sticky strips can be used to check insects that are in the garden. By checking the sticky strips on a regular basis (using magnifying glass), the grower knows what insects are present and whether or not the population is growing.

Some harmful insects that are usually seen in the garden are:

Spider Mites

These are tiny pinhead size microscopic bugs forming cobwebs. They can be red, yellow or green in colour. They feed on sap. They feed by piercing the leaf surface and extracting leaf cells and fluid. This manner of feeding causes tiny holes to be punched into the leaf surface resulting in mottling of leaves that eventually turn brown and fall off. Their presence is indicated by plants generally looking sick. Other symptoms include 'white spots on leaves', webbing and 'general yellowing of leaves'. The most common two-spot spider mite that has two dark spots on its shoulders is found throughout the world.

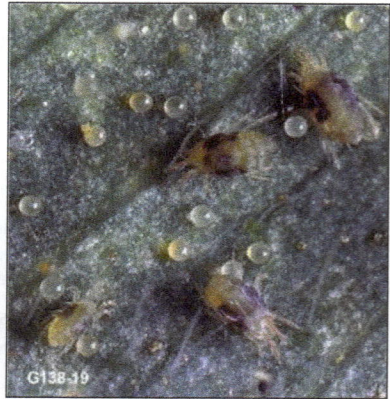

Spider Mites

☆ **Favourable conditions**: They have the ability to go dormant in winter, and then reappearing when it warms up again. They prefer lower humidity level, so raising the humidity helps control them.

☆ **Treatment**: Plants should be isolated; leaves should be washed with warm, soapy water and then spraying of the botanicals. Use insecticide only in severe infestation.

☆ **Natural enemies**: Spider mite predators that are similar in size to mites, eat them. Ladybugs are also seen to be effective in controlling these spider mites.

Aphids

They are slow moving insects and have tiny, light coloured, pear-shaped body and can be green, yellow or of other colour. They suck sap from leaves causing leaves to curl. Leaves become curled, puckered, and discoloured.

Aphids

They leave a sticky residue on the leaves.

☆ **Favourable conditions**: They prefer lower humidity level, so raising the humidity helps control them. They have the ability to go dormant in winter, and then return when it warms up again.

☆ **Treatment**: Remove damaged leaves, wash leaves with warm, soapy water for some days and spray with insecticidal soap.

☆ **Some natural enemies**: Ladybugs love to eat them. Lacewings, *Aphidius colemani* (in some species) and *Aphidoletes aphidimyza* are also effective.

Ladybug Eating Aphids

Fungus Gnat

Not too common in a hydroponics garden. These are similar to tiny fruit fly. It is ~2mm long and usually black with 2 wings and antennae. Their larvae, a small worm that lives in the top inch or two of the soil, feeds mostly on organic debris, fungus, algae, *etc*. While they're doing this, however, they can chew on the roots of young seedlings too. Its symptoms include: 1) small black flies located on top of the medium, 2) starts to show signs of nutrient deficiencies and root rot *i.e.* yellowing of leaves, and brown spots.

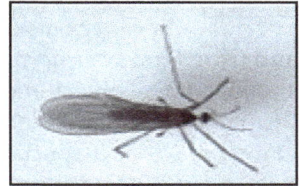

Fungus Gnat

☆ **Treatment:** Use bright yellow sticky strips. Discard any damaged plant and spray with insecticide.

☆ **Some natural enemies** are *Steinernema feltiae* and *Steinernema carpocapsae*.

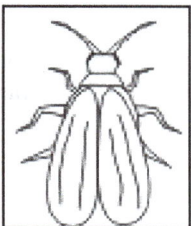

Whitefly

These are tiny white flying bugs and about ~2mm long having 4 wings and looks like a small white moth. They feed off plant sap causing plants to turn yellow and wilt. The whitefly sucks large quantities of sap from the plant and secretes the sugars as honeydew. This makes the leaves sticky and susceptible to fungal growth and rot. In a serious infestation, the fungus and

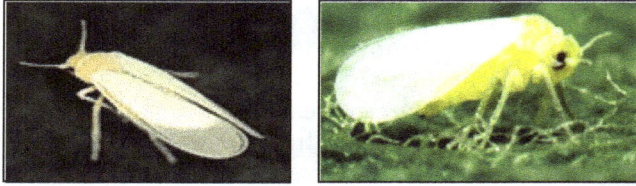

Whiteflies

rot associated with the honeydew can kill an entire crop. They rest on plant leaves and will quickly fly away when disturbed. Shiny, sticky leaves are signs of whitefly presence. Whitefly can pose a great threat to plant health because they are able to transmit many plant viruses.

☆ **Treatment**: Spray with insecticide or something stronger. Also spray all surrounding plants. Yellow sticky traps are also effective.

☆ **Some natural enemies**: Use of whitefly parasites is good for greenhouse whiteflies. Whitefly predators include: Green lacewings and Ladybugs. *Encarcia formosa* lays its eggs in the larvae of the whitefly. Parasitized larvae turn black and are easily recognized. Adult *Encarcia formosa* also feed on honeydew and the body fluids of whitefly larvae.

Thrips

These are very small (1-3mm long) and slender with hairy wings. Greenhouse thrips feed off plant sap leaving dark patches or streaks on the leaves, usually injure inner leaves

Thrips

and fruit. First symptoms are seen when leaves appear finely speckled with yellow spots. Later, a silvery-metallic looking sheen may cover leaf surfaces (not with all thrips, though), and black specks (thrips faecal material) may be scattered about. Only after close inspection is the real pest found.

Flower Thrips feed on buds, flowers, or leaves. The effect of their numerous but shallow punctures is to give the injured tissue a shrunken appearance. The **flower thrips** feed on the thick fleshy petals, pistils, and stamens of the flower, and then the affected parts turn brownish-yellow, blacken, shrivel up, and drop prematurely.

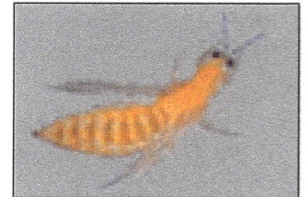

Flower Thrips

☆ **Favourable conditions**: Prefers lower humidity level and high temperature.

☆ **Treatment**: Remove bugs by hand, wash leaves with water and spray insecticide. Dust bulbs and corms to prevent **flower thrip** infestation on the following growing season.

☆ **Some natural enemies**: Predator nematodes can be used to kill immature thrips. Adults are controlled by thrips predatory mites (which work well in greenhouses with higher humidity levels), green lacewings and Pirate bugs. Ladybugs are also seen to be effective.

Scale

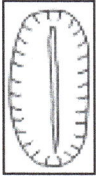

These are tiny, brown, circular to slightly oval and hard-bodied. They look like oyster shells attached to leaves and stems. Affected plant's symptoms include: stunted, yellowed and distorted. Often, the first symptom may appear as shiny honeydew covering the leaves. They suck sap from leaves and stems.

Scale

☆ **Treatment**: Scrape off and isolate plant, if possible. Wash leaves with warm soapy water and apply insecticide if needed.

☆ **Some natural enemies:** Green lacewings feed on the crawler stage of scales. Mealybug destroyers also feed on scales when mealybugs run low. Scale control with natural predators has been variable.

Mealybugs

They are small, soft bodied, coated with a fluffy, waxy coating and are very slow moving insects. These insects usually gather at joints in the plant forming powdery masses and suck on plant juices resulting in distorted leaves, weakened plants, covered with shiny honeydew, and finally a sooty mould grows, killing the plant.

☆ **Treatment**: Scrape off and spray with insecticide. Wash leaves with water or soapy water and apply insecticide if needed.

☆ **Some natural enemies:** Green lacewings.

Some Other Harmful Pests

Insect	Description	Best Controlled with
Earwigs	Small dark brown centipede-like insects with a pincer tail. They usually come out at night.	Pick off by hand.

Contd...

Contd...

Insect	Description	Best Controlled with
Cabbage worm 	Small velvety green caterpillars that eat plant leaves.	Spray with insecticidal soap.
Mexican bean beetle 	Small, 1.25 cm beetles will eat roots and leaves, leaving only the veins.	Wash with soap.
Caterpillars 	Most kinds of caterpillars are considered plant pests and will eat the leaves	Pick off by hand.
European corn borers 	Small, 2.5 cm long caterpillars that bore into plant stems and eat them from inside out.	Spray with insecticide.
Cutworms 	Small worm type insects that, when touched, curl up into a ring shape. They eat through plant roots and topple the plant.	Not too common in hydroponics, if found, spray with insecticide.
Colorado potato beetle 	2.5 cm long striped beetles and their larvae will eat everything on a plant.	Pick off by hand and optionally, spray with insecticide.
Tarnished plant bugs 	Small beetles about 6 mm long with tarnish-like markings on its back. They inject plants with a substance that deforms leaf tips and stem joining.	Clean off all nearby organic debris and spray with insecticide.

Contd...

Contd...

Insect	Description	Best Controlled with
Striped cucumber beetle	Small, very destructive, striped back beetles 6 mm long. As adults, these plant pests eat leaves, and as larvae eat roots.	Spray with insecticide.
Root maggots	Fly larvae that hatch at the base of the plants and eat the roots.	Not too common in hydroponics gardens. If found, spray with insecticide.
Leaf hoppers	Small, 3mm long, wedge shaped insects that suck the sap through the plant leaves.	Pick off by hand and apply insecticide if needed.

8.4 Insect-pest Control in Hydroponics

The best way to get rid of unwanted insects is their detection and identification. Once the insects in the garden are identified correctly, they can be controlled easily using one or more of the following options:

1. Bright Sticky Trap Strips

Sticky strips provide a safe method of trapping insects. The insects are attracted to the bright colour of the sticky strip and, once they land, they are stuck. When the strips are full, simply discard and replace with new ones. These sticky traps can be made at home as:

a) Spray paint a cardboard with bright yellow colour (for insects such as aphids, fungus gnats and whiteflies) or with bright blue colour (for insects such as thrips).

Bright Sticky Trap Strips

b) Let it dry.

c) Coat with Vaseline (petroleum jelly).

2. Physical Control Methods

a) Sometimes a strong blast of water from a hose may kill and wash away the pests.

b) Pests like caterpillars can be hand-picked and dropped into soapy water.

c) Potato beetles can be removed by shaking off the plants onto a sheet.

d) Aphids, cabbage worms and white flies are attracted to yellow colour so they can be trapped by covering a piece of yellow poster board with sticky glue and hanging it among the infested plants.

3. Predators and Parasites

A predator is an animal that kills and eats other animals. A parasite is a small animal or plant that lives on or inside a larger animal or plant and gets its food from it. Almost all insects have predators and/or parasites. These enemies of harmful pests can be purchased from the predators and parasites garden supply centres or insectaries (facilities that raise insects) that breed and sell beneficial insects. These beneficial insects can be introduced in the gardens to eradicate harmful pests. These beneficial insects are shipped as eggs, larvae or adults. Some such beneficial insects are:

a) **Ladybugs beetles**: They are small spotted insects that love to eat aphids and effective in controlling many other insect-pests (*e.g.* thrips, mites, scale insects and small caterpillars) in the larval stage as well as the adult stage.

Ladybugs Beetles

b) **Lacewings**: They are known for their voracious appetite for various insects [*e.g.* aphids, thrips, whiteflies, mealybugs, and scales (only if mealybugs run low)] in the larval stage as well as the adult stage.

Lacewings

c) **Praying mantis:** Effective in controlling various insects in the larval stage as well as the adult stage. It produces egg cases which contain approximately 200 babies. The babies are also deadly predators.

d) *Trichogramma* **wasp:** This wasp feeds on more than 200 types of insects (such as tomato hornworms, cabbageworms, fruitworms, corn earworms, codling moths, cutworms, armyworms, cabbage loopers, corn borers and cane borers) and does no harm to other beneficial insects or people.

e) **Predatory polistes wasp:** Controls bollworms or other caterpillars on a cotton plant.

f) **Larvae of many hoverfly species:** Principally feed upon greenfly and also eat fruit tree spider mites and small caterpillars. Adults feed on nectar and pollen, which they require for egg production.

g) *Encarsia formosa:* This tiny parasitic wasp is effective in some species of the whitefly (*e.g.* tobacco whitefly) and lays its eggs in their larvae. Parasitized larvae turn black and are easily recognized. Adult *Encarcia formosa* also feed on honeydew and the body fluids of whitefly larvae.

h) *Eretomocerus californicus:* The tobacco whitefly can be parasitized by *Encarsia,* but they are controlled better by *Eretomocerus* species.

i) *Ichneumonid* **wasps:** These are 5–10 mm parasitic wasps and prey mainly on caterpillars of butterflies and moths.

j) **Braconid wasps:** These are tiny parasitic wasps (up to 5 mm), attack caterpillars and a wide range of other insects including greenfly.

k) **Chalcid wasps:** These are among the smallest of insects (<3 mm). Parasitize eggs/larvae of greenfly, whitefly, cabbage caterpillars, scales and strawberry tortrix moth (*Acleris comariana*).

l) **Tachinid flies:** Parasitize a wide range of insects including caterpillars, adult and larval beetles, true bugs, and others.

m) *Aphidius colemani*: This parasitic wasp is effective against some species of aphids. Parasitized aphids form characteristic white mummies.

n) *Aphidoletes aphidimyza*: Effective on a wide range of aphid species and lays its eggs in aphid colonies. The orange larvae hatch from these eggs and feed voraciously on aphids.

o) **Predatory mite** (*Phytoseiulus persimilis*): Used for control of the two-spotted spider mite.

p) *Verticillium lecanii*: A common soil borne fungus which affects various kinds of insects. It is widespread in temperate and tropical areas, but does not harm birds, fish, mammals or plants. The fungal spore germinates and begins to grow on the honeydew secretion on the whitefly body and the whitefly dies from infection before the fungus even becomes visible. Since the fungus is not mobile and cannot seek its host, it is only effective in very high densities of whitefly and repeated applications are necessary.

q) **Parasitic nematode** (microscopic worms): They destroy harmful insects such as grubs, fleas, gnats, flies, cutworms, billbugs, ants, and Japanese beetles. They control harmful soil-dwelling insects by feeding on their larvae and do no harm to other beneficial insects.

   i) *Steinernema feltiae*: Seen to be effective in reducing fungus gnats.

   ii) *Steinernema carpocapsae*: Also used to control fungus gnats.

   iii) *Phasmarhabditis hermaphrodita*: It is a microscopic nematode that kills slugs, thereafter feeding and reproducing inside.

r) **Black and yellow garden spider**: They feed on juice of various harmful as well as beneficial insects.

4. Micro-organisms

Pathogenic micro-organisms such as bacteria, fungi, and viruses can be used to kill or weaken their host. These are relatively host-specific and can be introduced in the gardens. They occur naturally only when insect populations become denser.

a) Bacteria

Bacteria used for biological control infect insects *via* their digestive tracts, so insects with sucking mouth parts like aphids and scale insects are difficult to control with bacterial biological control. *Bacillus thuringiensis* is the most widely applied species of bacteria used for biological control. The bacteria are available in sachets of dried spores which are mixed with water and sprayed onto vulnerable plants such as brassicas and fruit trees. *Bacillus thuringiensis* has also been incorporated into crops, making them resistant to these pests and thus reducing the use of pesticides.

b) Fungi

Fungi that cause disease in insects are known as entomopathogenic fungi. An example is *Beauveria bassiana*, which is used to manage a wide variety of insect pests such as whiteflies, thrips, aphids and weevils. A remarkable additional feature of some fungi is their effect on plant fitness. *Trichoderma* species may enhance biomass production promoting root development, dissolving insoluble phosphate containing minerals.

5. Homemade Organic Pesticides

Homemade organic pesticides explained earlier can be used.

6. Mild Organic Pesticides

Horticultural oils, insecticidal soaps and botanical insecticides are all organic pesticides. Botanicals are non-toxic soaps and include plant compounds combined into efficient formulas for easy use. Most are safe to use on food crops because they break down quickly and have no long-lasting effect on the environment. However, these are not only toxic to the pests and diseases but can also be toxic to insects, birds and soil microorganisms that are part of a healthy garden.

7. Harsh Pesticides

Its should only be used if all the above methods are ineffective.

8.5 Beneficial Microbes or Beneficial Bio-inoculants

Beneficial fungi and bacteria can be added in the nutrient solution to enhance the growth of roots and keep them clean and healthy. Enzymes also work to break down dead roots. Using a combination of these three additives, results in a robust, expanded root zone that absorbs nutrients faster and more efficiently. For beneficial bacteria to survive in a hydroponic environment they will need ideal environmental conditions. Hydroponic nutrients lack organic carbon sources for beneficial bacteria to survive. Bacteria must obtain nutrient materials necessary for their metabolic processes and cell reproduction from their environment. Thus adequate food and oxygen must be present in solution and media for microbes to thrive in their environment. They can metabolize humic and fulvic extracts but one of the best sources of food for beneficial bacteria is molasses.

Inorganic substrates are more effectively colonized by bacteria, while organic substrates are more effectively colonized by fungi. For example, cocofibre contains a greater amount of *Trichoderma harzianum*, while the rockwool system contains the highest amount of fluorescent pseudomonads bacteria. Materials that are high in lignocellulose are the organic media (straw, woodbark, and coconut fibre). This makes cocofibre an ideal environment for *Trichoderma* colonization.

The use of beneficial microbes not only controls or eradicates pathogens, but also enhances yield through hormone stimulation, enzyme production and other mechanisms. The same cannot be said for sterilization methods such as UV, ozone, monochloramine, chlorine and hydrogen peroxide. Two main organic additives that act as microbial nutrient/stimulators and plant fertilizers are kelp and fish products. It is recommended that in water based systems (*e.g.* NFT and aeroponics), too much organic matter or additives should not be used, otherwise the proliferation of unwanted microbial life may potentially rob oxygen from the root zone.

Some beneficial fungi and bacteria are explained below:

Example	Description
Beneficial Fungi	
Trichoderma species ☆ *T. harzianum* ☆ *T. viride* ☆ *T. virens* ☆ *T. koningii* ☆ *T. hamatum* ☆ Other species	☆ These species are biological control agents against a wide range of pathogenic fungi such as *Rhizoctonia* spp., *Pythium* spp., *Botrytis cinerea*, and *Fusarium* spp. ☆ These are also effective against *Phytophthora palmivora*, *P. parasitica* and different other species. ☆ *Trichoderma harzianum* is reported to be most widely used as an effective bio-inoculant. *T. harzianum* controls or eliminates all manner of pathogens in both inorganic and organic media. ☆ *Trichoderma harzianum* is noted to increase the uptake and concentration of a variety of nutrients (copper, phosphorus, iron, manganese and sodium) in hydroponic culture. ☆ Cellulases (enzymes) produced by *Trichoderma* spp. is the most efficient enzyme system for the complete hydrolysis of cellulosic matter (*e.g.* decaying root matter) into glucose. *Trichoderma viride*, *T. reesei*, *T. harzianum* and *T. asperellum* produce high levels of cellulase enzymes. ☆ *Trichoderma* spp. possesses an inherent resistance to most agricultural chemicals, including fungicides, although strains differ in their resistance. ☆ The optimum temperature for rapid colonization and bioactivity of most of the commonly applied *Trichoderma* spp. is 25-30 ºC. Their colonization will be slow if temperature is too cold and or too warm and then they will gradually disappear. ☆ Optimum pH for *Trichoderma* fungi may vary between species, however, fungi thrive in semi-acidic conditions at pH 5.5–5.8.
Beneficial Bacteria	
☆ Bacillus spp. ☐ *Bacillus subtilis* ☆ *Pseudomonas* spp.	☆ These bacteria are applied to a wide range of agricultural crops to suppress pathogens. ☐ *Pseudomonas* spp. and *Bacillus* spp. have been noted to control over *Fusarium oxysporum* in hydroponic settings. ☆ They produce hydrolytic enzymes and antibiotics. *Bacillus subtilis* can produce great number of antibiotics, for example: lipopeptide antibiotics which exhibit wide spectrum antifungal activity. ☆ pH neutral environment is best for bacteria species. Optimum pH for *Bacillus subtillis* is 6.5–7.0. ☆ Optimum temperature for *Bacillus subtillis* is between 40–47ºC.

Contd...

Contd...

Example	Description
	☆ It will be better to always maintain optimum temperatures and pH for optimal plant growth in hydroponics systems.
	❑ pH: 5.5–6.0.
	❑ Nutrient/media temperature: 20–22 ºC.
	☆ Food for Bacteria
	❑ Bacteria thrive well in a high carbon environment.
	❑ Molasses has been shown to be a cost efficient source of providing this carbon.
	❑ Other sources of food are humates (fulvic and humic acids), kelp, and fish emulsion.

⑨
What Can be Grown Hydroponically?

A large variety of vegetables, fruits, herbs, and ornamental plants can be grown hydroponically. The main advantage of growing crops hydroponically is that they can be grown outside of their normal season. Since different crops have their own particular requirements (space, temperature, light, nutrients, *etc.*), therefore, special care is to be taken while selecting different species that are to be grown in the same hydroponics unit. Plants with different needs cannot be mixed in the same system. For example, roses need a larger amount of potassium than most other plants or flowers. Leafy crops generally need higher N, root crops need higher K, and fruit crops such as tomatoes or cucumbers should maintain relatively low N levels.

Plants with similar nutritional needs can be placed on one system. Given below is a brief idea for selecting the types of hydroponics systems for various crops.

Green Leafy Vegetables

Almost all leafy vegetables such as lettuce, chard, spinach and cabbage grow well in nearly all types of hydroponics systems. Short plants like lettuce grows well when floated on a styrofoam raft, in which holes are cut to allow the roots to grow, and a pump brings air to the roots. Leafy crops generally need higher nitrogen. Basil, chard, lettuce and spinach have similar needs and can be put on the same system. Spinach and lettuce can be harvested in small quantities or by taking the whole plant. Basil plants must be replaced after three to four months. There should be a steady supply of lettuce and spinach by seeding every few days.

Vine Vegetable Plants

Vine plants such as tomatoes, cucumbers, peppers and peas grow well in hydroponics systems with an ample support such as trellis, especially in case of tomatoes and cucumbers. Cucumbers, tomatoes and peppers have similar requirements, including light exposure. Pole type of beans, similar to peas, will also need a trellis support. In absence of sufficient support structure, these plants should be given enough room to grow. These plants can be wrapped around a vertically-tied string as they grow. Large crops with a long growing season, like tomatoes and cucumbers, need sufficient room for growth of their roots. Fruit crops such as tomatoes or cucumbers should maintain relatively low nitrogen levels.

Bushy Vegetables

Bush beans and peas also require some sort of support structure such as trellis and can be wrapped around to a vertically tied rope or string. Beans and peas can be grown on the same system and both have medium to high requirements for exposure to light. Broccoli, cabbage and cauliflower prefer cool temperatures and can be grown in the same system. However, cabbage has slightly lower light requirements than broccoli and cauliflower.

Vegetables that Grow Beneath the Soil

Vegetables such as beets, carrots, leeks, onions, parsnips, potatoes, radishes and yams will also grow well hydroponically, but may require extra care. Root crops such as beets, carrots and radishes require a large deep container for their roots to have enough room to grow. For larger crops such as potatoes, channels must be 20-25 cm deep, while for smaller crops such as carrots the depth of channels should be 7.5-10.0 cm. Root crops generally need higher potassium. However, succulent plants thrive well in dry conditions therefore these are not good choices for hydroponic gardening.

Fruit Yielding Plants

Small and light fruiting plants thrive well in a hydroponics system. Water-loving fruits such as watermelon, cantaloupe, tomatoes, strawberries, blueberries, blackberries, raspberries and grapes are the best choices to be grown hydroponically.

Fruit Trees

A small fruit trees such as dwarf banana, papaya or citrus, and lemons may be grown well in a hydroponic greenhouse. A large and heavy fruit tree can be started in a hydroponics system and when it outgrows its container, it can be transplanted to the soil. To grow a large tree hydroponically, it requires a very well built-up support system.

Herbs

Numerous herbs such as arugula, basil, chervil, chives, coriander, dill, lemon balm, mache, mint, oregano, rosemary, watercress, parsley, sage, sorrel, sweet marjoram, tarragon and thyme can be grown very well in hydroponics systems, even when planted quite densely. Herbs mature rapidly, require little care, have a wide variety of culinary and health uses, and do very well when grown hydroponically. Even a small hydroponics system can produce an impressive crop of herbs. The herbs that grow in wet conditions are found to be more successful in hydroponics systems as compared to those that thrive in drier environments, like sage.

Ornamental Plants

Seedlings for a variety of flowers can be developed in a hydroponics unit and when the plants mature they can be transplanted into other hydroponics unit or can even be planted in the soil. The ornamentals may be perennials but mostly the flowering is seasonal, so using hydroponics the flowers can be produced year-round in large numbers. The flowers such as *Alstroemeria, Amaryllis, Anthurium, Antirrhinum, Begonia*, Bromeliads, *Callistephus, Cymbidium* and many other orchids, *Dahlia, Dendranthema, Dianthus caryophyllus, Eustoma, Freesia, Gerbera, Gladiolus, Gypsophila, Heliconia, Hyacinthus, Iris* (bulbous), *Lilium, Limonium, Narcissus, Polianthes, Rosa, Strelitzia, Tulipa, Watsonia, Zantedeschia* and many others do very well when grown hydroponically. Most of the non-flowering ornamental plants such as *Aphelandra, Caladium, Dieffenbachia, Dracaena, Impatiens, Monstera, Philodendron,* palm trees and others are grown well in a soilless culture.

There is no end on the number of crops that can be grown hydroponically. Even crops such as corn, cacao, sugarcane, rice, tea, tobacco and cereal grains can be successfully grown hydroponically. In most cases these crops are started hydroponically and when the seedlings reach their desired size they are transplanted to the fields.

Given below is the list of some plants with their nutrient solution requirements:

Crops	pH	CF	EC	ppm
Vegetables				
Artichoke	6.5–7.5	8–18	0.8–1.8	560–1260
Asparagus	6.0–6.8	14–18	1.4–1.8	980–1260
Bean (common)	6.0	20–40	2–4	1400–2800
Beetroot	6.0–6.5	8–50	0.8–5	1260–3500
Broad bean	6.0–6.5	18–22	1.8–2.2	1260–1540
Broccoli	6.0–6.8	28–35	2.8–3.5	1960–2450
Brussells sprouts	6.5–7	25–30	2.5–3.0	1750–2100
Cabbage	6.5–7.0	25–30	2.5–3.0	1750–2100
Capsicum	6.0.6.5	18–22	1.8–2.2	1260–1540
Carrots	6.3	16–20	1.6–2.0	1120–1400
Cauliflower	6.0–7.0	5–20	0.5–2.0	1050–1400
Celery	6.5	18–24	1.8–2.4	1260–1680
Cucumber	5.5–6.0	17–25	1.7–2.5	1190–1750
Eggplant	5.5–6.5	25–35	2.5–3.5	1750–2450
Endive	5.5	20–24	2.0–2.4	1400–1680

Contd...

Contd...

Crops	pH	CF	EC	ppm
Garlic	6.0	14–18	1.4–1.8	980–1260
Leek	6.5–7.0	14–18	1.4–1.8	980–1260
Lettuce	5.5–6.5	8–12	0.8–1.2	560–840
Marrow	6.0	18–24	1.8–2.4	1260–1680
Okra	6.5	20–24	2.0–2.4	1400–1680
Onions	6.0–6.7	14–18	1.4–1.8	980–1260
Pak choi	7.0	15–20	1.5–2.0	1050–1400
Parsnip	6.0	14–18	1.4–1.8	980–1260
Peas	6.0–7.0	8–18	0.8–1.8	980–1260
Pepino	6.0–6.5	20–50	2.0–5.0	1400–3500
Potato	5.0–6.0	20–25	2.0–2.5	1400–1750
Pumpkin	5.5–7.5	18–24	1.8–2.4	1260–1680
Radish	6.0–7.0	16–22	1.6–2.2	840–1540
Spinach	5.5–6.6	18–23	1.8–2.3	1260–1610
Silverbeet	6.0–7.0	18–23	1.8–2.3	1260–1610
Sweet corn	6.0	16–24	1.6–2.4	840–1680
Sweet potato	5.5–6.0	20–25	2.0–2.5	1400–1750
Taro	5.0–5.5	25–30	2.5–3.0	1750–2100
Tomato	5.0–6.5	20–50	2.0–5.0	1400–3500
Turnips	6.0–6.5	18–24	1.8–2.4	1260–1680
Zucchini	6.0	18–24	1.8–2.4	1260–1680
Herbs				
Basil	5.5–6.5	10–16	1.0–1.6	700–1120
Chicory	5.5–6.0	20–24	2.0–2.4	1400–1600
Chives	6.0–6.5	18–22	1.8–2.2	1260–1540
Fennel	6.4–6.8	10–14	1.0–1.4	700–980
Lavender	6.4–6.8	10–14	1.0–1.4	700–980
Lemon balm	5.5–6.5	10–16	1.0–1.6	700–1120
Marjoram	6.0	16–20	1.6–2.0	1120–1400
Mint	5.5–6.0	20–24	2.0–2.4	1400–1680
Mustard cress	6.0–6.5	12–24	1.2–2.4	840–1680
Parsley	5.5–6.0	8–18	0.8–1.8	560–1260
Rosemary	5.5–6.0	10–16	1.0–1.6	700–1120
Sage	5.5–6.5	10–16	1.0–1.6	700–1120
Thyme	5.5–7.0	8–16	0.8–1.6	560–1120
Watercress	6.5–6.8	4–18	0.4–1.8	280–1260

Contd...

Contd...

Crops	pH	CF	EC	ppm
Flowers				
African Violet	6.0–7.0	12–15	1.2–1.5	840–1050
Anthuurium	5.0–6.0	16–20	1.6–2.0	1120–1400
Antirrhinim	6.5	16–20	1.6–2.0	1120–1400
Aphelandra	5.0–6.0	18–24	1.8–2.4	1260–1680
Aster	6.0–6.5	18–24	1.8–2.4	1260–1680
Begonia	6.5	14–18	1.4–1.8	980–1260
Bromeliads	5.0–7.5	8–12	0.8–1.2	560–840
Caladium	6.0–7.5	16–20	1.6–2.0	1120–1400
Canna	6.0	18–24	1.8–2.4	1260–1680
Carnation	6.0	20–35	2.0–3.5	1260–2450
Chrysanthemum	6.0–6.2	18–25	1.8–2.5	1400–1750
Cymbidium	5.5	6–10	0.6–2.5	420–560
Dahlia	6.0–7.0	15–20	1.5–2.0	1050–1400
Dieffenbachia	5.0–6.0	18–24	1.8–2.4	1400–1680
Dracaena	5.0–6.0	18–24	1.8–2.4	1400–1680
Fern	6.0	16–20	1.6–2.0	1120–1400
Ficus	5.5–6.0	16–24	1.6–2.4	1120–1680
Freesia	6.5	10–20	1.0–2.0	700–1400
Impatiens	5.5–6.5	18–20	1.8–2.0	1260–1400
Gerbera	5.0–6.5	20–25	2.0–2.5	1400–1750
Gladiolus	5.5–6.5	20–24	2.0–2.4	1400–1680
Monstera	5.0–6.0	18–24	1.8–2.4	1400–1680
Palm	6.0–7.5	16–20	1.6–2.0	1120–1400
Rose	5.5–7.0	15–25	1.5–2.5	1050–1750
Stock	6.0–7.0	16–20	1.6–2.0	1120–1400
Fruits				
Banana	5.5–6.5	18–22	1.8–2.2	1260–1540
Black Currant	6.0	14–18	1.4–1.8	980–1260
Blueberry	4.0–5.0	18–20	1.8–2.0	1260–1400
Melon	5.5–6.0	20–25	2.0–2.5	1400–1750
Passion fruit	6.5	16–24	1.6–2.4	840–1680
Paw–Paw	6.5	20–24	2.0–2.4	1400–1680
Pineapple	5.5–6.0	20–24	2.0–2.4	1400–1680
Red Currant	6.0	14–18	1.4–1.8	980–1260
Rhubarb	5.0–6.0	16–20	1.6–2.0	840–1400
Strawberries	5.5–6.5	18–22	1.8–2.2	1260–1540
Watermelon	5.8	15–24	1.5-2.4	1260-1680

Experimental Data

Lettuce	Tomatoes	Cucumbers	Strawberries
☆ NPK-8-15-36 hydroponic fertilizers	☆ NPK-4-18-38 hydroponic fertilizers	☆ NPK-8-16-36 hydroponic fertilizers	☆ NPK-10-8-22 hydroponic fertilizer
☆ pH for seedlings: 6.3-6.5	☆ pH: 6.2	☆ pH: 6.4 - 6.7	☆ pH: 5.6-6.5
☆ pH for mature plants: 5.8-6.0	☆ ppm: 1400-3500	☆ ppm: 1190-1750	☆ ppm: 1260-1540
☆ ppm: **560-840**	☆ Calcium and magnesium sulphate added to the main solution	☆ Calcium and magnesium sulphate addes separately	☆ Under direct sunlight
Note: Some varieties of lettuce are nitrogen sensitive and can get leaf tip burn.	☆ Tomatoes need a lot of light and we kept this system in out balcony where the plants can receive lots of sunlight.	☆ Do not like extreme heat	
☆ Calcium and magnesium sulphate added separately to the hydroponic solution	Note: The flower clusters/truss need to be shaken in absence of bees or the wind		
☆ Lettuce does not need a lot of strong light and we grow them very successfully under fluorescent grow lights (T5 or T8 bulbs)			
Result: High yield of good quality	**Result:** Moderate yield	**Result:** Good yield.	**Result:** Low yield

These specified values for pH and electrical conductivity are given as a broad range and may vary depending on season, regional climatic conditions and various stages of growth (growing, flowering and fruiting stages). Nutrient solutions (ppm) and other factors (pH, temperature, *etc.*) need to be adjusted during the growing cycle of the crop as different crops may have different requirements. As a general rule, the plant requires higher nutrients in cooler 'grow rooms' and needs lower levels of nutrients in the warmer rooms.

Plants tolerate a range of nutrients. It is the grower's responsibility to read the plant and if necessary modify the nutrient to suit the environment. The grower must regularly check the plants for nutrient deficiencies and toxicities symptom. A close observation should be made regularly for any symptoms of pests and diseases. Water quality, temperature, humidity, light intensity, EC and pH are some other factors that should be regularly monitored and managed. Hence, any nutrient solution formula used for any crop in one system may not necessarily be suitable for the same crop in another system and may require some modifications for that system.

9.1 Sample Data for some Plants

Different varieties of plants may require different nutrients, temperatures and pH for better growth. We (Sanjay Misra and Sanyat Misra) have successfully grown lettuce, peppers, tomatoes, cucumbers (self pollinating varieties) and strawberries hydroponically. The hydroponic fertilizers are added into the water in the reservoir as instructed on the bottles. The Table on previous page shows the NPK and pH that we have maintained in our system in Allahabad.

⑩
Hydroponics
Step by Step

10.1 Selection of Plants

Many types of vegetables and ornamental plants can be successfully grown in hydroponic systems. The first step for a hydroponic system is selection of the plants that are to be grown hydroponically. The selection of plants depends on factors such as space needed (larger plants take more space so they should not be crowded), profitability, potential market, and whether the hydroponic system is to be implemented indoor or outdoor.

Note: Almost anything can be grown indoor, but this is not the case for growing outdoors as there are heat and cold weathers to deal with. Some Hydroponic plants also require high light levels and some require lower light levels.

After selecting the plants to be grown hydroponically, their requirements such as nutrient strengths, pH, nutrient solution temperature, light levels, deficiency

symptoms, diseases and pathogens that may attack these plants must be recorded. Given below is the list of some of the plants that are well suited for hydroponic gardening.

10.1.1 Vegetables

Artichoke, asparagus, beans, beetroot, broccoli, brussels sprout, cabbage, capsicum, carrot, cauliflower, celery, corn, cucumber, eggplants, garlic, leeks, lettuce, melons, onions, parsnips, peas, potato, pumpkin, radishes, rhubarb, spinach, squash, sweet potato, tomato, yams and many others.

10.1.2 Fruits

Banana, blueberries, raspberries, red currants, strawberries, and many others.

10.1.3 Flowers

Amaryllis, asters, begonias, chrysanthemums, dahlias, freesia, gerbera, gladiolus, iris, orchids, roses, and others.

10.1.4 Non Flowering Ornamentals

Aphelandra, bromeliads, caladiums, dieffenbachia, dracaena, impatiens, monstera, philodendron, palm trees and others.

10.1.5 Herbs

Basils, lemon balm, mints, parsley, rosemary, sage, thymes, and others.

10.2 Steps

10.2.1 Starting Seedlings

To obtain high quality and economic yields, production of vigorous seedlings or planting material of high yielding varieties is an essential step of soil-less culture. Although the plants for hydroponic system can be bought from the garden centre or the soil plants can be transplanted to a hydroponic system, it is very simple to start healthy seedlings at home. The hydroponic seedlings are usually placed in the dark at about 21°C until they sprout. After sprouting they are put under adequate light (5500 K or 6500 K fluorescent bulbs can be used) for healthy hydroponic seedlings.

If the grower wants to transplant the plants from soil into a hydroponics system, before transplantation, the roots of the plant should be gently and completely cleaned by removing away the soil. After removing the soil, the root zone should be washed by dipping and shaking it into a bucket of pure water.

10.2.1.1 Growing Medium for Seedlings

The growing medium must provide satisfactory conditions for seed germination and to raise pest and disease free seedlings. The medium to be selected for seed germination or rooting the planting materials should be moderately fertile, well drained yet have sufficient water holding capacity and good aeration and free of pests and disease causing organisms. Some media that can be used to raise seedlings or to root planting materials are 'old coir-dust', 'carbonised rice husk', 'fine sand', rockwool, peat, perlite, vermiculture, *etc*. The medium should be properly sterilized before use. For coir-dust, hydrated lime can be added to bring its pH to neutral (for a 05 kg coir-dust block, about 100–250 g hydrated lime is needed).

Most growers usually use rockwool to start seedlings which is made from fibres of molten rock and then spun and compressed into cubes, blocks, loose material, and slabs. It has a good water and nutrient holding capability making watering time much less frequent. The major drawback of rockwool is that it has a high and unstable pH.

Note: Rockwool is not very environment friendly and does not degrade quickly over time. At the same time, it is bad for lungs if breathed and can irritate the skin.

Given below are the pictures of root cubes and plug trays that can be used for starting seed. When root cubes are used and when seedlings are ready to be moved to their final location the cubes can just be snapped off from each other and planted into the hydroponic system. Plug trays can be used over and over again and when the seedlings become ready to be transplanted to the hydroponic system, the plants can be simply popped out and can be transplanted in the hydroponic system after washing it using chlorine free water.

Root Cubes   **Plug Tray**

The growing mix for starting the seeds can be easily prepared at home by mixing equal amounts of fine peat moss, perlite and vermiculture in a bucket. To neutralize the acidity of the peat moss a $1/4^{th}$ teaspoon of lime can be added. The ratio of peat moss can be increased to increase the moisture holding capability of the mixture. Note that the ratio of lime is directly proportional to that of fine peat moss. The sand is also a good medium for seedlings. In some cases the seeds can be directly placed in the grow media.

10.2.1.2 Sponge Nursery Technique

Sponge pieces can be used as nursery medium instead of the above mentioned media materials. 2.5 cm cube sponge blocks can be used for this purpose. The seeds can be placed at the centre in a cut made on the topside of the sponge block. Sponge nursery is maintained as other nurseries. Nutrient supply must begin when the first true leaf begin to unfold. Depending on the cultivation method, the seedlings can be planted in hydroponics system with the sponge block intact. The sponge block may be removed with minimum damage to roots when plants begin to grow.

10.2.1.3 Seedling Containers/Trays

Growers must select the seedling containers/trays that will provide the suitable conditions for seed germination and also according to crop and cultivation method. The seedling containers must be properly sterilized before used.

10.2.1.4 Seed Germination

If the seedlings are started in a plug tray, a sterile fine growing medium appropriate for seed starting should be used and it should be light enough to let the seedlings push their roots and tops through and it should stay just moist but not wet. It also needs to drain quickly so as not to promote damping off of the seeds. It also needs to come off the roots easily when rinsed them for transplanting into the hydroponic system.

The bottom of the root cubes/plug tray must touch the water to ensure that they are receiving moisture. Once the roots have begun to grow and extend beyond the planting cones, the water levels only need to be kept at a level at which they touch the bottom of the plastic cones.

Grower should place one seed per block (filled with growing medium) at the correct depth in the pots or trays and provide appropriate conditions for seed germination (moisture, temperature, humidity). Germination trays can be covered with wet papers or cloth to provide adequate temperature for germination until the seeds sprout and these papers should be removed at the time of seedling emergence. The grower must maintain the moisture level of the medium at correct level for uniform germination.

10.2.1.5 Planting Material Production

Vegetative parts separated from mother plants can also be rooted and used as planting materials. Individual containers or trays filled with growing medium are used for rooting these vegetative parts. Only those vegetative parts that are free from pests, disease causing organisms and nematodes should be selected for propagation.

For example, following materials can be used for propagation.

Crop	Vegetative Parts
Strawberry	Plantlets or runners separated from mother plants
Gerbera	Plantlets separated from mother plants
Mint	10–12 cm long semi-hardwood stems cuttings

Propagation

Propagation of plants refers to producing more plants by dividing, grafting or taking cuttings from existing plants. Unlike seeds, cuttings and divisions of plants will result in an identical plant (clone) that will reproduce the same beautiful flowers, blossom *etc.* as that of parent plant. Most plants can be cloned, although it takes different methods to do so.

Propagation of plants can be sexual (through seeds) or asexual (using leaf, stem and root cuttings).

☆ **Sexual propagation**: Sexual propagation is used for increasing plant numbers by germinating seed and growing the seedlings to maturity. This is probably the most widely used method by the majority of growers, including many agricultural grain crops.

There are many different ways to pre-treat seed for attempting germination in the artificial setting of a greenhouse. These are processes used in an effort to overcome seed dormancy and reach successful germination on a human driven schedule. Mother Nature does a wonderful job of overcoming dormancy, so seed will germinate at the safest time for the seedlings to emerge into a growth friendly environment.

☆ **Asexual propagation (cloning):** Asexual propagation is used for processes that involve using the leaf, stem, and root cuttings. All of these asexual methods develop plants that are essentially clones of the original specimens.

1. Dividing clumps of plants, such as 'hostas', when they get very large is another form of asexual propagation
2. Planting bits of a rhizome root will give rise to new plant growth and is another form of cutting
3. The practice of grafting which is often used with roses and fruit trees is a type of asexual propagation
4. Stem cuttings

Propagation by Division

☆ Dig the plant up when the flowers have faded and clean the roots.

☆ Break the plant into several pieces.

☆ The divisions should follow fairly natural points on the plant (at nodes, or between leaves etc.).

☆ The important thing is to make sure each divided piece has shoots and roots on it.

☆ Replant each piece in good medium in a container (in shade)

☆ Water thoroughly.

Propagation by Cutting

Growers can either get the cuttings from the healthy plants at their gardens or can purchase rooted or unrooted cuttings from specialist propagators/breeders. Specialist propagators maintain mother stock that is true to type, use culture-virus indexing to eliminate diseases, and grow and harvest large areas of stock to supply the market.

The kind of cloning performed most often in greenhouse situations is to take cuttings. Taking cuttings from plants is a viable way of propagating plants, without the expense of buying seeds or cuttings. Grower should always take cutting from the healthiest plants and from the youngest leaf set. Parent plant should be at least two months old and it should still be in the stages of vegetative growth. Cuttings need to have only 2-3 leaves on them for best success. Extremely sharp, sterile razor blade should be used for cutting off leaves and foliage. For best results, some kind of propagation/seedling box and a heating mat should be used, so that the humidity and the root zone temperature can be controlled. The seedling box is designed with two trays and a vented lid. The vents in the lid allow controlling the humidity and airflow in the box.

The cut should be at a 45 degree angle and it should be dipped into a cloning gel or powder (rooting hormone or solution). The excess cloning gel or powder should be removed so that only a thin film of gel is placed on the "wound" as too much cloning gel can actually inhibit the roots from developing. Best practice technique for cuttings is to always make sure that whatever gel is being used should be poured into a separate container or lid and then disposed off once used. Cuttings should never be dipped straight into the bottle as this will contaminate the whole bottle making it less effective for any future cuttings. Once the gel or powder is applied, it should be placed into the moist rockwool cube. Once planted, the rockwool cubes should be placed on the seedling tray and must be covered with the vented lid making sure the vents are closed. The next step is to place seedling tray in a warm well lit area. Ideally, in winter a heating pad can be used, but in summer this is not always necessary. Provide enough light, and leave the vents closed for the first week. The moisture of the cube should be checked periodically and the cubes must not be allowed to dry out. Also, over watering should not be done. The cube must be kept moist but not saturated.

After a week, or when the roots start developing, the vents must be opened slightly and the airflow should be gradually increased through the box. Continue opening the vents over the next few days until they are fully open. Once the roots get fully developed, the lid can be removed and gradually acclimatise the cuttings to what will be their new environment (outside or in a grow-room). It is very important that the transition from cutting stage to transplanting is done gradually to ensure best success. It is no good taking a newly developed cutting and placing them in direct sunlight or artificial light.

Tips

☆ Some people like to dip the ends of cuttings into a "rooting hormone". This adds a bit more expense, but its growth promoting elements give the cutting a better start. This should be available at gardening stores.

☆ Plants can be propagated from cuttings or by dividing at any time of the year but be aware that roots form much more slowly during winter months. For a very cold climate, indoor propagation is the only suitable method during the depths of winter.

☆ Avoid woody or hard plant pieces when cutting or dividing. These are unlikely to strike. The same goes for very soft and wilted pieces of a plant.

Warnings

☆ Don't let the cuttings dry out, get too hot or too wet. Keep them in blotchy shade or indoors until established.

☆ Do not over-water otherwise there would be a risk of encouraging fungal growth or rotting.

☆ The cutter (knife or razor) used to make a cutting must be clean, to avoid any possibility of transferring disease to the plant.

10.2.1.6 Nutrient Supply

Only clean water without any nutrient should be applied until the emergence of first two true leaves. However, when they unfold, nutrient supply must begin gradually as the growing medium contains very little plant nutrients. The fertilizer mixture meant for hydroponics plants could also be used for nursery plants. Diluted nutrient solution can be applied every day. At the early stage, the trays or pots should be placed in shallow containers that are filled with nutrient solution in such a way that the tray's or pots' lower portion is submerged in the solution. The nutrient solution will reach the media through the holes at the bottom of the pots or nursery trays by capillary action. Vegetative parts for propagation planted in individual containers or trays should also be placed in shallow nutrient solution containers as seedling trays.

Once the seedlings or planting materials reach the correct size for planting, they can be planted with the medium. Vegetative parts can sometimes be directly established in the hydroponics system.

10.2.1.7 Light Requirement for Indoor Cloning and Seed Starting

Seedlings and clones require bright light for healthy growth. Most growers use special spectrum fluorescent lights for these early stages of plant growth. They will result in a much healthier start for the plants. Fluorescent lights should not be placed above 16 cm from the container. Some growers choose to use H.I.D. lights, but these should be hung up higher from the plants so that they do not burn (3, 4 feet away should do the trick). Most clones and seedlings benefit from 16 to 18 hours of light.

10.2.1.8 Temperature Requirement for Clones/Seedlings

The temperatures for seedlings should be about 21 to 24°C in the air around the plants, but the bottom heat from the propagation mat should supply heat of about 26 to 27°C. Bottom heat encourages root growth.

10.2.1.9 Humidity Level for Clones/Seedlings

The humidity level should be kept at or near 90 per cent. These conditions encourage compact, bushy, vigorous growth while minimizing disease.

10.2.2 Select or Build the Best Suitable Hydroponics System

After selecting the type of plants that are to be grown hydroponically, select the hydroponic system (ebb and flow, NFT, *etc.*) that is suitable for the selected plants and the available space for the system. There are many types of hydroponic systems

Images of some Types of Hydroponics Systems

available either for retail purchase or can be built yourself as most hydroponic systems are quite simple to build and can be very cost effective. The choice of any hydroponic system would be highly dependent on what is being grown, where it is being grown and how much it is being grown. Any water culture or medium culture hydroponic system discussed previously can be selected according to the requirement.

10.2.2.1 Building a Simple Ebb and Flow System

☆ **Select the containers**

☐ Select the growing container. First select an appropriate planting container of almost any size, rectangular or square that has a flat bottom and should be deep enough for the plants that are to be planted. The size of the container depends on what and how much plants are to be grown. The size of the growing container will need to correspond with the nutrient tank so it will have enough water volume to fill the container. For example, a tub container that is 60 cm x 40 cm x 15 cm deep. The selection of the growing container should be according to whether the pots (nursery pots or net pots) filled with substrate (grow medium) are to be placed in the growing container or grow media is to be put directly in the container.

☐ Select the nutrient reservoir. The nutrient reservoir can be any bucket of the size that is suitable for the selected container. For example, for a 24″ x 16″ x 6″ deep growing container, a 3-5 gallon nutrient bucket would be appropriate.

☐ Plan on the types of drain setups. It can be fitted at the bottom or to the side of the container. Setting the drains at the side of the container may eliminate the problems of placing the container on the shelves and leaking of underwater fittings.

☆ **Plan where the nutrient reservoir should be placed:** This will help in knowing the side of the container where the holes are to be drilled.

☆ **Now decide the maximum water level that should be based on the pots in which plants will be grown**

☐ The maximum water level should be about three quarters of the way up the pot. A good way to do this is to place a pot in the container and mark the location on the side of the container.

☐ The top fill hole can be 2.5 to 5.0 cm above this.

☐ Now drill two holes (approximately 5/8") at these marks. These holes are used to install **fill** (in upper hole) and **drain** (in lower hole) flow tubes, which will go to the nutrient tank placed below.

☐ Attach the hoses to these holes. Here, a little bit of plumbing is required.

☐ The same tube can be used to flood the grow bed and to drain back the solution to the reservoir. When the solution is flooded in the tray and the roots get briefly submerged, the pump is made inactive using a switch (timer operated or manual) and the solution is drained back to the reservoir using the same tube.

☐ Read the previous discussion on ebb and flow hydroponic system.

☆ **Build a nutrient solution reservoir:** The nutrient reservoir can be any strong and waterproof container of any size or shape. It should be large enough to supply the nutrient solution to the plants in the grow bed (growing container) and should be located below the growing container. It should be accessible for filling and stirring of the nutrient solution. Also, the nutrient tank must not get empty before the growing container is full. Since water will evaporate on daily basis, therefore, the growing tank should be larger in water volume as compared to growing container. Another reason for nutrient tank to be larger than the growing container is that the more water volume in the nutrient tank helps with the buffering action of the hydroponic nutrient solution so that the pH does not drift radically. For example, if the hydroponic growing container holds 2 gallons of water, the nutrient tank should hold about double or more gallons of water otherwise, the system will require the frequent refilling and pH adjustments of the nutrient tank. The nutrient tank should be covered using a lid to slow down the evaporation and to keep dust and debris out of the nutrient solution tank.

Note: *A transparent nutrient tank should be covered or painted to prevent algae growth.*

☆ **Installation of Water Pump**: Water pump can be a submersible pump that can be placed in the nutrient solution tank or it can be an external pump. Attach the pump supply hose (pipe) to the pump and bring one end of the drain hose in the nutrient tank (if fill and drain have

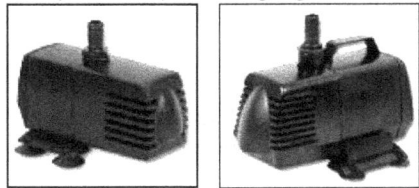

Water Pumps

separate paths). Make sure there are no kinks in the hoses. The water pump should be controlled by a timer which can be set to any number of flood times. The flood time depends on the growing medium, plant types and stages of plant developments.

☆ **Installation of air stone and air pump:** The air stone and air pumps can be purchased from any aquarium store. These are added in the nutrient tank to oxygenate nutrient solution which will improve the growth of the plants.

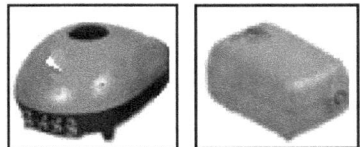

Air Pumps

10.2.3 Prepare and Add Nutrient Solution to the Base Tank

The nutrients for the hydroponic system can be prepared using fertilizer salts or pre-prepared hydroponic nutrients can be purchased from the store. During the different stages of plant life, different nutrient levels are needed to optimize the growth, flowering and fruiting stages. The product label for any regular fertilizer has a list of ingredients in percentage of each ingredient in each container. The macro elements (usually N-P-K) and minor elements are listed in the label. The main point is that the nutrient solution should have all the essential nutrients for the plants that are to be grown in this hydroponic system. The following points are to be considered while preparing nutrient solution.

10.2.3.1 Check the Water Quality

The selection of fertilizers and their concentration in the hydroponic nutrient solution depends greatly on the quality of the raw water as the raw water for hydroponic nutrient solution may contain minerals. Therefore, it is necessary to test the raw water prior to deciding on a fertilizer formula. Also, the water for nutrient solution should be free from pathogens.

10.2.3.2 Electrical Conductivity

The nutrient solution should be checked regularly for its electrical conductivity (EC). Proper nutrient management would enhance the growth of plants in a hydroponic system.

10.2.3.3 The pH of Nutrient Solution

All Hydroponic nutrient solutions need to be kept at the proper pH if they are to be used by the plants. A major factor in determining a plant's ability to uptake hydroponic nutrients is the relative acidity or pH of the hydroponic growing medium or hydroponic solution from which they feed.

10.2.3.4 Temperature of Nutrient Solution

For healthy growth of the plants, a proper temperature of nutrient solution is necessary. An **aquarium heater** can be used to warm nutrient solution in the winter and **chillers** can be used to cool nutrient solution in the summer (if high nutrient temperature becomes a problem).

Plants tolerate a range of nutrients. It is the grower's responsibility to read the plant and if necessary modify the nutrient to suit the environment. The grower must regularly check the plants for nutrient deficiencies and toxicity symptoms. A close observation should be made regularly for any symptom of pests and diseases. Water quality, temperature, humidity, light intensity, EC and pH are some other factors that should be regularly monitored and managed. Hence, any nutrient solution formula used for any crop in one system may not necessarily be suitable for the same crop in another system and may require some modifications for that system.

10.2.4 Hydrogen Peroxide (H_2O_2)

Hydrogen peroxide is water with an extra oxygen atom. The weak hydrogen peroxide is usually used by many hydroponic gardeners to add extra oxygen to the hydroponic nutrient solution. It enhances the plant's root development and can be used to treat root rot and other diseases. Apart from adding extra oxygen to the nutrient solution, hydrogen peroxide can also be used to sterilize the growing media to prevent bacteria and some harmful pathogens (that thrive in low oxygen solution) from growing in the system. It can also be used to remove chlorine from the water. Hydrogen peroxide also helps in increasing the plant's ability to uptake more nutrients from the nutrient solution resulting in healthier plants. The percentage of hydrogen peroxide that is to be added in the nutrient solution depends on the hydroponic system and plants that are being grown. The instruction given on the hydrogen peroxide product should be read carefully before using it. Normally 3 per cent of hydrogen peroxide is added in the nutrient solution.

Some important points

☆ If some burning or yellowing of the leaves is seen in the plants, the amount or the strength being used may be too strong and should be reduced.

☆ Hydrogen peroxide should be mixed in a separate container with gloves on, before adding it to the nutrient tank.

☆ Hydrogen peroxide acts only as root disease preventatives, and once root disease is present in the crop they are ineffective and other means for controlling the disease should be sorted out.

☆ Hydrogen peroxide is not suitable for use where organic media (*e.g.* coco substrate) or organic additives are used. These products are oxidants and oxidants break down organic matter.

10.2.5 Hydroponic Growing Media

Hydroponic growing media provide support for the plants and distribute appropriate oxygen and carbon dioxide. Therefore, it is very important to select the proper growing media for the hydroponic system being developed. Growing media is not needed in 'solution culture hydroponic systems'. Hydroponic growing media should allow maximum water availability to the roots; it should have good nutrient and moisture holding capabilities until the next feeding. It must also be eco-friendly and pH Stable. Some popular growing media are hydroton (expanded clay balls

Expanded Clay Aggregate

Brick Shards

that have been heated at temperatures of 1200 degrees or more and have very high capillary action), cocofibre or cocopeat, rockwool, lava rock, perlite and vermiculite.

Note: The sharp edged growing media may cut the plant roots which can be harmful to the plants. The poor quality growing media are another cause of unstable pH.

10.2.6 Grow Pots and Net Pots

Although the growing media can be directly added into the growing bed but most of the hydroponic gardeners find it convenient to add the growing media into pots or net pots and then they place these pots in the grow tank. They found it easy to maintain and monitor the individual pots rather than the whole of the container filled with the growing media. Another reason for using pots is that there is rarely any chance of blockage in the drain or fill tube and if any blockage occurs, it can be easily corrected.

Net pots in hydroponic system allow very fast flooding and filling with nutrient solution. If net pots are being used to grow plants in a hydroponic system, the growing media of smaller size (such as **perlite** and **sand)** should not be used as these may fall through the holes. With net pots, roots usually spread out of the pots and this may sometimes cause a problem when moving and relocating the hydroponic plants to other locations. Net pots can be used in NFT, top Irrigation, aeroponics and ebb and flow hydroponic systems and in most other hydroponic systems.

Net Pot   Net Pot Filled with Medium   Roots Emerging from Net Pot

10.2.7 Grow Lights

Light is necessary for the overall development of the plants because plants use carbon dioxide, water and light along with many minerals to produce food for themselves. There is no need to install artificial grow light for an outdoor hydroponic system or even for indoor hydroponic system if it is kept at a place where sufficient sunlight is available. If the hydroponic system does not get enough sunlight, the grow lights can be used. The choice of grow lights for the hydroponic system depends on the plant type (variety), plant size, plant stage (flowering and fruiting), type and size of hydroponic system being used, place (indoor or outdoor), and season of the year (such as summer or winter). There are many types of grow lights that are specifically designed for indoor hydroponic system. Some of them are fluorescent hydroponic grow lights, metal halide hydroponic grow lights, high pressure sodium grow lights and special LED grow lights.

10.2.8 Light and Nutrient Timers

Timers for nutrient pump and grow lights are used to make hydroponic system automatic and thus hydroponic gardening becomes much easier. The feed times and duration along with the grow lights coming on and going off becomes fully automatic after the installation of good timers. Alarms can also be installed to monitor pH and ppm of the nutrient reservoir.

Timer

10.3 Example

Growing lettuce in a non-circulating (static) hydroponic system.

1. **Preparing the seedlings (starter plants):** Prepare the starter plants 12-16 days before planting.

   a. Place the lettuce seeds in the **root cubes** (one seed in one cube) and place these cubes in water in a dish and let them sprout into little seedlings. Make sure that the base of cubes should be slightly immersed in water so that the cubes never get dry.

   Root Cubes in a Dish

2. **Select and prepare the planting container:** Use a dark five gallon plastic bucket or other container. If the container is transparent, either paint it with a dark colour, or cover it using a dark plastic or using a tin foil.

Five Gallon Bucket with Cover

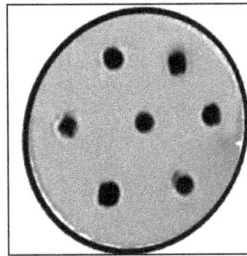

Container Cover with Evenly Spaced Holes

3. **Prepare a cover for the container:** A cover for the container can be made from plastic, plywood board or foam board. This cover should be able to completely cover the surface area of the container

   a. Make evenly spaced holes around the cover slightly away from the rim on the container. These holes will be used to suspend the plastic planting cones (tubes) in which plants will be grown to maturity.

4. **Plastic tube (preferably cone shaped):** Suspend a plastic tube in each hole on the cover of the container.

5. **Prepare nutrient solution:** Dissolve and add the necessary nutrients to the water in the container.

   a. Dissolve each chemical with water in the cup.

      i. 2 teaspoons of water soluble hydroponic fertilizer 8-15-36 (N-P-K percentage)

      ii. 1 teaspoon of epsom salt (magnesium sulphate)

      iii. 2 teaspoons of calcium nitrate

   b. Add the dissolved chemicals to the 5 gallon water in the container.

6. Place the cover containing the planting tubes on the container and use a duct tape to fix the cover.

7. Transplant the seedlings

   a. Place a root cube with the seedling into the planting cones.

8. Keep a watch on the plants for insects (grasshoppers, caterpillars, slugs, *etc.*).

9. Since the lettuce prefers about equal amounts of direct and indirect sunlight, therefore this hydroponic system should be kept in an area where plants would get lots of sunlight and at the same time it should be safe from rain.

<p style="text-align:center">

⑪

Growing Tomatoes in Hydroponics System

pH			CF	EC	ppm
Min	Max	Optimum			
>=5.0	<=6.5	5.5-6.0	20-50	2.0-5.0	1400-3500

11.1 Introduction

Tomato is botanically classified as fruit but generally prepared and served like other vegetables. Not only tomatoes but green beans, eggplants, cucumbers, and squashes of all kinds (such as zucchini and pumpkins) are all botanically fruits, yet cooked as vegetables. The tomato plant *Lycopersicon esculentum,* typically grow to 1–3 metres (3–10 ft) in height and have a weak stem that often sprawls over the ground and vines over other plants. It is a perennial in its native habitat, although often grown outdoors in temperate climates as an annual. There are thousands of tomato varieties that vary in shape, size, and colour.

Tomatoes can be commonly classified as **determinate, indeterminate** or **vigorous determinate** or **semi-determinate**. The majority of heirloom tomatoes are indeterminate, although some determinate heirlooms exist.

1. **Determinate tomatoes**: Determinate tomatoes, or "bush" tomatoes, are varieties that grow to a compact height (generally 3 - 4'). They stop growing when fruit sets on the top bud. All the tomatoes from the plant ripen at approximately the same time (usually over period of 1- 2 weeks). They require a limited amount of staking for support and are usually good choices for container growing. They are preferred by commercial growers who wish to harvest a whole field at one time.

2. **Indeterminate:** These varieties develop into vines and continue producing fruits until killed by frost. The vegetative growth of the plant of indeterminate varieties is continual and does not stop once flowering begins. This creates long tomato vines which must be trained up strings hanging from the greenhouse structures to maximize space and manage the crop. They are preferred by those farmers who want ripe fruit throughout the season.

3. **Vigorous determinate** or **semi-determinate:** These are like determinates, but produce a second crop after the initial crop.

11.2 Hydroponic Tomatoes

There are thousands of tomato varieties and almost all of them can be successfully grown in hydroponics system. Several tomato varieties have been specifically developed for hydroponic production in controlled environments. Tomato plants are quite large and require an ample support such as trellis, or can be wrapped around a vertically-tied string as they grow. In absence of sufficient support structure, these plants should be given enough room to grow. Large crops with a long growing season, like tomatoes and cucumbers, need sufficient room for their roots to grow.

11.3 Propagation

11.3.1 Plant Selection

The first step is selection of the tomato variety (**determinate or indeterminate**) that is to be grown hydroponically. The selection of plants depends on factors such as space needed, profitability, potential market, and whether the hydroponic system is to be implemented indoor or outdoor.

11.3.2 Starting Media (Seedling Media)

See, "10.2.1 Starting seedlings - Chapter 10"

11.3.3 Sowing Seeds

Prepare the tomato **seedlings** (starter plants) 12-16 days before planting. Tomato seeds should be sown 0.6 to 1 cm deep. Sprinkle a thin layer of vermiculite over the seeds or cover the germination cubes or pots with a large piece of clear plastic to conserve moisture at the surface. Avoid the use of plastic if the cubes receive direct sunlight, as the temperature may get too hot for good germination. The plastic must be removed as soon as emergence begins.

11.3.4 Watering Seedlings

Overhead watering or flood and drain (ebb and flow) systems can be used for germinating seedlings. When watering, the water must be sprinkled uniformly over all seedlings to avoid uneven growth. In flood and drain systems a shallow tray containing the sown cubes or pots is flooded with nutrient solution or water, providing even moisture from the bottom by capillary action.

For more detail on Flood and Drain, see "Passive hydroponics (Medium Culture Hydroponics)–Chapter 3"

11.3.5 Transplanting

Once true leaves appear (during post-emergence), seedlings should be transferred into larger growing blocks (pots) and should be evenly spaced to maximize light to each plant. The transplants must be spaced so as not to touch one another, and may need to be spread several times during their growth. A good transplant is one that is as wide as it is tall. Tomato plants readily grow adventitious roots from the stems if given the opportunity, producing a stronger plant with more roots. Adventitious roots will grow from the bent stem inside the block.

Transplanting into the final growing media should be done before any flowering and before transplanting the growing media should be properly leached and moistened, and should be at the proper temperatures. Plants should be irrigated with nutrient solution immediately after moving. The spacing of tomatoes in hydroponic systems can be much denser than in soil (approximately 60 square cm per plant). In lower light condition, wider spacing should be applied.

11.4 Light

Photosynthesis plays an important role in the growth of plant. If photosynthesis is decreased (due to low light conditions, high humidity or water stress), then the production of sugars will decline and will affect the fruit quality and size. If tomato plant does not get sufficient light for a long period, its leaves become low in sugars, and may become pale and thin. Excess nitrogen at that time can be damaging. Supplementary artificial light, from cool white, high output fluorescent or high intensity discharge sodium vapour lamps is beneficial to plants when sunlight is unavailable but it is not a complete substitution of the sunlight.

Artificial lighting is generally not economical for vegetable crops with the exception of seed production. During winter months, especially in Northern latitudes, supplemental light may be required for strong growth of seedlings. The lights should operate 14 to 18 hours per day.

Sometimes it may be necessary to provide shade to tomato plants especially in the areas of high summer temperatures and humidity, because direct sunlight on fruits can cause yellow or green shoulders and cracking. Therefore, shading using a 30 per cent shade cloth keeps the temperature within reasonable range and can improve fruit quality.

11.5 Temperature

The day temperatures for a tomato plant should be around 21 -26 C and night temperatures around 16 -18.5 C. Some new varieties can do best with little difference between day and night temperatures. The recommended growing temperatures can be known from the seed company. For seedlings, the temperatures should be constant, 20 -22 C, then slowly the plants can be acclimated to the diurnal temperatures before transplanting.

To maintain the temperature of the greenhouse to optimum level, an increase in temperatures can be brought using electrical furnaces or solar energy and a decrease in temperature can be done using fan, evaporative pad or by fogging.

11.6 Humidity

The ideal humidity for tomatoes during day is 80–90 per cent and during night is 65-75 per cent. Under higher humidity conditions, plants are unable to transpire and as a result they are unable to draw water or nutrients into the roots, resulting in nutrient deficiency symptoms, such as **blossom end rot** (calcium deficiency). The tomato yields, fruit quality and leaf size gets reduced and flower and fruit abortion increases significantly. High atmospheric humidity can cause **glassiness** and **gold fleck** in tomato fruit. Under low humidity, water and nutrients are pumped too quickly through the plants due to which nutrient ions are deposited in the leaves rather than in the fruits. This too results in nutrient deficiencies.

Misting and fogging systems can be used to increase humidity and decrease temperatures. Exhaust fans and proper ventilation helps in reducing humidity.

11.7 Preparing Nutrient Solution

As already mentioned, the selection of fertilizers and their concentration in the hydroponic nutrient solution depends greatly on the quality of the raw water because it may contain pathogens, pesticides and minerals. Therefore the raw water should be properly tested and treated accordingly, and if possible RO water should be used. It is necessary to test the raw water prior to deciding on a fertilizer formula.

The concentration of macronutrients in nutrient solution for tomatoes progresses with the maturity of the crop, whereas the concentration of micronutrients remains the same throughout the growth cycle. The following Tables (adapted from Jensen and Malter, 1995) shows the concentration of macronutrients in the first stage of growth (seedlings from the first true leaf until the plants are 60 cm tall, when initial fruit is 6-12 mm in diameter) and the second growth stage (fruit set to harvest).

Table 1: Nutrient Requirements for Tomatoes
(Adapted from Jensen and Malter, 1995)

Nutrient	Seedlings to First Fruit Set (ppm or mg/l)	Fruit Set to Harvest (ppm or mg/l)
Macronutrients		
Magnesium (Mg)	50	50
Potassium (K)	199	199
Phosphorus (P)	62	62
Nitrogen (N)	113	144
Calcium (Ca)	122	165
Micronutrients		
Iron (Fe)	2.5	2.5
Boron (B)	0.44	0.44
Copper (Cu)	0.05	0.05
Chlorine (Cl)	0.85	0.85
Manganese (Ma)	0.62	0.62
Molybdenum (Mo)	0.06	0.06
Zinc (Zn)	0.09	0.09

Merle H. Jensen, Alan J. Malter World Bank Publications, 1995 - Agricultura - 157 pages.

Table 2: Preparation of Macronutrient and Iron Solutions for Tomato
(Adapted from Jensen and Malter, 1995)

Chemical Compound (Fertilizer Grade)	Seedlings to First Fruit Set (g/1000 litres)	Fruit Set to Harvest (g/1000 litres)
Magnesium sulphate (epsom salts)	500	500
Monopotassium phosphate (0-22.5-28)	270	270
Potassium nitrate (13.75-0-36.9)	200	200
Potassium sulphate (0-0-43.3)	100	100
Calcium nitrate (15.5-0-0)	500	680
Chelated iron	25	25

Merle H. Jensen, Alan J. Malter World Bank Publications, 1995 - Agricultura - 157 pages.

The formula in the above table has been standard for many years; some new tomato varieties may require much higher nitrogen and potassium. Therefore it is advisable to consult the seed company for the recommended nutrient formulas for the grown tomato variety.

As the crop progresses towards maturity and as the available light and day length changes, it becomes necessary to optimize the N: K ratio. Plants use more nitrogen (N) under high light conditions and high potassium (K) under low light conditions.

Table 3: Preparation of Micronutrient Stock Solution for Tomatoes

Fertilizer Salt	Grammes of Chemical in 450 ml Stock Solution
Boric acid	7.50
Manganous chloride	6.75
Cupric chloride	0.37
Molybdenum trioxide	0.15
Zinc sulphate	1.18

Use 250 mL of this micronutrient stock in each 1000 litres of nutrient solution from Table 2, above (Adapted from Jensen and Malter, 1995).

Note: If a concentrated stock solution is used for the macronutrients, then the calcium salts should be kept apart from the other salts in a separate solution.

11.8 Symptoms of Nutrient Deficiencies and Toxicities

Table 4: Common Nutrient Disorders in Tomatoes

Element	Deficiency	Toxicity
Nitrogen	☆ Older leaves become chlorotic (yellow). ☆ New growth is restricted and spindly. ☆ Small fruit. **Treatment**: Use of foliar spray of 0.25 per cent to 0.5 per cent solution of urea.	Plants dark green with abundant foliage but reduced root growth or fruit production, flower drop.
Phosphorus	☆ Plants stunted. ☆ Delayed maturity. ☆ Restricted root growth. ☆ Fruit-set delayed. ☆ Red/purple colours appear (usually on underside of leaves).	No recognizable symptoms, however Cu and Zn deficiencies may occur in presence of excess P
Potassium	☆ Older leaves chlorotic, with scattered dead spots. ☆ Uneven ripening in fruit (blotchy). **Treatment**: use foliar spray of 2 per cent potassium sulphate	Not usually absorbed by plants in excessive amounts, but high levels may lead to deficiencies in Mg, Mn, Zn, or Fe
Sulphur	☆ Very rare. ☆ Some yellowing in young leaves. ☆ Upper leaves become stiff and curl downward. ☆ Stems, veins, and petioles turn purple.	Stunted growth, may see interveinal yellowing or leaf burning.
Magnesium	☆ Interveinal chlorosis on older leaves. **Treatment**: use foliar spray with 10 per cent magnesium sulphate	No visual symptoms
Calcium	☆ Blossom end rot on fruit. ☆ Yellowing on margins of young leaves. ☆ Undersides turning purple. ☆ Curling of leaves.	No visual symptoms

Contd...

Table 4–*Contd...*

Element	Deficiency	Toxicity
	☆ Growing tip and root tip death ☆ Thick woody stems. Can be caused by boron deficiency. **Treatment**: foliar spray of 0.75 to 1.0 per cent calcium nitrate solution or 0.4 per cent calcium chloride.	
Iron	☆ Pronounced interveinal chlorosis on young leaves, starting at margins and spreading through entire leaf. ☆ Stunted growth and aborted flowers. ☆ High pH can lead to iron deficiency. ☆ Low pH can lead to preferential uptake of aluminum, restricting iron absorption. **Treatment**: Foliar spray with 0.2 to 0.5 per cent iron chelate every 3 to 4 days	Not usually a problem
Chlorine	☆ Rarely occurs. ☆ Manifests as wilted leaves. ☆ Chlorotic with a bronze colour. ☆ Stunted root growth.	Burning of leaf tips, bronzing or yellowing, leaf drop and stunted growth
Manganese	☆ Interveinal chlorosis on older leaves. ☆ Light green leaves with dead patches ringed in yellow. ☆ Few flowers or fruit **Treatment**: Foliar spray using 1 per cent solution of manganese sulphate	Chlorosis, stunted growth
Boron	☆ Growing points wither and die. ☆ Interveinal chlorosis of upper leaves. ☆ Brittle leaves. ☆ Boron deficiency can lead to calcium deficiency. **Treatment**: foliar spray of 0.1 to 0.25% borax	Yellowing of leaf tips, leading to browning
Zinc	☆ Reduction of internodal length. ☆ Puckered margins on leaves. ☆ Brown spots on petioles. ☆ Small leaves, sometimes long and narrow **Treatment**: foliar spray with 0.1 to 0.5 per cent solution of zinc sulphate	Commonly accompanied by Fe chlorosis
Copper	☆ Young leaves dark green and misshapen. ☆ Curling into a tube. ☆ Petioles bent downward. ☆ Few or no flowers. **Treatment**: use foliar spray with 0.1 to 0.2 per cent solution of copper sulphate to which 0.5 per cent hydrated lime has been added.	Reduced growth, symptoms of Fe chlorosis
Molybdenum	☆ Interveinal chlorosis on older leaves. ☆ Margins of leaves curl up upward. **Treatment**: foliar spray with 0.07 to 0.1 per cent solution of ammonium or sodium molybdate	Tomato leaves turn golden yellow

Source: Adapted from Resh, 1995.

On seeing any deficiency symptoms, nutrient solution should be changed with the increased concentration of the deficient element (25 to 30 per cent) and when the deficiency is rectified; the concentration should be lowered back down to slightly higher than normal levels. Foliar sprays can be applied for a faster response, however, burning of the plants may result.

11.9 The pH

The desirable pH range of hydroponics nutrient solution for tomatoes is 5.5 to 6.5. At pH below 5.0, some nutrients get excessively absorbed by the plants resulting in toxicity of those elements. At pH above 7.0, some nutrients may become unavailable to the plants resulting in nutrient deficiency. The pH of the nutrient solution can be lowered with nitric or phosphoric acid.

11.10 Carbon Dioxide

Carbon dioxide is necessary for growth of plants and plants can deplete the CO_2 at a very high rate in a closed greenhouse or growroom. The optimal levels of CO_2 for tomatoes may be 2 to 5 times the normal atmospheric levels. CO_2 level of 1100–1500 ppm during the light phase will show a great increase in tomato yields. It should be injected during daylight (light phase) where it is used by plants for photosynthesis. When CO_2 is being injected, the greenhouse should remain sealed with the inlet and outlet fans turned off. If the fans are not turned off it will remove the CO_2 from the greenhouse before it is absorbed by the plants.

Bottled CO_2 with a regulator and solenoid valve can be used to inject CO_2 into a greenhouse. A specialized CO_2 controller can be used to monitor the CO_2 levels in the greenhouse. After a set time period, to allow for absorption of the CO_2, the fans in the greenhouse can again be switched on.

11.11 Diseases and Pests

Although diseases and pests in hydroponics system can be prevented to a great extent by proper sanitation and observation, these cannot be completely eliminated. Some of the common diseases that may infect tomato are:

- ☆ **Fungal disease:** Gray Mold (*Botrytis*), various forms of mildews, various forms of blight, *Verticillium* wilt, *Fusarium* wilt.
- ☆ **Nematodes**
- ☆ **Bacteria:** Bacterial canker (*Clavibacter michiganensis*), bacterial spot (*Xanthomonas vesicatoria*), bacterial wilt (*Burkholderia solanacerum*, also called *Pseudomonas solanacearum* and *Ralstonia solancearum*).
- ☆ **Virus:** Tobacco mosaic virus, tomato yellow leaf curl virus (TYLCV), common mosaic, aucuba mosaic, cucumber mosaic
- ☆ **Insects:**
  - ☐ Whiteflies
    - ✦ The greenhouse whitefly (*Trialeurodes vaporariorum*)

◆ Tobacco whitefly, *Bemisia tabaci*, also known as the sweetpotato, silverleaf, or cotton whitefly

❒ Tomato fruit worm (*Heliothis armigera*), leaf miner, tomato pinworm, cabbage looper, two-spotted spider mites, stink bugs, cutworms, tomato hornworms, tobacco hornworms, aphids, cabbage loopers, flea beetles, red spider mite, slugs, colorado potato beetles

11.12 Pollination

Tomato flowers are naturally pollinated by wind but for hydroponics system a grower cannot completely rely on natural air movements for good pollination. He has to opt for other pollinating options such as artificial wind, shaking and vibrating each flower cluster or maintaining hives of bumblebees in greenhouse. Bumblebees-pollination is not suggested for locations like porch or for small hydroponics systems at home. Mechanical pollination such as shaking or vibrating each flower cluster should be done at least every two days when humidity and temperature conditions are best. Pollination should be attempted even if conditions are not ideal. Although tapping or shaking the entire vine will move some pollen, it will be better to use an electric vibrator on each truss. Commercially available pollinators can also be purchased.

Using bumblebees is the best approach for pollinating large greenhouses of tomatoes where pesticides are not being used on crops. One hive will work for approximately 0.2 hectare of tomatoes.

⑫
Growing Cucumber in Hydroponics System

	pH		CF	EC	ppm	Temperature	Humidity
Min	Max	Optimum					
>=5.0	<=6.0	5.5-6.0	17-25	1.7-2.5	1190-1750	26.7-29.5ºC	75–85 per cent

EC and Light intensity

☆ Preferable EC during low light intensity: 2.2

☆ Preferable EC as light intensity increases: 2.5

12.1 Introduction

Cucumber (*Cucumis sativus* L.) is an annual plant from Cucurbitaceae family. Cucumbers, like tomatoes, are botanically classified as fruits but are generally prepared and served like other vegetables. The cucumber is a warm season crop and its plant is a creeping vine that grows up trellises or other supporting frames, wrapping around supports with thin, spiralling tendrils which tend to catch any support they can find to direct the plant in the upper direction allowing it to get maximum sun and warmth. The stem is covered with white hair,

and the plants can be 2 metres long. Some varieties form hermaphrodite flowers, those flowers form round fruits.

Although conventional cucumber varieties can be successfully grown in hydroponics, the **parthenocarpic** cucumbers (produce fruits without pollination) are the best choices for the growers all around the world. If pollination does occur in parthenocarpic cucumbers, the fruits will form seeds, the fruit shape will be distorted and a bitter tasting fruit will develop.

Cucumbers varieties can be:

☆ Profiled and hybrid

☆ Bee-pollinated, partly self-pollinated and self-pollinated parthenocarpic fruits

☆ Bunch-forming (short-vined) and long-vined

☆ With normal ovaries and clustered ovaries (bucketed)

12.2 Hydroponic Cucumbers

There are many cucumber varieties and almost all of them can be successfully grown in hydroponics system. Self-pollinated (parthenocarpic) varieties are best suited for hydroponic production in controlled environments. Cucumber plants are quite large and require an ample support such as trellis, or can be wrapped around a vertically-tied string as they grow. In absence of sufficient support structure, these plants should be given enough room to grow. Large crops with a long growing season, like tomatoes and cucumbers, need sufficient room for their roots to grow.

12.3 Propagation

12.3.1 Plant Selection

The first step is selection of the cucumber that is to be grown hydroponically. The selection of plants depends on factors such as space needed, profitability, potential market, and whether the hydroponic system is to be implemented indoor or outdoor.

12.3.2 Starting Media (Seedling Media)

See, "10.2.1 Starting seedlings - Chapter 10"

12.3.3 Sowing Seeds

It has been seen that cucumber seed germinates rapidly (2 to 3 days) at an optimum germinating temperature of 29°C (84°F) in the germination room. At 21°C (71°F), it takes 5-6 days for germination and at 16°C (60°F), 9-10 days are required for seedlings to emerge. Once the seed has germinated, the temperature should be raised to 25°C (77°F). During the seeding and transplant production stage, the plants must never become stressed for water or nutrients.

12.3.4 Watering Seedlings

See "Watering Seedlings" - Chapter 11"

12.3.5 Transplanting

If rockwool is to be used as the growing system, it should be soaked for a day before transplanting the cucumber seedlings. If other growing medium such as perlite is to be used, it should be properly sterilized, soaked and drained before seedlings are transplanted. Some growers have successfully direct seeded cucumber into soilless media, such as perlite, coconut fibre, or composted pine bark, and at the same time seeded a few transplants to use as replacement plants.

Cucumber Vines Trained on Trellises

A cucumber plant with three or four true leaves is ready to be transplanted. The transplants should maintain an upright growth habit (no curves in the stem) to aid in successful transplanting. There should be plenty of space between cucumber plants. Spacing instructions on the seed packets should be followed. Hydroponics/ greenhouse cucumber plants have very large leaves, grow vigorously, and require large amounts of sunlight. Under good sunlight conditions, each plant should be provided 2 metres of greenhouse space. More space is often required where light conditions are poor. Exact spacing between rows and between plants within a row will depend upon the type of training and production system to be used.

Cucumber vines can be trained on trellises to save space. This also improves fruit quality and yield.

12.4 Light

A fully grown cucumber crop benefits from any increase in natural light intensity, provided that the plants have sufficient water, nutrients, and carbon dioxide and of course optimum air temperature. Insufficient lighting during early stages of plant development leads to extending and weakening of the plant. Indoor growing cucumber requires at least 14 to 16 hours of light every day, so artificial grow lights will be needed.

12.5 Temperature

The greenhouse cucumbers (especially seedless cucumbers) are more sensitive to low temperatures than tomatoes. Minimum temperatures should not go below 18°C (65°F) for sustained maximum production. Prolonged temperatures above 35°C (95°F) should be avoided because fruit production and quality get reduced at extremely high temperatures.

Cucumbers thrive in warm conditions with air temperatures of 26°C to 30°C and plenty of sunlight. If possible, the night temperature should be at least 20°C (68° F), 22-23°C is preferable. The optimum day temperature is between 23–26° C (75° and 78° F). Some cucumber varieties may do best with little difference between day and night temperatures. Cucumbers are very sensitive to cold. Some seedless (parthenocarpic) varieties grown under the following controlled temperatures have shown good production.

- ☆ Although seeds of some varieties start germinating only at 15 C and higher, optimal temperature for germination is 25-30 C (preferably 28°C).
- ☆ Stem extension and leaf development is linearly dependent on a mean air temperature between 18°C and 27°C. The growth of the plant reduces (not growing at optimum rate) at temperatures above 28°C.
- ☆ The optimal temperature for flowering and pollination is 18-21 C.
- ☆ Recommended temperature for maximum fruit production is around 20-22 C day temperature and around 19-20°C night temperature during fruit bearing stage.
- ☆ The optimum temperature for nutrient solution is 25-30°C. Plant roots will not survive at solution temperature below 16°C.

Other experiment that we did with same varieties has also shown good results. In this experiment the temperature was kept between 23.8° C (75°F) and 25°C (77°F) during the day and 21.1°C (70°F) at night until the first picking. The temperature was reduced by 2° F when picking started. After picking started, the night temperatures were reduced to 2°F per night gradually to 63°F temporarily (for 2-4 days) to stimulate growth. Also the maximum temperatures exceeded temporarily to cause some flower abortion so that fruit-vine balance is maintained.

Optimum growth has also been achieved in greenhouses when the air temperatures are maintained between 20–26°C with nutrient solution temperatures between 22°C and 25°C.

To maintain the temperature of the greenhouse to optimum level, an increase in temperatures can be brought using electrical furnaces or solar energy and a decrease in temperature can be done using fan, evaporative pad or by fogging.

Note: Cool and cloudy growing season may produce bitter fruits.

12.6 Humidity

Cucumbers do best in a humid environment. The ideal humidity for cucumbers during day is 75–85 per cent and during night is 60-75 per cent. Optimal substrate humidity depends on vegetation stage. Maintaining the following humidity levels at various stages have shown good results:

☆ During seedlings growing - 50-70 per cent

☆ From planting of seedlings to fruit-forming–70-80 per cent

☆ During mass-flowering - substrate humidity can be lowered to 55-60 per cent

☆ From fruit-forming to first harvest–75-85 per cent

☆ From first harvest to the end of vegetation–85-95 per cent

If the humidity remains higher than 95 per cent for few days (7-10), first symptoms of ascochyta-leaf spot might appear.

Misting and fogging systems can be used to increase humidity and decrease temperatures. Exhaust fans and proper ventilation helps in reducing humidity.

12.7 Preparing Nutrient Solution

The selection of fertilizers and their concentration in the hydroponic nutrient solution depends greatly on the quality of the raw water because it may contain pathogens, pesticides and minerals. Therefore the raw water should be properly tested and treated accordingly, and if possible RO water should be used. It is necessary to test the raw water prior to deciding on a fertilizer formula.

The concentration of nutrients may vary for different varieties and growth stage. Usually during the initial stage of plants when their root system is not well developed, the concentration of the nutrient solution should be kept low. The concentration must be raised according to the stages of growing and fruit-bearing of the plant. It is advisable to consult the seed company for the recommended nutrient formulas for the cucumber variety that is to be grown hydroponically.

Experiments performed with the following nutrient solution composition have shown good results.

Component	Minimum	Optimum	Maximum
$N–NO_3$	120	160	190
$N–NH_4$	-	7	20
P	30	38	50
K	200	240	270
Ca	120	140	190
Mg	20	25	60
S	25	35	60
Fe	0.4	0.6	2.0

Contd...

Contd...

Component	Minimum	Optimum	Maximum
Mn	0.3	0.6	1.0
Zn	0.1	0.3	1.0
B	0.1	0.2	0.5
Cu	0.02	0.03	0.04
Mo	0.03	0.05	0.08
EC	1.5	2.0	2.5

Apart from cucumber variety and growth stages, the concentration of the nutrient solution also depends greatly on the growing conditions such as light intensity, pH, temperature and humidity.

The main idea is that the grower must keep observing the plants for deficiency or toxicity and accordingly increase or decrease the concentration of nutrient solution. On seeing any deficiency symptoms, nutrient solution should be changed with the increased concentration of the deficient element (25 to 30 per cent) and when the deficiency is rectified; the concentration should be lowered back down to slightly higher than normal levels. Foliar sprays can be applied for a faster response, however burning of the plants may result.

12.8 pH

The desirable pH range of hydroponics nutrient solution for cucumbers is 5.5 to 6.0. At pH below 5.0, some nutrients get excessively absorbed by the plants resulting in toxicity of those elements. At pH above 6.5, some nutrients may become unavailable to the plants resulting in nutrient deficiency. The pH of the nutrient solution can be lowered with nitric or phosphoric acid.

12.9 Carbon Dioxide

Carbon dioxide is necessary for growth of plants and plants can deplete the CO_2 at a very high rate in a closed greenhouse or growroom. The cucumber needs higher level of carbon dioxide in the air, and raising it to 2-5 times in protected environment of greenhouses, hydroponics has shown great increase in productivity.

Bottled CO_2 with a regulator and solenoid valve can be used to inject CO_2 into a greenhouse. A specialized CO_2 controller can be used to monitor the CO_2 levels in the greenhouse. After a set time period, to allow for absorption of the CO_2, the fans in the greenhouse can again be switched on.

12.10 Diseases and Pests

High temperatures and humidity in greenhouse environment are ideal for diseases to develop. Several diseases, insects, and nematodes can potentially be pests of greenhouse cucumbers. These can be controlled by selecting resistant varieties, greenhouse sanitation, and well-timed applications of properly selected pesticides. Most fungus and virus diseases are prevented by sanitation and sterilization of hydroponic media, containers, and equipments. They can be controlled with fungicides, *etc.*

Most wide-spread problems are **anthracnose, cucumber mosaic virus, grey rot (*Botrytis*), gummosis, bitter fruits, withering of ovaries,** *etc.*

Growing Lettuce in Hydroponics System

Lettuce is one of the most widely hydroponically grown (especially in greenhouses) vegetable crops in the world. Some of the varieties of lettuce are 'Romaine' ('Parris Island - Green' and 'Freckles - red'), 'Oakleaf' ('Cocarde', 'Berenice - Green', 'Oscarde', 'Dano–red'), red curly varieties such as 'Lolla Rosa', 'Ruby', 'Red Sails', 'New Red Fire', 'Brunia'.

The majority of hydroponic lettuce is bibb types. During the longer days of summer months these lettuce will mature within 40 to 48 days. Seedlings are grown for 12 to 18 days as transplants and then 28 to 32 days to maturity.

Green vegetables like lettuce, spinach, watercress and some herbs grow well in water culture such as NFT and raft culture.

13.1 Starting the Plants

13.1.1 Seedling Production

The growing system to which the lettuce seedlings will be transplanted determines what method of sowing is best. Substrates used include, rockwool cubes, Oasis cubes, and multipacks (celled trays) with vermiculite medium. The small size of lettuce seed makes it difficult to sow by hand. To make this task easier "**pelletized**" lettuce seed can be used. It is more expensive, but is a lot easier to sow by hand. Each seed is encapsulated in an inert clay-like material making it about 3 mm in diameter. To sow thousands of seeds daily, the grower must go for an automatic seeding system.

13.1.2 Germination

Lettuce seed loses viability quickly with age, so the grower should be careful that all seed is dated. The seed must be stored in a refrigerator providing a cold, dry atmosphere. Under such conditions it can be stored for up to 6 months or longer. When vermiculite or perlite in multi-celled trays is being used, grower should sow the seeds first then moisten the substrate. When using rockwool or Oasis cubes, they must be thoroughly soaked with a dilute nutrient solution of 0.5mS EC prior to sowing. For rockwool, a pH of 5.2-5.4 should be used to lower the pH of the rockwool that initially is 7.5 or greater. After germination a solution of EC 1.2 should be used.

Lettuce requires cool temperatures to germinate. The seeded trays or cubes can be stacked and placed in a cooler at a temperature of 4.5° C for 1 to 2 days to allow initiation of germination. Once the seeds crack and begin to grow, they should be immediately placed in the greenhouse at temperatures between 15 - 18° C.

13.1.3 Seedlings

Keep the seedlings at temperatures from 18–21° C during the day and 13 to 16° C at night in the greenhouse. Carbon dioxide enrichment should be maintained at 1000 ppm during the day. The optimum pH of the nutrient solution is between 5.5 and 6.0 and the EC should be from 1.0 to 2.3 mS depending upon the light. Lower the EC during bright sunny days. Maintain relative humidity (RH) from 60 per cent to 80 per cent. In northerly latitudes during the winter months supplementary artificial lighting is applied for 14 to 16 hours during cloudy days and to extend the day length. In high solar light regions sunlight can be reduced in the seedling area of the greenhouse by a 35 per cent to 40 per cent shade curtain. High temperatures can cause burn on the tips and margins of the leaves (tip burn). As mentioned earlier some varieties have resistance to tip burn.

Seedlings are generally grown 14 to 21 days before transplanting to the hydroponic production area of the greenhouse. A lettuce operation, regardless of its size, must produce lettuce daily for the market. Planting schedules are altered according to the length of the growing period (longer in the winter months under shorter days). To achieve continuous production, the operation must include daily sowing of seed, transplanting, harvesting and clean up of the growing system.

13.1.4 Transplanting

The lettuce seedlings are transplanted when they reach 2 to 3 leaf stage (from 14 to 21 days). Multicubes are broken apart into individual cubes during transplanting. The cube with its seedling is placed into the hydroponic system of NFT channels or raft culture boards depending upon the system used. The grower should be careful not to damage plant roots during transplanting, as such damage predisposes the plant to disease infection.

It is better to transplant in the late afternoon to avoid the plants getting stressed during the heat of the day under high solar conditions. The transplant will start to adjust to the new location during the night and roots will begin to grow into the solution below. When transplanting, position the base of the plants so that they touch the flow of nutrient solution below. In this way, the solution will be absorbed by the base of the cube to keep it moist. Within several days the roots will extend out of the cubes into the solution and the plants will grow vigorously.

13.2 Nutrient Solution

A complete nutrient formulation provides all essential elements to the plants. Lettuce seedlings are fed a half-strength solution until they are transplanted. A half-strength solution contains about one-half of the concentration of macro-elements, but the full concentration of micro-elements. The specific formulation to use is dependent upon temperature, day length and sunlight, as during summer conditions with high sunlight and long days the plants can be forced to grow faster by use of higher nitrogen levels. Under low light levels the potassium and nitrogen should be reduced. A typical nutrient formulation can be as:

Nutrient	Seedlings to First Fruit Set (ppm or mg/l)
Macronutrients	
Magnesium (Mg)	40-50
Potassium (K)	200-210
Phosphorus (P)	50-60
Nitrogen (N)	140-160
Calcium (Ca)	175-195
Micronutrients	
Iron (Fe)	2.5-3
Boron (B)	0.44–0.5
Copper (Cu)	0.1
Manganese (Mn)	0.5 - 0.62
Molybdenum (Mo)	0.05 - 0.06
Zinc (Zn)	0.09–0.1

Best grade, highly soluble fertilizers of highest purity should be used.

The optimum pH for lettuce is between 5.5 and 5.8. The EC of most lettuce formulations will be between 1.5 and 2.0.

13.3 Nutritional Disorders

The most common nutritional disorder is tip burn. It is caused by excessive water loss from the leaves accompanied by inadequate water uptake by the roots.

Control is through proper nutrition, adequate water supply and healthy roots. Sufficient oxygen in the nutrient solution is important to maintain healthy roots.

Note: The EC should not be too high.

13.4 Environmental Disorders

13.4.1 Relative Humidity

Sometimes high relative humidity (RH) in excess of 70 per cent within the head of the lettuce may cause tip burn that may cause necrotic areas on the leaf margins. Therefore the air RH should be kept at about 60 per cent. Calcium deficiency too causes tip burn.

13.4.2 Temperatures

There should not be much fluctuation in air and water temperatures and of course in pH. The day temperatures should be kept under 26.5°C.

13.4.3 Light

Lettuce needs 12-16 hours of daylight. During short days in winter months supplementary artificial lighting is beneficial to shorten the cropping period. Metal halide (MH) lighting is best for lettuce as it is a leafy crop.

13.4.4 Pests and Diseases

Most of the pests that attack other crops will also infest lettuce. Thrips, whiteflies and larvae from moths and butterflies are the most common. These should be treated with biological agents or homemade organic sprays. The worst disease of lettuce is *Pythium*.

(14)

Growing Capsicum in Hydroponics System

Temperature Requirements (Until the end of crop)

Nutrient Solution			Air		
Min	Max	Ideal	Min. Night	Max. Night	Min. Day
18ºC	28ºC	22-25ºC	15ºC	28ºC	18ºC

Nutrient Solution's pH and CF Requirements

pH			CF		
Min	Max	Optimum	CF at Planting	CF at Cropping	ppm
>=6.0	<=6.5	6.2-6.3	15-18	18-22	1260-1540

Capsicum varieties occur in many shapes, sizes and colours, and their plants are bushy, about 60–80 cm high. Capsicums are semi-perennials that are usually grown as annuals in cultivation. *Capsicum* species can be consumed fresh or dried, whole or ground, alone, or in combination with a variety of other flavouring agents. Capsicums are very versatile and are commonly used in salads, baked dishes, stuffed dishes, salsa, pizzas, *etc.*

Capsicum can be grown in almost all types of hydroponics systems such as NFT or in media-based systems. The nutrient solution should be well oxygenated and its temperature, CF and pH should be mainteined at desirable levels.

In NFT systems, the gullies should not exceed 15 metres in length and should have a fall of at least 1 in 40 down the 15 metres. Nutrient flow rate down the gully should be 1 litre of nutrient per minute.

In media-based systems, capsicums can be grown in a wide variety of different media such as rockwool, perlite, pumice, *etc*. Almost any inert sterile material of a grain size that allows air to permeate down to the roots can be used. We have used composted pine bark in our hydroponics systems under open shade structure in Allahabad.

14.1 Open Shade Structure

Capsicum production in an **open shade structure** would be much higher than yields for most open field production systems, but the growing season in the shade could be 3-4 months longer. Also, an open shade structure would help to successfully extend the harvest season [*e.g.* from June to frost (November)]. An added benefit is that this system will produce large, high-quality fruit for the entire season.

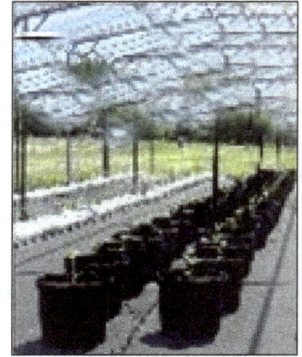

Open Shade Structure

14.2 Propagation

Capsicum plants can be propagated in any suitable media, such as perlite/vermiculite. Capsicum seeds can also be sown directly into the growbeds or into the netpots in growbeds. Propagation media temperatures should be 21 to 25°C. It will take around 2 to 3 weeks for seeds to germinate. Clean water should be used to maintain a damp medium, until the seeds get germinated. At the two leaf stage they should then be fed with a balanced nutrient at 5 to 10 CF until 200 mm tall. Then the CF should be increased to 12 - 15 until 30 to 35 cm tall when they can be planted onto the system. It should be planted at about 2.5 to 3 plants per square metre.

Propagation Environment

Water Quality			Media Type	Media Temp.	Plants per Square Metre
Before Germination	At the Two Leaf Stage (until 20 cm tall)	Until 30 to 35 cm Tall			
Clean water to maintain a damp media should be used until the seed has germinated.	Should be fed with a balanced nutrient at 5-10 CF	When plants can be planted onto the system, the CF should be increased to 12-15 and then to 15-18.	Any suitable medium, such as rockwool/ perlite/ vermiculite	21-25ºC	2.5-3 plants per square metre

14.3 Plant Training Systems

Most of the growers usually use a single string to overhead wire for young plants. Once the plant is established it is usually stopped at the head and allowed with 4 lateral branches, these are then attached to strings to the top wire.

We have used 1.52-1.55 m long bamboo poles in each pot and then tied the plants to the poles. Additionally, a metal fence post was driven into the ground down the middle of the double row every 3.0 m. The bamboo poles were attached to the metal posts with horizontal polypropylene strings. As the plants grew taller, additional horizontal strings were added to support the weight of the branches.

14.4 Pollination

Some varieties require assistance and others are pollinated by insects and air movement. The pollination technique should be asked from the seed sellers. Minimum air temperatures are required to set good fruit.

14.5 Diseases

Common diseases are powdery mildew, botrytis, and leaf spotting. The diseases can be controlled to some extent by maintaining good air movement and keeping the humidity down on cool damp days. See "Plant Enemies (Chapter-8)"

14.5.1 Insects

White fly and red spider mites are the main threat. Other common insect pests were armyworms, corn earworms, aphids, and stinkbugs. Insect-pests and diseases should be treated as needed. Predators can be used with a totally integrated pest management programme. See "Plant Enemies (Chapter-8)"

14.5.2 Root Problems

Clean water free from pests, should be used to reduce the introduction of root diseases such as *Phytophthora* and *Pythium*. See "Plant Enemies (Chapter-8)".

14.6 pH of Nutrient Solution

The pH of nutrient solution should always be between 6.0 and 6.5, and there should not be a major pH drift. All pH correctors should be added in a very weak form and full strength acid or alkali should never be used.

The pH is usually lowered with phosphoric acid, but if required, phosphoric/ nitric acid mixes or nitric acid alone can be used.

The pH is usually raised using potassium hydroxide (caustic potash).

14.7 Nutritional Requirements

Capsicum crops require nearly all the mineral nutrients (nitrogen, potassium, phosphorous, calcium, magnesium, sulphur, iron, manganese, boron, copper, zinc, molybdenum). Growth promoters can be added to the nutrient solution to produce quality growth during periods of poor conditions, *e.g.* in winter under low light.

Nutrient formulations are blended to suit the water supply, so a water mineral analysis is essential for optimum formulation.

Hydroponic fertilizer solution concentrations that can be used to grow shadehouse capsicums in a composted pine bark media:

Nutrient Concentration (ppm mg/l)

Nutrient	Transplant to First Flower	After First Flower
N	75	125
P	45	45
K	110	180
Ca	140	140
Mg	35	35
S	45	50
Fe	2.6	2.6
Cu	0.2	0.2
Mn	0.7	0.7
Zn	0.3	0.3
B	0.6	0.6
Mo	0.05	0.05

Note: Capsicums are particularly susceptible to over-feeding. Although they don't actually die, they just look ridiculous.

⑮
Chrysanthemum
(Dendranthema)

Chrysanthemums are available in a wide range of flower colours, flower types, and plant sizes. Some of these are more suited for cut flower production and others for outdoor planting, often called hardy mums or fall mums. The flower heads occur in various forms, and can be daisy-like or decorative, like pompons or buttons. This genus contains many hybrids and thousands of cultivars developed for horticultural purposes. Important cut flower ones are 'Barliz' (spider), 'Dark Splendour' (red), 'Greeno' (green), 'Leman's' (red), 'Paragon' (white), 'Reagan Cherry' (red), 'Reagan Lemon' (yellow)', 'Reagan Orange (orange), 'Reballet' (pink), 'Rosalis' (red), 'Snowdon White' (white), 'Sunny' (yellow), ' Tigro' (light orange), 'Tikro' (white), 'Vesuvio' (spider), 'Yellow Paragon' (yellow), 'Yellow Venn', etc. Important cut flower varieties for export are 'Ajay', 'Anne', 'Birbal Sahni', 'Bright Golden', 'Chandrama', 'Lehmans', 'Nanako', 'Mountaineer', 'Snow Ball', 'Snowdon White', 'Sonar Bangla', 'Sonali Tara', *etc*. NBRI, Lucknow has developed chrysanthemum cultivars for thoughout the year sequential flowering though usually chrysanthemum flowers during November-December. These cultivars are 'Himanshu', 'Jwala' and 'Jyoti' for April-June flowering, 'Phuhar' for July-August, 'Ajay' and 'Sharda' for September-October, 'Makhmal', 'Megami', 'Mohini' and

'Sharad Bahar' for October-November, and 'Maghi' for flowering in February-March.

15.1 Propagation

Chrysanthemum (*Dendranthema grandiflora*), national flower of Japan, is grown for cut flowers, loose flowers, as potmums, for beddings, as border plants and for hanging baskets. It is a typical short day plant requiring critical day length of less than 9½ hours. It is popular for its wide range of beautiful flower colours and sizes and excellent vase life. It has various classes such as Single, Anemone, Korean, Double, Decorative, Pompon, Incurved, Incurving, Reflexed, Reflexing, Quilled, Spoon, Spider, *etc.* Although chrysanthemums can be propagated through seeds (only the bedding types and a few pompons), plant divisions, cuttings or tissue culture, commercial growers usually use cuttings for its propagation. Chrysanthemum growers can either get the cuttings from the healthy plants at their gardens or can purchase rooted or unrooted cuttings from specialist propagators/breeders. Specialist propagators maintain mother stock that is true to type, use culture-virus indexing to eliminate diseases, and grow and harvest large areas of stock to supply the market. So the propagation choice for commercial cultivation of chrysanthemum crops becomes:

☆ **Rooted cuttings**: It is the easiest method for small growers with no or limited propagation space. The cost per cutting may be higher than other methods and the grower has less control over cutting quality. The only facilities required is an area of reduced light where the humidity can be raised with periodic mist (often by hand) for 2-3 days until root growth begins. Rooted cuttings from the propagator should be inspected upon arrival and transplanted into final containers as soon as possible. If this is not possible, the cuttings can be stored in a refrigerator at 0.6–4.5° C for several days.

☆ **Unrooted cuttings**: The unrooted cuttings are to be directly placed in the final container or propagate in specialized cells, cubes, or trays.

15.1.1 Propagation Environment

The relative humidity in the propagation area should be maintained close to 100 per cent. To maintain this, the propagation area should be isolated and be large enough to accommodate about 3 weeks of production. Larger, unused areas make the environment more difficult to control.

☆ **Sanitation**–Everything that may come in contact with the cuttings should be disinfected. Bench surfaces and floors should be easy to clean and free of weeds.

☆ **Temperature**–Air temperature should be maintained between 21°C and 29.5°C. Temperature in winter should not drop below 20°C. Bottom heat to keep the propagation medium 22°C–24°C dramatically speeds rooting.

☆ **Light** - Light intensities should be 3200-3800 fc. Shading in the summer can be used to reduce the light intensity as well as to control heat. Night-break incandescent lighting to simulate long days is mandatory. HID supplemental light is useful during low-light periods.

☆ **Mist**–Uniform mist is necessary to keep cuttings turgid during rooting. Excessive mist will leach nutrients from the medium and leaves, over-saturate the medium reducing aeration, and stretch the cuttings. Inadequate mist causes wilting, necrotic leaf margins, and hard cuttings that do not root uniformly. Thus, a continuous film of moisture should cover the leaves until roots form. The frequency (how often the mist turns on) and duration (length of time the mist is on) of mist depends on many environmental factors and varies from season to season and throughout the day. This is usually controlled by programmable automatic systems such as an automatic mist controller. The mist frequency depends on the environment and the mist duration should be long enough to cover the foliage with a film of water. Mist may be required at night to keep cuttings turgid during the first 3-4 days. Afterward, mist is applied from 1 hour before sunrise to 1 hour after sunset. Tests of the quality of the water supplied to the mist system should be done to determine alkalinity. High alkalinity can increase the pH of the propagation medium and cause chlorosis of the foliage during propagation.

☆ **Media** - Almost any well-drained, course medium can be used to root chrysanthemums. This may be cells, flats, strips, or pots containing rockwool, foams, or other artificial media. Cuttings are usually stuck 2.5 cm apart in rows and 2.5-5.0 cm between the rows.

☆ **Rooting hormone** - Rooting hormones containing IBA or NAA may not speed rooting but increase rooting uniformity. Liquids or talcs containing 1500 ppm IBA should be applied to the lower end of the cutting.

☆ **Growth retardants** - This is used on certain vigorous cultivars during warm times of the years to prevent stretch and produce a compact plant. Cuttings are dipped into B-Nine at 1000 ppm and placed in plastic bags in a cooler overnight, then stuck the next morning.

☆ **Nutrition** - Many growers begin fertilizing chrysanthemum cutting as soon as callus forms (4-7 days) using a balanced fertilizer at 200-250 ppm N. Some growers apply a dilute solution of potassium nitrate through the mist system for the first several days.

15.1.2 Direct Stick

Many growers stick unrooted cuttings directly in the final container. The method reduces labour costs by eliminating the transplant step and can produce a more uniform product. Production time is also often reduced. However, a larger, more controlled propagation area is required. Death of one cutting in a pot can ruin the marketability of a 15 cm pot. Also, cutting must be graded to the same size in each pot for a uniform final product.

Containers for direct sticking should be filled close to the top with medium and watered before sticking. One or two lower leaves should be removed from the base of the cuttings so they may be stuck 1.25-2.5 cm deep.

15.2 CO_2 Enrichment

CO_2 enrichment to a level of 900 ppm will accelerate growth rates.

15.3 Temperature

Research has shown that once the root system has become established, the optimal temperature level during day is 22-26° C and during night is 18-22°C. Lower temperature of 18°C during the day and 14°C during night will prolong the time to flowering. High temperatures of 30°C day and 24°C night significantly reduce flower quality and yields.

15.4 Nutrients Requirements

Chrysanthemums have a high requirement for nutrients, particularly nitrogen and potassium. High N is required for the first few months of growth but should be reduced as buds are formed. Potassium should be increased when buds are formed. The recommended EC is 16-22 for media-based systems.

Nutrients	ppm
Nitrogen	150-160
Potassium	120-140
Phosphorus	60-72
Magnesium	50-60
Calcium	170-180
Sulphur	70-80

Trace elements are required in standard ranges.

16

Roses

Roses (*Rosa* sp.) are one of the most common and beautiful flowers in the world. They are available in wide range of colours and sizes, and are fairly easy to grow. Rose bushes must be given at least 0.36 m² (4 ft²) of growing room because they may grow in big sizes. This will allow enough light to reach the roses and also prevents branches from becoming entangled and harming each other. Further, rose branches tend to grow in a haywire direction and will need some kind of support. In hydroponics, the rose plant can be fixed in a container with pebbles or coco peat as a growing medium.

Roses need a lot of water to grow properly and need constant moisture. This doesn't mean that the roots are soaked in hydroponics nutrient solution for long, as this will only harm the roots. It will be better to use 'ebb and flow' hydroponics system.

16.1 Propagation

16.1.1 Propagation Environment

1. **Humidity**: The humidity level should be kept at or near 90 per cent. Cuttings need to remain in a humid location so that they can take moisture in through their leaves and stem since they don't have any roots yet. However, the medium should not be too wet otherwise the rose cuttings

will rot and die within a few weeks. If possible, create a 'greenhouse effect' using a dome, greenhouse, gallon bag or other moisture barrier. This will help keep the moisture in the air. If it is not possible to create a greenhouse effect, then mist with a spray bottle 2-3 times a day. Make sure that the soil remains moist at all times but not soggy wet.

2. **Sanitation**: Everything that may come in contact with the cuttings should be disinfected. Bench surfaces and floors should be easy to clean, and the pots should be free of weeds.

3. **Temperature**: Air temperature should be maintained between 21.0 °C and 23.5 °C. Bottom heat in the propagation medium at 25 °C–27 °C will speed up rooting dramatically.

4. **Light**: At least 6 hours of sunlight should be provided. Shading in the summer can be used to reduce the light intensity as well as to control heat. HID supplemental light is useful during low-light periods.

5. **Mist**: Uniform mist is necessary to keep cuttings turgid during rooting. Excessive mist will cause leaching down of nutrients from the medium and leaves, will reduce aeration and will cause cutting elongation. Inadequate mist causes wilting, necrotic leaf margins, and hard cuttings that do not root uniformly. The frequency (how often the mist turns on) and duration (length of time the mist is on) of mist depends on many environmental factors and varies from season to season and throughout the day. This is usually controlled by programmable automatic systems such as an automatic mist controller. The mist frequency depends on the environment, and mist duration should be long enough to cover the foliage with a film of water. Mist may be required at night to keep cuttings turgid during the first 3-4 days. Afterwards, mist is applied from 1 hour before sunrise to 1 hour after sunset. Tests of the quality of the water supplied to the mist system should be done to determine alkalinity. High alkalinity can increase the pH of the propagation medium and may cause chlorosis of the foliage during propagation.

16.1.2 Propagation Methods

16.1.2.1 Sexual (Seeds)

Like most other plants, roses too can be propagated through seeds for production of new varieties. **Seed** propagation is generally followed by breeders as through this method usually seedlings show variation in colour though there are many species which breed true such as *R. glauca*. Seed propagation is also used to multiply *R. canina*, very popular rootstock in Europe. Seed is harvested from matured and ripe hips normally during autumn. The seeds of rose contain an immature embryo which results into dormancy, *vis-à-vis* a very hard seed coat. Therefore, it is essential to soften it through acid scarification so its dormancy is broken. The seeds after extraction are stored in a dry, dark and airy place at 1-5 °C until their use. However, for raising them, these are first soaked in water for 24 hours and then placed in media like sphagnum moss or moist peat + sand, and wet-stratified at 1.6-4.4 °C for 2-6 months depending on the genotype. The treated seeds are sown to their own depth with sand or grit and placed at an airy place or in raised beds at 5 cm apart or in plugs in the month of October- November in fine and lightly pressed compost (having only a little ~ roughly one-quarter of the fertilizer contents of the potting compost and a fungicide to prevent damping off disease) which is capable of drawing water up to seeds placed near its surface. However, sowing during other months will require placing of the trays in cold frames. These may take about one year to germinate. After formation of the first true leaves, the seedlings being highly fragile are pricked out gently in individual containers. Up to their establishment, the seedlings are retained in the cold frame, and then these are hardened off gradually by moving them out of the cold frame during day time. After full acclimatization these seedlings are kept in the open, potted on as and when necessary and cared as other rose plants.

16.1.2.2 Vegetative

16.1.2.2.1 Layering and Inarching

Vegetatively, the roses are multiplied through cuttings, layering, inarching (grafting) and through budding, but now except budding which is the commercial way of propagation, all others have become completely redundant. Earlier and best rooting in layering may be induced through application of 25 ppm 2, 4-D, 500-2,000 ppm IBA, and 200 ppm NAA or 500 ppm IAA. These promoters may be variety or species specific. Rooting may occur within a month and then after a fortnight the layers may be severed from the parent plant and planted as the usual plant which will flower similar to that of its mother plant. In case of air-layering, the bark to a length of some 2.5 cm is removed without damaging the cambium layer and the wounded portion with a little adjoining bark is wrapped with moist sphagnum moss and then with polythene securely over the wound to both the sides, however, certain rooting hormones may also be added with the sphagnum moss for better results. Tissue culturing is at present only of academic interest.

16.1.2.2.2 Rootstocks, Cuttting and Budding (The Commercial Methods)

Commercially the roses are propagated only through cuttings for raising rootstocks for budding the desired scion varieties onto them. The growers can either get the cuttings from the healthy plants at their gardens or can purchase them from specialist propagators/breeders. IBA 500-2,000 ppm (Singh *et al.*, 2003) is used for rooting of cuttings. Cuttings taken during monsoon and spring respond well (Bose and Mukherjee, 1977). The length of hardwood cuttings should be 20-25 cm (in case of miniatures only 5-10 cm) and softwood (semi-ripe) cuttings 10-15 cm severed above a bud where the shoot is turning woody but the soft tips are trimmed off with the basal cut just below a node, at least with 3 nodes and devoid of leaves at least the lower ones, out of which 2/3rd of the length is pushed down in the soil for rooting and 1/3rd above. When there is dearth of material, even those cuttings with 1-2 nodes can also be used, especially during high humid atmospheric conditions or under mist. However, miniature flowering pot-roses are exclusively propagated by single leaf-node stem cuttings (2 cm of stem above and below the node), best being from shoots from the second cutback or pinch of plants already in production. These cuttings can be stored moist in plastic bags at 2 to 5 °C for up to 7 days (Dole and Wilkins, 1999). The upper cut should be slant to distinguish it from the bottom cut so that while planting there may not occur any polarity mistake and also to facilitate rain water not stagnating on the wound, and trickling down easily. Cuttings are always planted in slanting position with upper cut facing downward to avoid any pathogenic infection. Rooting media are fresh sand, sand + peat moss, vermiculite or vermiculite + peat moss, perlite or perlite + peat moss or good garden soil. Media should be porous and should have good water holding capacity. The rooting and further growth of cuttings are influenced by type of cuttings (the age and the physiological condition) and their treatment, *vis-à-vis* environmental conditions. All the roses can be propagated through soft-wood, semi-hard wood or hardwood cuttings taken from current season's growth under mist and with treatment of IBA. Miniatures under mist respond well through soft-wood or semi-hardwood cuttings, though in case of rootstocks, and strong-growing climbers, ramblers and polyanthas hardwood cuttings are most suitable (Harstmann and Kester, 1972), though large-flowered bush roses and that too of complex parentage do not root so readily and when root also the growth may be very poor. However, to avoid damping off caused by *Cylindrocladium scoparium*, it would be better to steam-sterilize the media. During course of rooting and even afterwards, there should be proper illumination.

Roses are commercially propagated by 'T' budding or shield budding. In this case the plant material from two different roses are united to combine the virtues of both, one that is bud-grafted is known as scion and is meant for producing desired quality of flowers on a desired stock on which the scion is budded. Usually the stock (rootstock or understock) is chosen on the basis of its suitability to a particular agro-climatic condition, its compatibility with the scion, *vis-a-vis* vigorosity. The dormant bud from the scion is taken and inserted under the bark of the rootstock and then scion with the stock is securely wrapped and tied tightly around leaving only the portion of the bud for growth. In case of shield budding, the bud along with the rectangular bark around is gently removed and inserted through the incision made out on the stock at a desired height but below a node but in case of 'T' budding, an inverted 'T' cut is made onto the stock where the bud with a little portion of bark from desired scion is inserted. There are various stocks to suit varying conditions. Climate, compatibility, soil, pests and diseases, and utility play an important role in the selection of rootstocks. *Rosa indica* var. *odorata* is recommended for northern plains. This can withstand dry or wet conditions and salinity. This is also tolerant to powdery mildew and insect-pests. *R. indica major* (syn. *R. chinensis major, R. × odorata*) is easily propagated through cuttings and this bears pink, semi-double and fragrant flowers. This is most suitable rootstock for hot conditions, and especially for the 'Baccara' roses. *Rosa bourboniana* is widely used in North India for producing standard roses as it throws vigorous and strong shoots though it did not gain prominence due to being susceptible to powdery mildew. *Rosa multiflora* is widely used as rootstock in South and Eastern India, and in the temperate regions of the country where winters are very cold. It is very vigorous all through the season from summer to autumn and is most suited to poor and light soils though scions budded onto this are not long-lived. It is resistant to nematodes. *Rosa canina* and *R. c. inermis* are commonly used rootstocks in Europe as in latter case the plant remains less thorny, that is why, it has been named *inermis* meaning thornless, however, *R. c. pollmeriana* rootstock possesses coloured green shoots. *Rosa canina* produces hardy plants and is popular where winters are severe or on heavy soils but it suckers heavily. It is propagated by seed and is resistant to drought and alkaline conditions. 'Dr. Huey' (syn. 'Shafter') is most suitable with areas having short-lived dormancy in roses. *R. ceriifolia froebelii* (syn. *R. dumetorum laxa*, popularly known as 'Laxa') rootstock, a moderately thorny or almost thorn-less producing no suckers, is highly vigorous and most suitable for budding the red varieties. This is reliable on most soils and climates and is replacing most other commercial rootstocks. *R. eglanteria* (syn. *R. rubiginosa*) as rootstock is very thorny, and slow-growing, hence is most suitable for pot-roses and miniatures. *R. manetti* (*R.* × *noisettiana*) and Natal Briar are also used as rootstocks in other countries. It is *R.* × *noisettiana* which is used in North America while *R. indica major* (*R. chinensis*) and *R. canina* are commonly used in Europe and Israel. Colombia, Ecuador and Mexico growers generally use *R.* × *noisettiana*, *R. chinensis* and *R. canina* as rootstocks. *R. canina* and *R. manetti* perform well in India also (Singh *et al.*, 2003). Under Delhi condition when various rootstocks (*R. bourboniana* strain, *R. canina*, *R. indica odorata*, *R. laxa*, *R. manetti*, 'Dr. Huey') were tried, the performance of *R. indica odorata* was recorded best (Singh *et al.*, 2003). Natal Briar is gaining prominence as rootstock for greenhouse roses. The ideal time for budding in North India is from January to March. In case of areas with mild climate, it can be done throughout the year and in temperate areas from mid- to late summer when air is quite humid. When new shoots grow after bud take, the shoots are cut back 20-25 cm above the graft union and then for 4 weeks the plants are exposed to -0.5 to 0 °C and high humidity which will stimulate axillary bud activity after planting (Dole and Wilkins, 1999). Anitha (1989) when tried the rootstock *Rosa indica* for three scion varieties, *viz.* 'Ambassador', 'Pink Panther' and 'Princess' during different seasons, recorded 82-98 per cent success from second fortnight of August to first fortnight of October when there is downpour and atmospheric humidity is high, however, from first fortnight of February to second fortnight of March was found least favourable due to hot climate and high sunshine, and stated that critical periods for success in budding was recorded as the current fortnight, one preceding and the one succeeding. Swelling initiation of the bud at the union shows the success of the bud take, and when shoots grow out sufficiently from this bud, growth of the rootstock above the union is cut off.

Growers can follow the following steps to grow roses using cuttings:

1. Before proceeding to choose rose cuttings, the most important step is to choose the spot where these cuttings are to be rooted. It should be in indirect light and the better rooting medium would be coco-fibre as it has great drainage and it virtually cannot be over-watered, besides the roots are even sturdier in this substrate.

2. Rooting container (10 cm size container will be most suited) are prepared and filled with sterile rooting medium such as coco-fibre. It is now watered thoroughly and is allowed to drain out the excess water.

3. Right stem is chosen for cuttings. These should be from healthy plant, *i.e.* free from insect-pests and plant pathogens. The cuttings should be from softwood, a stem from a rose that flowered this season. It should be about as thick as a pencil. It should be 10-20 cm long or even more, with a couple of leaf pairs on. This will insure that the grower will get a rose that will be a bloomer as well as a stem that is mature enough to root. Stems that are too old or too young will not produce root properly.

4. Stems are cut just above a node (the place on the stem where leaves emerge) and at an angle (45°) using a sharp and clean knife, or pruning shears and place them immediately in water. The leaves on the lower part of the stem have to be removed to make the joints visible. The roots will emerge from these joints.

Pruning Shear

5. Now with a sharp knife, about 2.5 cm of the outside of the stem is cut off (shaving a thin layer off on one side of the stem). This will leave an open area for more roots to form.

6. Optionally root hormone can be used. The ends of the cuttings are moistened in water and then inserted as a quick dip in the powdered or gel-form rooting hormone. Excess hormone is removed with a little shake. Always remember, only the area of the cut end of the stem that would be under the soil should be dipped in rooting hormone.

Rose Cuttings Covered with Plastic Cover to Create a Greenhouse Effect.

7. A hole in the medium in rooting container is now made with a pencil and the cuttings are placed therein (if the cane has two nodes, place one under the medium and if it has three, two nodes can be placed under the growth medium). Now the medium is gently firmed up around the cuttings with finger pressure to make sure there is no air pocket left.

8. Light watering is done to help settle the medium around the stems. Over-watering is always avoided as it may make the medium non-aerobic and the rooting hormone may be also washed off.

9. Cuttings are kept consistently moist (not wet) and for that either these should be in the mist chamber or water can be misted over them several times a day or even a transparent plastic bottle or transparent plastic sheet can be used to cover them creating the greenhouse effect (bottom of the bottle should be removed with a scissors before using it).

Rose Cuttings Covered with Plastic Bottle to Create a Greenhouse Effect.

10. The roots may start developing after about 1-3 weeks or sometimes may even take more time. Never pull the rose stem cuttings out to check and see if it has roots. The best way to tell if cuttings have rooted is when new leaves appear and new growth can be seen.

11. Just like anything else, roses grown from cuttings need to be hardened off. Allow the roots to become as long as possible so that they can grow stronger before being transplanted.

12. After the cuttings have been well established for a few months and are starting to grow well, they can be considered for planting.

   (a) Harden-off the roses by placing them outside (if not already) for a few hours a day in the warm sunshine. Be sure not to let them stay out too long or they will sunburn. Each day add a little more sunshine to their stay.

   (b) They can also be placed in a shady spot and allowed to get the sun they need as the sun moves east to west.

   (c) Do not leave out in the cold, in the wind or bad storm. These plants are too fragile for this. Once they are hardened off for a few weeks, they can be planted. Be sure to keep them well watered. Some growers choose to leave them in their pots for the first year so that they can be moved in and protected on their first winter.

13. Remove the rooted cuttings carefully from the rooting container and place them in grow-container of the hydroponics system.

Keeping the cuttings in a zip-bag with a little moist medium until they form roots is another easy cutting propagation method, the only inconvenience is that it may be more difficult to transplant since those roots are very fragile and may get broken during transplantation.

There are more elaborate propagation methods such as application of heat underneath the potted plants as it stimulates root formation but this method needs an extra effort, perfect skilling and equipment.

Note: Cutting propagation is not suitable for all rose varieties. Miniature and climbing have the greatest success rate with this method.

16.1.2.2.3 Direct Sticking or Stenting

Many growers stick non-rooted cuttings directly in the final container. The method reduces labour costs by eliminating the transplant step and can produce a more uniform product. Production time is also often reduced. However, a larger, more controlled propagation area is required. Death of one cutting in a pot can ruin the marketability of a 15 cm pot. Also, cutting must be graded to the same size in each pot for a uniform final product. Containers for direct sticking should be filled close to the top with medium and watered before sticking. One or two lower leaves should be removed from the base of the cuttings so that they may be stuck 1.25-2.5 cm deep.

Stenting is a quick method of propagation of roses based on whip grafting. In stenting method, a selected cultivar budded on an unrooted cutting of a rootstock results in a complete plant in 3-4 weeks. A piece of stem with a 5 leaflet-leaf and a single dominant axillary bud is used as a scion. A piece of stem consisting of a single internodes without buds and/or leaves is used as rootstock. Scion and stock are held together with the help of a peg. The basal end of a rootstock is dipped in a talc with IBA at 0.4 per cent. The rooting medium is a mixture of peat and perlite in 1:1 ratio. The relative humidity is maintained at 100 per cent with intermittent misting and the temperature is maintained around 25 °C.

16.1.2.2.4 Division of Clumps

There are roses that grow on their own roots and send out suckers which when root out should be **divided** and planted separately when these are dormant (Brickell, 1992). The roses such as some of the Gallicas and the cultivars of *R. pimpinellifolia* and *R. rugosa* can easily be propagated through this technique. Roses grown through cuttings and also through seeds can be propagated through this method.

16.1.2.2.5 Root Cuttings

Hartmann and Kester (1972) stated that certain species such as *Rosa blanda, R. nitida* and *R. virginiana* can be propagated even through **root-cuttings**. Proximal (uppermost) ends of the root-cuttings should always be up and at the level of rooting media to maintain the polarity.

16.1.2.2.6 Micropropagation

For rapid multiplication, roses can also be propagated through **tissue culturing** the terminal buds, immature floral buds and petals. Uma (1991) studied chemical mutagenesis in rose cv. 'Alliance' under *in vitro* culture, and recorded 0.08 per cent mercuric chloride treatment for 15 minutes to be the best for surface sterilization of axillary buds, BAP 2 mg l^{-1} + 2, 4-D 1 mg l^{-1} for culture establishment, BAP 2 mg l^{-1} + GA_3 1 mg l^{-1} for shoot proliferation and the full strength MS medium when supplemented with 2 mg l^{-1} IAA for maximum rooting. For EMS treatment, 0.125 and 0.250 per cent were found to be the best for time taken for bud take, multiple shoot production and rooting for the buds excised at the time of flower harvest though 0.375 and 0.500 per cent EMS curbed multiple shoot production. Under *in vitro* studies with the cv. 'Folklore' (Wilson, 1993), mercuric chloride 0.08 per cent for 12 minutes for shoot and axillary bud explants and 0.06 per cent for 12 minutes for internodal segments and leaf disc explants minimized the contamination maximally. Axillary buds excised 4 days after flower opening had the best response in culture establishment. The MS medium supplemented with kinetin 2.0 mg l^{-1} + GA_3 1.0 mg l^{-1} recorded early multiple shoot induction and highest number of shoots per culture, though addition of BAP 2.0 mg l^{-1} + GA_3 0.75 mg l^{-1} proved best with respect to highest percentage of cultures with multiple shoots, however, floral bud initiation was observed under BAP 2.0 mg l^{-1} + GA_3 0.50 mg l^{-1}. *In vitro* rooting was best when the medium was supplemented with IAA and NAA each at 1.0 mg l^{-1} + activated charcoal 500 mg l^{-1}, however, successful hardening and *ex vitro* establishment of plantlets were achieved by surface inoculation of germinated spores of VAM mycorrhizae in liquid suspension. Callus induction was found best under BAP 0.5 mg l^{-1} + NAA 2.0 mg l^{-1} +2, 4-D 0.5 mg l^{-1} though callus proliferation was recorded best under BAP 0.5 mg l^{-1} + NAA 0.1 mg l^{-1} + ascorbic acid 5 mg l^{-1}.

Pre-culture dipping of explants for 3 hours in 2,000 mg/l carbendazim and 200 mg/l 8-HQC, and afterwards with 500 mg/l gentamycin to the establishment medium control the contamination to a greater extent. More than 90 per cent explant survival and its sprouting with dark green colouration is obtained in MS medium where sprouting occurs within 8 days, and addition of 3 mg/l BAP + 0.1 mg/l NAA and 1 mg/l GA_3 causes more than 95 per cent sprouting within 4 days producing more than 3 shoots per explant (Singh *et al.*, 2003). A combination of 0.2 mg/l BAP, 0.2 mg/l NAA and 0.5 mg/l GA was found for cvs 'Priyadarshni', 'Raktagandha' and 'Raktima'; and 1.0 mg/l GA instead of 0.5 mg/l in case of cv. 'Arjun' quite favourable for initial establishment through shoot tip culture though for maximum proliferation it was 1.5 mg/l BAP and 0.2 mg/l NAA in case of 'Priyadarshni' and 'Raktagandha' and 2.0 mg/l BAP and 0.2 mg/l NAA in case of 'Arjun' and 'Raktima' (Singh *et al.*, 2003).

16.2 Protected Cultivation

Soilless cultivation of roses can be done only when there is provision to protect them from the adversities of the environment. Greenhouse roses are usually day-neutral with recurrent flowering occurring throughout the year to sustain regularity in supply to the market or directly to the consumers so that reliability of the market or consumer is fully ensured. Zieslin and Moe (1985) stated that garden roses include both recurrent and non-recurrent flowering cultivars. Market reliability is lost when only the roses are supplied to the market periodically. Therefore, it is necessary to grow them in controlled atmosphere where optimum conditions for their growing are met, and it is possible only when the grower is equipped with greenhouses. **Greenhouse** is a glass or transparent plastic structure, often on a wooden or metal frame, for growing high value plants by providing required heat and light (intensity and duration), *vis-à-vis* protection from adverse weather conditions. The structures for roses are, in fact, warm greenhouses with 13-18 °C temperatures as these continue blooming throughout the year at 15/27 °C night/day temperatures. In temperate areas, glasshouses are required only during winter months when temperature goes very low and plants tend to become dormant. So, to create an environment for growing plants in winter such greenhouses are erected though summers are quite congenial for growing plants outdoors without any protection. During sunny days of May-June when temperature is anticipated to go beyond 30 °C, roses may require to be roofed with some semi-transparent material to provide only filtered lighting from 10.30 a.m. to 2.30 p.m. so that the plants are protected from scorching sunlight and uncongenial temperatures, however, the sides are left free for proper air circulation. During winters, *i.e.* from third week of December to second week of March, the greenhouses prepared from polyethylene will be damaged due to weight accumulation of snows on their roofs in snowfall areas, especially higher reaches of Uttarakhand, Kashmir and Himachal Pradesh. Hence in such areas only glass-made greenhouses will work though cost will be prohibitive. In the sub-tropical areas, the polyhouses are required from mid-March to mid-October (7 months) to reduce the temperature levels as roses are successfully grown at 15-27 °C temperature range though here electricity cost becomes prohibitive, however, from mid-November to mid-February the polyhouses without electricity device for temperature control will be effective as the erection of such structures will increase the temperature levels from 8-12 °C which will facilitate rose growing as usual temperature in December-January in sub-tropical region of Delhi remains in the range of 5/15 °C (night/day). In quite tropical areas, the polyhouse will work throughout the year to maintain the congenial temperatures for its growing, except in winter. It may be either (i) **conventional greenhouses**, *viz.* 'traditional span', 'Dutch light', 'three-quarter span', 'lean-to', and 'Mansard' or 'curvilinear', or (ii) **special greenhouses**, *viz.* 'dome-shaped', 'polygonal', 'alpine house', 'conservation greenhouse', 'mini-greenhouse' or 'polytunnel'. All of these may have aluminium or timber framework, and glass walls or part-solid walls, except Dutch light greenhouses. Glasshouses, now days are quite expensive but polytunnels are much cheaper and have added benefits. Greenhouse culture leads to 10-15 times higher yield, blooming consistently throughout the year, than that of outdoor cultivation where neither quality is good nor blooming is perpetual, though greenhouse cultivation depends upon the greenhouse design, availability of environment control facilities, cropping systems, greenhouse management and crop type. A modern greenhouse has three major components–structure, covering and environment control system (temperature, light and humidity). Of various designs, gutter connected house covered with heavy-duty transparent plastic sheets having roof ventilation is most efficient for wide range of conditions. These types of greenhouses are cheaper and most feasible to automate the single consolidated space inside a gutter connected greenhouse than the multiple equivalent space in Quonset greenhouses. Management of materials and products into and out of the greenhouse requires less labour in a single large space than in numerous small spaces. The heating cost is less in multispan (gutter) greenhouses, because there is a less expanded area. The height of the gutter above the ground is increasing over the years to accommodate the growing evolution of climatic control equipment and automation devices. The original gutter connected greenhouse typically has a 2.4 m gutter height, but today 4.3 m is becoming very common. Gutter may be constructed from galvanized sheet of aluminium. The distance between gutter rows depends on the greenhouse brand purchased. The distance ranges from 3.2 to 12.2 m. Greenhouse with spacing between gutters of 3.7, 5.2, 6.7 and 9.1 m can be covered by film plastic sheets 4.3, 7.3, 7.6 and 11 m wide. Greenhouses are now offered that have roll up side curtains and can be installed on two or all the four walls. Greenhouses with retractable roof are becoming very popular in USA and Canada. The purpose of side curtain plus roof-ventilator system or the retractable roof with or without roll-up curtains is to replace high energy consuming 'fan and pad cooling systems'. These passive cooling systems work well in hot or cold climates. The greenhouse frames should preferably be covered with double layering covering 0.10-0.18 mm thick UV stabilized plastic. Today, polythene film as well as rigid FRP, acrylic and polycarbonate panels are available with an antilog surfactant built into the film or panel. It is advisable to use an antilog product because in addition to water dropping, the condensation also reduces light intensity within the greenhouse.

Fresh flower crops are grown in either ground beds or raised benches in the greenhouses. Such beds are 1.1 or 1.2 m wide and 20 cm deep, but 30 cm is best for rose beds. Fresh flower beds are oriented along with the length of the greenhouse with 45 cm aisles between them. This arrangement of beds allows for 67-70 per cent of floor space for growing. Soil is the natural medium for growing greenhouse crops, though roses and gerbera do not respond well in soil. The **cocopeat** available in plenty in South India can be used to get manifold yield with better quality. This substrate provides better root spread with minimum resistance, sufficient porosity, maximum nutrient uptake, better drainage of excess water, *vis-à-vis* minimum pathogens and scarce algal growth, as the medium is quite independent of the soil. Cocopeat, a powdered coconut fibre, is decomposed for 3-4 months in a tank filled with water which takes out the excess salts from the dust, and weathers it to the extent it becomes effective for crop growing. When roses inside polyhouse are planted in pots, a grower has almost a different mixture than others but a mixture containing coarse turfy loam 180 parts: wellrotten farmyard manure 100 parts:dry bonfire ash 40 parts:old soot 5 parts:bone meal 1 part, all by volume, should be mixed thoroughly and stored inside the store room where it should have been turned over several times, and this at the time of planting should be filled in 25 cm pots leaving only 4 cm at the top and then the plants are planted firmly pressing the roots all around and watered. Temperate as well as sub-tropical climates of North India face heavy chilly weather, requiring heating to sustain crop growth for flower regulation on special events like Christmas, New Year, Easter, Mother's Day and Valentine Day. Greenhouses can be **heated** with the oil burners, hot water or steam and electric heaters. A central heating system can be more efficient than unit heaters. In this system, two or more large boilers are in single location from where heat is transferred in the form of hot water or steam pipe mains to growing areas. Their heat is exchanged from hot water in a pipe coil located in plant zone or through overhead pipe. **Cooling** in the greenhouse is very essential where outside temperature goes above 30°C. Cooling system generally consists of fan at one side and pad at the other where principle of evaporation cooling is facilitated by running water stream over pad and consequent withdrawal of air through it by fans on the opposite side. A reduction of maximum 10-15 °C difference in temperature could be achieved depending upon the system and outside climate. Greenhouse should be airtight during the running of fan-pad system and care must be taken to periodically clean the pads from salts and algae. Other active way of cooling alternative is fog cooling. Control can be achieved through analog machines or by computers through aspirated chambers. Top and side ventilation also adds in cooling along with maintaining relative humidity.

The flower production and its quality may deteriorate due to very high **light intensity** during summer. **Shading** reduces light intensity and cools the microclimate inside the greenhouse. Shade paints (lime or Redusol or Vari clear), agro-shade nets or retractable thermal screens are generally used or operated manually or through automatic devices. Several plant species flower only when they are exposed to specific light duration. Yield and quality of flower crops could be increased with artificial lighting during night hours. **Fertigation** varies from single broadcasting of fertilizers to use of soluble grade fertilizers over different operating systems. One of the most modern technologies is currently offered by Priva-Phillips Nutriflux or Van Vliet Midi Aqua Flexilene System. Both the systems have a nutrient recycling method translating plant demand of nutrients in relation to EC/pH of the media, temperature, RH, light intensity, crop growth, mineral deficiency, etc. **Water quality** is very important though often overlooked. Total salt content levels, alkalinity levels, the balance of Ca and Mg, and levels of individual ions such as boron and fluoride can all have serious bearing on crop success. The water source should be tested before a greenhouse is established. **Electrical conductivity** level should be 0.75-1.50 dS/m and a **pH** of 6-7. Automatic watering system through drips or over-head foggers are generally used. A manual or semi-automatic control system is less capital intensive but requires a lot of attention and care. Now computerized control systems are available which can integrate temperature, light intensity, RH, CO_2, plant moisture, nutrient requirement, and plant protection measures. There are certain unheated structures in the gardens which are used to advance or extend the growing season such as cloches used *in situ* over plants in the garden itself, and **cold frames** also used for storing 'resting plants' and for hardening off plants propagated in the greenhouse.

In polyhouses, rose is planted on raised beds having one metre width, 30 cm height and length as per the structure. Roses are planted in double lines at 30 15-20 cm, accommodating 7 to 13 plants per square metre. The cocopeat may be use as the medium along with perlite and sand stones. Most growers use soil which is sterilized with formaldehyde and thoroughly mixed with FYM, phosphatic and potassic fertilizers but in soilless culture nothing of the sort is required as fertigation is given looking into the requirement of the nutrients and water by the plants. Sterilized farmyard manure at the rate of 100 t/ha should be thoroughly incorporated in the soil if the roses are to be planted in the flat beds. Usually these are planted in strips one metre wide with medium filled in double-layered polythene sheets. Medium used is usually cocopeat whose depth may be from 20 to 30 cm. Initially the plants are supported to stand properly. Roses grown at a minimum **temperature** of 12 °C with 2,000 lux supplementary **lighting** (total light period of 16 h/day) from HPS lamps during November to February increases flower yield. There may be species that requires cold for dormancy release. There are certain cut flower roses that produce fewer leaves under low irradiation levels than under high, and there are facultative LD cultivars which flower even with fewer leaves under long days (Moe, 1970).

Rose plants respond to **CO_2** enrichment (1,000 to 2,000 ppm) of the greenhouse atmosphere (Mastalerz, 1987a, b) with a decrease in the rate of lower bud abortion, an increase in lateral sprouting and an increase in the fresh weight of flower stems (Mortensen and Moe, 1983). CO_2 (700 to 1,000 ppm) in miniatures produce increased number of flowering shoots and hasten flowering though shoot lengths are increased (Clark *et al.*, 1993), whereas in bare-rooted potted plants, Mortensen and Moe (1983) recorded 19 per cent increased floral shoots in floribunda 'Garnet' roses at 950 ppm CO_2 level. Prevention of flower bud abortion can be a result of alteration in partitioning of assimilates between upper buds. An additional possibility is that the effect of the CO_2 on the flower bud retention is due to anti-ethylene action of CO_2. The increase in number of flowers and quality may be the result of CO_2 on photosynthetic process, partitioning of assimilates and the effect of CO_2 on ethylene. Increasing CO_2 levels during day time to 800-1,000 ppm are beneficial for rose production due to increased dry matter production.

16.3 Varieties to Choose From

There are **glasshouse roses** for long-lasting cut flower use out of which the **long-stemmed** ones are 'Aalsmeer Gold', 'Bellona', 'Bianca', 'Black Magic', 'Carona', 'Carambole', 'Cora', 'Corvetti', 'Dallas', 'Diplomat', 'Dr. Verhage', 'Esacada', 'Femma', 'Film Stone', 'First Red', 'Gabriella', 'Golden Gate', 'Grand Gala' (almost thornless), 'Grand Prix', 'Ilona', 'Jacaranda', 'Kiss', 'Konfetti', 'Lambada', 'Laser', 'Magic Moment', 'Maybella', 'Monte Carlo', 'Movies Star', 'Nicole', 'Noblesse', 'Osiana', 'Papillion', 'Parea', 'Passion', 'Pavrotte', 'Pink Sensation', 'Prophyta' 'Ravel', 'Red Velvet', 'Rodeo', 'Rossini', 'Sacha', 'Samura', 'Sandra', 'Sandy Femma', 'Sangria', 'Skyline', 'Soledo', 'Sonia', 'Spinks', 'Star Light', 'Sublime', 'Susanne', 'Temptation', 'Texas', 'Tineke', 'Virginia', 'Vivaldi', 'Zéphirine Drouhin', etc.; the **medium-stemmed** ones are 'Aalsmeer Gold', 'Bellinda', 'Europa', 'Flirt', 'Frisco',

'Golden Times', 'Jack Frost', 'Jaguar', 'Kardinal', 'Kiss', 'Lambada', 'Mercedes', 'Motrea', 'Nordia', 'Ohio', 'Pink Prophyta', 'Prophyta', 'Sandokan', 'Souvenir', 'Tina', 'Vanilla', 'White Success', etc.; and the **small-flowered Floribundas** are 'Cecile Brunner', 'Garnette', 'Marimba', 'Sweetheart Rose', etc. The leading export cultivars having long and strong stems, tapering and pointed buds with slow opening but with more number of flowers, and attractive colour with more longevity are 'Cora', 'Corvetti', 'Diplomat', 'Femma', 'First Red', 'Golden Gate', 'Grand Gala', 'Kiss', 'Konfetti', 'Lambada', 'Laser', 'Nicole', 'Noblesse', 'Osiana', 'Papillon', 'Parea', 'Passion', 'Pavrotte', 'Prophyta', 'Rodeo', 'Rossini', 'Sacha', 'Samura', 'Sandy', 'Sangaria', 'Soledo', 'Susanne', 'Texas', 'Tineke', 'Vivaldi', *etc.*

16.4 Light

Planting sites may have full sun to partial shade; however, roses do best with 6 hours or more of direct sun with 6,000-8,000 f.c. light. In a hydroponic setting, sodium halide lamps or compact fluorescent bulbs produce favourable results. Lighting must be consistent and strictly controlled during the flowering stage. Whether the grower had better success with 18 hours of light and 6 hours of darkness, or a 12/12 photoperiod, the lights should be turned off and back on at exactly the same time each day. A digital timer that makes such precision possible can be used.

16.5 Temperature

Ideal temperatures for rose growing are between 18 and 24°C, however, these tolerate from 15 to 28 °C for flowering. Many varieties of roses are able to withstand cold winters, even then gardeners still need to protect them to ensure survival and minimize stress on the plants. Although the greenhouse will take care of nearly all the environmental conditions for the roses, growers may want to place their hydroponics systems outside in their gardens so as to add beautification to their garden. If the hydroponics system is placed outdoor, wrapping roses in gauze or a burlap mesh before cold weather arrives is a good idea, as is planting them in a sheltered spot. Some growers mound mulch around their rose bushes, and hold it in place with chicken wire. This can protect roots and stalks. Wait until just before the onset of serious frost before applying winter protection. If protection is applied too early, the winter hardiness of the plant is jeopardized. Before wrapping or mulching, the fall clean up should be done by removing plant debris and diseased plant parts. Spring pruning will remove tip dieback. In very hot climates, the roses must be protected from intense afternoon sun.

16.6 pH Requirements

Roses prefer well-drained medium with a pH value of 5.5 to 7.0 (optimum being 6.0 to 6.5) and EC less than 1.0. The likelihood of micronutrient deficiencies becomes greater as pH increases, especially for pH values above 7.5.

16.7 Nutrient Requirements

Roses are not salt tolerant, so EC values which measure the level of salt ions in the nutrient solution, should be less than 2.0 dS/m.

Suggested EC and nutrient levels for growing roses.

Soil Characteristics	Unit	Low	High
EC	dS/m	0.5	2.0
NO₃-N (nitrate-N)	ppm	35	150
NH₄-N (ammoniacal-N)	ppm	0	20
P	ppm	5	50
K	ppm	50	300
Ca	ppm	40	200
Mg	ppm	20	100
B	ppm	0.1	0.75
Fe	ppm	0.3	3.0
Mn	ppm	0.2	3.0
Cu	ppm	0.001	0.5
Zn	ppm	0.03	3.0
Mo	ppm	0.01	0.10

Tissue tests may be used to provide information on the current nutritional status of the rose plant. Suggested nutrient levels are presented below.

Suggested Values for Nutrient Levels in Rose Tissue

Nutrient (Unit)	Low	High
N (per cent)	3.0	5.0
P (per cent)	0.2	0.3
K (per cent)	2.0	3.0
Ca (per cent)	1.0	1.5
Mg (per cent)	0.25	0.35
Zn (ppm)	15	50
Mn (ppm)	30	250
Fe (ppm)	50	150
Cu (ppm)	5	15
B (ppm)	30	60

Roses express deficiency symptoms quickly though return to normalcy slowly. Generally, the basic **nutrient** concentration of N in ammonium form (5:1 nitrate:ammonium in summer and 10:1 in the winter when soil or water has high pH) at 150 to 200 ppm is recommended. Jørgensen (1992) stated that low EC (1.5 ds/m) is critical and he advocated N:P:K levels at 220:30:195 ppm fertigation. A fertilizer dose of 520 kg N, 868 kg P_2O_5 and 694 kg K_2O/ha per year has been recommended for Delhi conditions for high density planting (30 30 cm) for 'Super Star' roses under open field conditions (Bhattacharjee and Damke, 1994). Potassium application rates affect the number and quality of flowering stems and reduce the disease incidence.

Potassium closer to 50 g/m^2 per year is quite effective to obtain maximum quantity of commercial stems regardless of the cultivar. Heavy dose of organic manure like FYM at the rate of 50-100 t/ha is recommended in rose. Bhattacharjee (1994) observed that the soil application of sulphur at 10 kg/ha, magnesium sulphate at 50 kg/ha and calcium sulphate at 50kg/ha significantly improves flower yield and quality in cv. 'Raktagandha'. Ushakumari (1986) worked out effect of N, P and K nutrition on bud take, vegetative growth and flowering of roses under Kerala conditions. The fertilizer application improved the girth of stock at the time of budding. A fertilizer combination of 0.75 g each of N and P$_2$O$_5$ and 0.25 g K$_2$O, per plant has been found effective in promoting early bud take; a dose of 1.0 g N, 0.75 g P$_2$O$_5$ and 0.25 g K$_2$O for early emergence of leaves and for maximum height of the plants; however, the best combination was 0.75 g each of N, P$_2$O$_5$ and K$_2$O. She further reported that nitrogen and potassium alone and in combination were effective in early induction of first floral bud but phosphorus was found effective only when combined with nitrogen and potassium. Nirmala George (1989) recorded maximum number of flowers with 10.0 g N, 30.0 g P$_2$O$_5$ and 10.0 g K$_2$O per plant when applied at 30 day's interval; the longest floral shoot with 20.0 g N, 30.0 g P$_2$O$_5$ and 5.0 g K$_2$O applied at 45 day's interval; maximum floral diameter with 20.0 g N, 15.0 g P$_2$O$_5$ and 10.0 g K$_2$O applied at 15 day's interval; and maximum floral life with 10.0 g N, 45.0 g P$_2$O$_5$ and 15.0 g K$_2$O applied at 45 day's interval.

16.8 Weed Control

Rajamani *et al.* (1992) recorded better control of monocot **weeds** with glyphosate 1.0 kg a.i/ha and dicot weeds with oxyfluorfen 0.5 kg a.i/ha. The weeds can be kept under check through mulching with black polythene strips and through a thick layer of straw or colossal garden waste which after rotting will also provide nutrients to the plants.

16.9 Insect-Pests and Diseases

There are a wide variety of insects that affect roses though in soilless culture soil is not present hence breeding atmosphere for most of the pathogens and insect-pests is eliminated. Even then many of the insect-pests and diseases get introduced in such environments due to frequent opening of doors and vents and human interference. Insect-pests degrade the ornamental value of roses in the garden and decrease productivity in commercial rose ventures. In case of severe infestation, they can even cause death of plants. Aphids (*Macrosiphum rosae*), thrips (*Rhipiphorothrips syriacus, Scirtothrips dorsalis*), leafhoppers (*Edwardsiana rosae*), scales (*Aulacaspis rosae, Aonidiella aurantii*), the mites (*Twtranychus cinnabarinus, T. neocaledonicus*), mealy bugs (*Pseudococcus* spp.), whiteflies (*Trialeurodes vaporariorum, Bemisia tabaci*), termites (*Odontotermus obesus*) and the leaf feeding chafer (*Oxycetonia versicolor, Adoretus* spp.) and ash beetles (*Myllocerus* spp.) are the common pests of roses when grown in open. Lady bird beetles feed on the aphids, and pressure water spraying kills the aphids. Chemical insecticides such as Acephate, Dimethoate, Imidacloprid, Malathion or Monocrotophos 0.1 per cent spraying will control these pests in case when live-stocks are not kept inside. Thrips which attack new rose-growths generally after pruning which can be controlled with Chloripyriphos, Dimethoate, Metasystox or Phosphomidon at 0.05 per cent, or Fenitrothion at 0.075 per cent. Spinosad is most effective in the control of thrips. Leafhopper nymphs and adults, both feed on the leaf-undersides and other tender parts and these can be controlled by spraying Carbaryl. Imidacloprid, Malathion, or Permethrin at 0.2 per cent spraying will control this pest. The covers of scale insects scale often blend in with the bark of the plant. These suck the sap from the plant with their thread-like sucking mouth parts. The infested parts can be pruned to prevent further spread. They can be effectively controlled by spraying insecticides like Dimethoate, Malathion, Rogor or Dimethoate at 0.2 per cent. **Spider mites** being so minute are usually not visible to the naked eye. They feed on the underside of the leaves by sucking the fluid from plant cells and in severe infestations even the plants are killed. Populations of plant-feeding mites are often kept in check by naturally occurring predatory mites and other predators. Outbreaks of spider mites often occur following insecticide treatments targeted against other pests because these treatments destroy the predatory mites. Foliar applications of Acephate, Carbaryl or Pyrethroid insecticides have a tendency to trigger mite outbreaks. Mites are favoured by hot-dry weather, especially if accompanied by dusty conditions. Keeping plants well watered during periods of drought helps reduce the potential for mite outbreaks. Washing foliage, especially the undersides with a water spray can also help control or prevent mites. Mites can be controlled by spraying miticides like Bifenzate, Dicofol, Dimethoate, Malathion, and Milbemectin from 0.5 to 1 ml per litre. Apollo (Clofentezine 0.6 ml/l), Cascade (Flufenoxuron 1 ml/l) or Vertimec (Abamectin 0.5 ml/l) controls the mites completely in one fortnight of application. Mealy bugs damage the plants by sucking sap from the shoots. They can be effectively controlled by spraying Monocrotophos and Dimethoate at 0.2 per cent spraying. Whiteflies are small insects that are covered with a white waxy powder. They most often occur on the undersides of leaves but clouds of adults will fly around infested plants when disturbed. Immature whiteflies are immobile, scale-like insects that feed on the undersides of leaves. Like aphids, whiteflies suck plant sap through piercing-sucking mouth parts. They can be effectively controlled by spraying Acephate, Imidacloprid or Malathion at 0.2 per cent or neem oil (20 ml/l). Termites cause damage to the rose plants even before they are fully established. They destroy the underground parts. They are very difficult to control as they colonize under the soil. One of the most difficult thing with termites is that by the time they are evident much of the damage has already been done to the plant. They can be controlled by soil application of Carbofuran granules at 5 g/m² followed by watering or Chlopyriphos at 0.5 per cent soil application. Chafer beetles (*Oxycetonia versicolor, Adoretus* spp.) and ash beetles (*Myllocerus* spp.) feed on the growing points making irregular holes and punches on the leaves. The grubs feed on the roots. The insecticides like Acephate, Carbaryl or Chlorpyriphos at 0.2 per cent controls these beetles. The larvae (**caterpillars**) of various moths or butterflies have been found infesting roses through their chewing type of mouth parts and feeding on the leaves, resulting in leaf skeletonization or defoliation. They can be effectively controlled through 0.2 per cent Acephate, Carbaryl, Dichlorvos or Spinosad sprayings and also through use of *Bacillus thuringiensis*. Digger wasps (*Crabro* sp.) damage the plants after pruning. They dig into the pruned stem through cut ends. Their digging also facilitates the entry of weak pathogens that cause die back. The pruned ends should be effectively sealed with a proprietary fungicidal paste to prevent the damage. In case of small gardens a few drops of insecticides like Dimethoate may be dropped into the tunnel. Leaf cutting bees (*Megachile* spp.) build their nests in hollow stems, reeds or pipes. They stuff the nests with pieces of leaves cut from various plants and feed them to developing larvae. Roses are one of the preferred plants for cutting leaves. Adult bees cut semi-circular holes in rose leaves, usually affecting only a few leaves on any one plant. Their damage is minor. They need not be controlled as their control may be harmful to the bee population.

Roses are also infected with many of the pathogens such as *Diplodia rosarum* (die back), *Diplocarpon rosae* (black spot), *Sphaerotheca pannosa* (powdery mildew), *Peronospora sparsa* (downy mildew) and *Phragmidium mucronatum* (rust). Most common disease is **die back** which starts from top to downwards, especially through wounds or pruned ends. In its infection, initially the affected stem-part becomes black but afterwards whole plant dies. The affected part should be cut immediately in slanting position with a sharp knife in one attempt and burnt, *vis-à-vis* cut ends should be sealed with Bordeaux paste or copper oxychloride. Slant cut will avoid water stagnation at the point of cut. Carbendazim 0.2 per cent spraying also controls this problem. **Black spot** disease appears as dark brown circular spots with fringed borders which after sometime expand and coalesce. Leaves also turn brown and start falling. The spread of the disease is checked by burning the plant debris and through spraying with 0.2 per cent Bayleton (Tridumefon), Captan, Chlorothalonil, Fenarimol or Benlate or 0.5 per cent Tilt (Propicenazole). *Rosa laevigata, R. multiflora, R. roxburghii, R. rugosa, R. virginiana* and *R. wichuriana* are highly resistant to black spot. **Powdery mildew** appears as powder-greying of entire plant, especially on the abaxial surface of young leaves and floral buds which ultimately causes unopening of the flower. The conidia spreads through wind and the disease is favoured by humid or excessive dry weather conditions and also when days are warm and nights are cool. Sanitation of the plant debris in the surrounding will prevent the spread of the disease. It can be controlled through spraying with 0.05 per cent Karathane followed by 0.2 per cent Sulfex or Topaz (Penconazole 0.05 per cent). **Downy mildew** is quite prevalent in polyhouse roses where its infection is noticed on young plants with purplish-red to dark irregular spots on leaves during extended period of cool and humid weather in early to mid-spring. Leaf falling is most common and mostly in its severe infection the entire plant defoliates completely. Downy mildew is controlled through 0.20-0.25 per cent spraying with Chlorothalonil, Fosetyl-aluminium, Ridomil, Cuman-L, Aliette or Aliette 0.25 per cent + mancozeb 0.20 per cent. **Rust** is evident by reddish-orange pustules on broken part of stems, petioles and on leaflets only in the temperate regions during rainy seasons, and these pustules become blackish at later stage. Defoliation of the plant occurs only in its severe infection. This disease is controlled through use of sulphur and oxycarboxin.

16.10 Pruning and Bending

Automated hydroponic systems will monitor light, temperature, humidity, CO_2 and other critical variables, but pruning still has to be done by hand.

Pruning starts right from planting, and (i) if any root is broken that is removed and the cane height is also reduced leaving only 3-4 dormant axillary buds, (ii) when new shoots start appearing and the floral bud is either smaller than pea-size, its removal is termed as 'soft pinch' and sometimes it is repeated twice or in case when bud size is larger than a pea-size it is termed as 'hard pinch'. In both these cases, either of the two is practiced, and these new flowering shoots are removed down to the second 5-leaflet leaf from the main stem though depends upon the cultivar and the season. Through soft pinch, axillary buds grow soon though from hard pinch it takes more time and flower later. This pinching structures the plant in proper architectural form to increase future quality production (Langhans, 1987). At the time of flower harvesting it is pruning No. (iii) when we should be sure whether one, two or three nodes on the stem are to be retained or it should be cut below the knuckle (the junction of the bud to the stem) and then afterwards the height of the plant can be controlled or long-stalked flowers can be obtained by cutting the flowers at their knuckle though further flowering will be delayed, and the No. (iv) pruning, *i.e.,* 4-5 weeks (depending upon the cultivar) prior to date of requirement of the flowers, developing shoots are pinched, and No. (v) pruning is drastic when demand of flowers is almost nil or little, maintaining the height of the plant from 45 to 90 cm and this is known as annual pruning. Rose requires annual pruning to revitalize the plant before it produces a new flush of blooms in the season. Pruning encourages new growth from bud union. This removal of old growth sometimes acts as a form of dormancy in warmer climates which is otherwise lacking. Pruning slows down the plant processes for a short period of time. Pruning in rose is done (i) to re-invigorate an old and unproductive plant, (ii) to remove old and diseased shoots, (iii) to shape the plant, (iv) to encourage the plants producing fewer but larger blooms or many but smaller blooms, (v) to manipulate the blooming time to avoid glut in the market and as per demand in the market, and (vi) to deadhead and to encourage repeat blossoms. **Hard pruning** is done to rejuvenate sickly and neglected plants, and in case of newly planted bush roses in HT, Grandifloras and Floribundas, and to produce show blooms for exhibition though it is not preferable for established plants. Here the shoots are cut back to three or four buds from the base or bud union to encourage short and sturdy shoots. **Light pruning** is usually not recommended as it produces spindly bushes, and its repetition year after year will result in an early bloom with poor quality flowers. However, only in very special cases such as very vigorous hybrid teas, climbers and shrub roses, this type of pruning is considered. **High pruning** is cutting back of shoots to about $2/3^{rd}$ of their length. This means that after removal of unwanted wood the remaining stems are merely tipped. **Moderate** or **intermediate pruning** is usually done in established plants, most suitable in case of HT, Floribundas and Grandifloras where shoots are normally removed to its half though weaker shoots more depending on their location on the bush. **Hybrid Teas** require hard pruning to build up a strong root system and to stimulate fresh and sturdy shoots close to the base of the bush in newly planted roses. In established HTs only 4-5 shoots are retained through moderate pruning, each shoot with 3-5 buds, the uppermost bud in each shoot facing outside. However, hard pruning in established roses is required for exhibition blooms. In case of 4-5 years old bushes, annually the oldest branch along with the thinning of the congested ones should be carried out. **Floribunda** and **Polyantha** roses should be cut back similar to HTs when newly planted but pruned hard the second year leaving only 1-2 buds (only top half of all strong stems) and moderate with 5-7 cm of the flowered stems and the tips as these bloom in clusters on shorter stems, *vis-à-vis* this ensures a long period of continuous blooming at varying heights. **Standard** roses are, in fact, HTs, Floribundas or Polyanthas budded 60-120 cm above the ground onto *Rosa canina, R. bourboniana* or *R. rugosa* stems and so the pruning should conform to the scion type budded. In newly **standard** roses hard pruning but in established standrads only moderate to induce plenty of blooms on a properly formed head, and on the top around 20-25 cm length of the stem should be retained. Established standards do not like hard pruning as this encourages vigorous canes that affect the shape of the plants adversely. **Miniatures** and **climbers** usually do not require pruning as such for blooming but to eliminate any unwanted shoot which is sick or broken and that too through scissors and not the pruning shears. The tips in the climbers with the spent blooms should also be removed. **Ramblers** as flower best on previous season's strong stems so pruning in this case is done to induce as many strong and 1-year old stems as possible. After initial planting these are cut back to 30 cm, next year every stem at 30 cm or up to the base and this practice is adopted year after year but only after flowering. After pruning, the cuts should be sealed with Bourdeaux mixture or Blitox paste, and just before sprouting of the buds an effective pesticide spray should be applied for healthy growth. Fertilizer can be applied only after three weeks of pruning.

Bending, which originated in Japan, in fact, is a modification of pruning to induce more flowering shoots with longer stems. This way one annual cut is prevented and instead the shoots are frequently bent whenever there is a need and to maintain the height at working level. In this case all weak shoots are bent down to fill any area void of foliage and thus attaining a desirable leaf area index to optimize photosynthetic potential and to facilitate the sugar transport to the developing shoots. Following knuckle cuts, as the weak shoots are bent the remaining shoots thus arising on dormant buds are normally more vigorous due to less competition for food and better exposure to light so these produce superior grade flowers.

Aquaculture

Aquaculture, also known as **aquafarming**, is the farming of aquatic organisms such as fish, mollusks, crustaceans and aquatic plants under controlled conditions for human consumption or for producing ornamental species, and other products. It involves some sort of intervention in the rearing process to enhance production, such as regular stocking, feeding and protection from predators. Aquaculture covers a wide range of aquatic species and methods. Depending on the species being farmed, aquaculture can be carried out in **freshwater, brackish water** or **marine water**. Aquaculture practiced in marine environments and in underwater habitats is referred to as **mariculture**.

17.1 Aquatic Species

Aquaculture ranges from production of aquatic organisms (such as fish) in naturally occurring ponds in rural areas to the highly intensive commercial closed re-circulating systems under fully controlled conditions. Particular kinds of aquaculture include fish farming, shrimp farming, oyster farming, algaculture (such as seaweed farming), and the cultivation of ornamental fish. Various species of aquatic organisms are:

17.1.1 Aquatic Plants

17.1.1.1 Microalgae

Microalgae, also referred to as phytoplankton, microphytes, or planktonic algae constitute the majority of cultivated algae.

17.1.1.2 Macroalgae

Maroalgae, commonly known as seaweed, also have many commercial and industrial uses, but due to their size and specific requirements, they are not easily cultivated on a large scale and are most often taken in the wild.

17.1.2 Aquatic Animals

17.1.2.1 Fish

The farming of fish is the most common form of aquaculture. The most important fish species used in fish farming in order are carp, salmon, tilapia and catfish. It involves raising fish commercially in tanks, ponds, or ocean enclosures, usually for food. The various stages of aquaculture operations include:

☆ A hatchery operation where fertilized eggs, larvae or fingerlings are produced

☆ A nursery operation which nurses small larvae to fingerlings or juveniles

☆ A grow-out operation which farms fingerlings or juveniles to harvestable sizes

17.1.2.2 Crustaceans and Molluscs

Other groups include aquatic reptiles, amphibians, miscellaneous invertebrates, such as echinoderms and jellyfish. These do not contribute enough volume to the aquaculture food production.

17.2 Culture Fish Species

17.2.1 Freshwater Species

☆ Commonly raised species in freshwater ponds are the carps, trout, tilapia, catfish, snakehead, eel, goldfish, gouramy, pike, tench, salmonids, palaemonids, and the giant freshwater prawn **Macrobrachium**.

☆ Habitats with high natural productivity for catfish, eels, mullets, *etc.* (*e.g.*, lakes, oxbow lakes, swamps, mining pools, rivers, and reservoirs).

☆ Habitats with low natural productivity: *Leptobarbus*, *Clarias batrachus*, *Oxyeleotris*, and *Macrobrachium*.

17.2.2 Brackish Water Species

Common species include sea bass, milkfish (*Chanos chanos*), mullet (*Mugil* sp.) and the different penaeid shrimps (*Penaeus monodon*, *P. orientalis*, *P. merguiensis*, *P. penicillatus*, *P. semisulcatus*, *P. japonicus*, and *M. ensis*).

17.2.3 Marine Water Species

Common species are the sea bass, sea bream, grouper, red sea bream, yellowtail, rabbitfish, and marine shrimps.

17.3 Aquaculture Environment

The three types of environment for a great variety of culture organisms are freshwater, brackishwater and marine. Freshwater aquaculture is carried out either in fish ponds, fish pens, fish cages or, on a limited scale, in rice paddies. Brackishwater aquaculture is done mainly in fish ponds located in coastal areas. Marine culture employs either fish cages or substrates for molluscs and seaweeds such as stakes, ropes, and rafts.

Aquaculture practices include:

☆ Freshwater pond culture

☆ Rice-fish culture or integrated fish farming

☆ Brackishwater finfish culture

☆ Mariculture involving extensive culture and producing fish/shellfish (*e.g.*, oysters, mussels, cockles) which are sold in rural and urban markets at relatively low prices.

17.4 Aquaculture Systems and Technologies

The systems and technology used in aquaculture have developed rapidly in the last fifty years, varying from very simple facilities (*e.g.* family ponds for domestic consumption) to high technology systems (*e.g.* intensive closed systems for export production).

17.4.1 Aquaculture Technologies

Much of the technology used in aquaculture is based on small modifications that improve the growth and survival rates of the target species, *e.g.* improving food, seeds, oxygen levels and protection from predators. Today, closed re-circulated systems have been developed due to advancements in technology and better understanding of complex interactions between nutrients, bacteria and cultured organisms, together with advances in hydrodynamics applied to pond and tank design. These closed systems have the advantage of isolating the aquaculture systems from natural aquatic systems, thus minimizing the risk of disease or genetic impacts on the external systems.

Aquaculture in coastal areas, in ponds, or in floating cages, has existed a long time for freshwater and anadromous species. With the advancements in technology, there is the potential to develop systems for rearing fish in the open ocean, either in sturdy cages or by so called **ranching**, where young fish are released to the wild and then collected by normal fishing or by training them to respond to specially generated sounds. Major progress has also being made in the aquafeeds technology, combining a large number of ingredients into very small pellets.

17.4.2 Feeds Technology

A very wide range of ingredients are used to prepare aquafeeds ranging from single component feeds available on-farm such as grass or rice bran to farm-made formulated feeds and commercial feeds. They include aquatic and terrestrial plants (duckweeds, azolla, water hyacinth *etc.*), aquatic animals (snails, clams, *etc.*) and terrestrial-based live feeds (silkworm larvae, maggots, *etc.*), plant processing products (de-oiled cakes and meals, beans, grains and brans) and animal-processing by-products (blood and feather meal, bone meal, *etc.*). Formulated commercial feeds are composed of several ingredients, mixed in various proportions to complement each other and form a nutritionally complete diet.

17.5 Aquaculture Systems

The systems used for aquaculture include but are not limited to ponds, cages, fibreglass, or concrete tanks. Depending on the level of input and output per farming area, stocking density of the culture organisms and the degree of management, aquacultural farming method can be categorized as **intensive, semi-intensive** or **extensive**. Intensive aquaculture involves intervention in the growing process, such as with supplemental feeding and water aeration, and/or artificial lightening and/or temperature control, *etc.* Extensive aquaculture allows the stock to grow on its own, using natural food sources and conditions.

Extensive Aquaculture

Extensive systems use moderate stocking densities with very large water bodies (*e.g.* large ponds) which may be shallow and may or may not be fully cleaned or well laid out. It includes no supplemental feeding, although fertilization may be done to stimulate the growth and production of natural food in the water. Water change is effected through tidal means, *i.e.*, new water is let in only during high tide and the pond can be drained only at low tide. Species used in this type of system may be monoculture or polyculture. Examples include: seaweed culture, coastal bivalve, culture (mussels, oysters, clams, cockles), coastal fishponds (mullets, milkfish, shrimps, tilapias), pen and cage culture in eutrophic waters and/or rich benthos (carps, catfish, milkfish, tilapias).

Semi-intensive

Semi-intensive systems use stocking densities higher than extensive culture with manageable size fully cleaned units with provisions for effective water management using pumps and aerators. Feeding of the stock is done at regular intervals during the day with high quality feeds. Formulated feeds may be used partially or totally depending on stocking density used. Fertilizers are usually used regularly with lime. Species used in this type of system are usually monoculture. Examples include: fresh and brackishwater pond (shrimps and prawns, carps, catfish, milkfish, mullets, tilapias), integrated agriculture-aquaculture (rice-fish; live stock/poultry-fish; vegetables-fish and all combinations of these), sewage-fish culture (waste treatment ponds; latrine wastes and septage used as pond inputs; fish cages in wastewater channels), cage and pen culture, especially in eutrophic waters or on rich benthos (carps, catfish, milkfish, tilapias).

Intensive

Intensive systems use maximum stocking densities in small fully cleaned ponds which are very well engineered systems with pumps and aerators to control water quality and quantity. Feeding of the stock is done at regular intervals with high-quality feeds (largely fry and feeds). Fertilizers are almost completely eliminated. Species used in this type of system are usually monoculture. Examples include: freshwater, brackishwater and marine ponds (shrimps; fish, especially carnivores - catfish, snakeheads, groupers, sea bass, *etc.*), freshwater, brackishwater and marine cage and pen culture (finfish, especially carnivores -groupers, sea bass, *etc.* - but also some omnivores such as common carp), other- raceways, silos, tanks, *etc.*

Note: *Financial returns in semi-intensive and intensive culture are much more attractive than those from extensive culture. Studies have shown that the return on investment from semi-intensive culture is better than from intensive culture due to the high cost of inputs (largely fry and feeds) used in intensive culture.*

17.6 Aquaculture Methods

Pond Culture

The breeding and rearing of fish in natural or artificial ponds are used for a wide variety of culture organisms in freshwater, brackishwater, and marine environments. It is carried out mostly using **stagnant waters** but can also be used in **running waters** especially in highland sites with flowing water. Outdoor freshwater fish farming is practiced commercially in many countries, including in the developed nations. In commercial farming, the density of stocking may be much higher with proper monitoring and control of fish health, water quality and fish feeding, but even in intensively operated systems, the key parameters that need to be monitored and balanced for success are similar to those in the traditional household pond system.

The stagnant water pond will typically be fed by natural rainfall or groundwater, or sometimes by diverting a nearby stream or irrigation canal. Some stocking occurs naturally, but more commonly the farmer will buy young fish from a breeder and stock them in his pond, often after holding them for a time in a small floating net to check they are in good condition before release to the pond. The production will improve if the farmer can stock a good balance of different species so that the varied kinds of food available in the pond can be made best use of. Typically the fish will be fed with household or agricultural by-products. Stocking and harvesting is often continuous, with the best stocking level being learnt from experience and the pond not being drained for many years. For most such farmers, fish production is a secondary activity, a useful additional source of protein to add to the supply or income from his main agricultural or commercial activities.

Pond Layout

The key parameters for the pond system's layout are the species for culture, and the shape and size of the area. These parameters in turn determine the number and sizes of the pond and the position of the water canals and gates. In a properly planned fish farm all the water control structures, canals, and the different pond compartments mutually go together in a balanced way. A fish farm is said to be complete if it has nursery and grow-out ponds and, in some instances, transition ponds for intermediate-sized fish/shrimp, all of which are properly proportioned and positioned within.

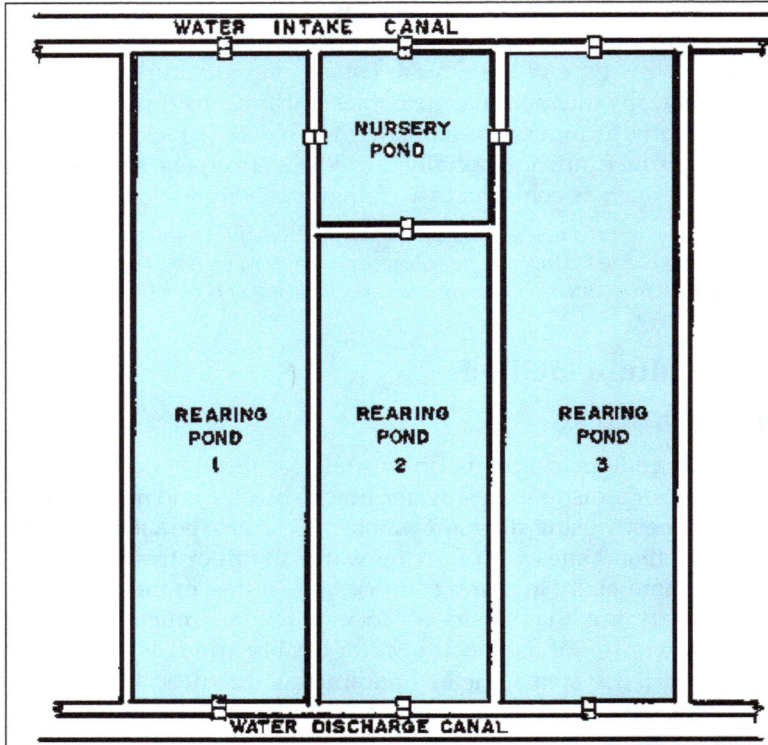

Pond Layout with One Nursery Pond and Three Rearing Ponds
(from ASEAN/SCSP, 1978)

Indoor Rearing

Holding of fish in controlled conditions indoors has been important in the development of seed production (hatcheries). Also, as knowledge has increased of nutrient cycles, bacterial action and water chemistry, it has become possible to rear fish, both for food and for ornament, in closed indoor systems, where the water is re-circulated. Passing the water through filters that use bacteria to naturally break down and recycle the waste products produced by the fish, allows a production system to be run in almost total isolation from the outside. Such systems have been important in the control of disease and also in maintaining the stable conditions that some species need to flourish and reproduce.

Outdoor Rearing

Nearly twenty percent of current world aquaculture production is of plants, mainly seaweeds that are usually grown on the seabed or on raft or racks in shallow coastal waters and used directly for food or for the production of alginate or carageenan (agar agar). This sector of aquaculture is largely run by small scale growers, mainly in Asia, who attach seedlings of marine plants to simple structures,

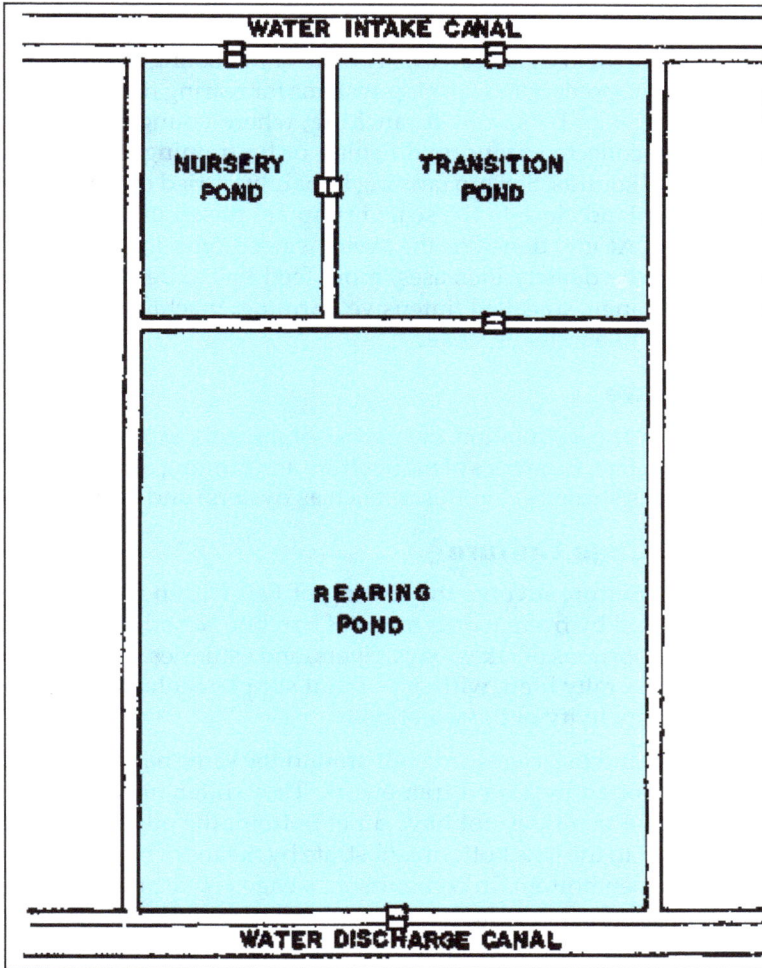

**Pond Layout with One Nursery Pond, One Transition Pond, and One Rearing Pond
(from ASEAN/SCSP, 1978)**

built close to shore and constructed from natural materials and then harvest the plants once they have grown.

Majority of production of molluscs such as oysters and mussels is done by small scale farmers and by coastal people who often combine fishing activities with farming. Oysters and mussels are mainly grown on structures built above the seabed - poles or racks on the shore, or ropes suspended from rafts or floating lines. The farmer's role is to supply suitable places for seed to settle or to add to natural production by bringing seed from a hatchery - and then to maintain the conditions of water flow and freedom from predators that the shellfish need to grow. Most of the commercially grown molluscs feed on microscopic algae floating in the water, so the farmer does not need to provide any feed.

Coastal Aquaculture

Fish are also grown in coastal areas, in ponds or in floating cages. As technology advances, there is the potential to develop systems for rearing fish in the open ocean, either in sturdy cages or by so called **ranching**, where young fish are released to the wild and then collected by normal fishing or by training them to respond to specially generated sounds. Shrimp farming is mainly carried out in earth bottomed ponds built on flat land close to the sea. Shrimp are raised in ponds at a range of different densities. At low densities the systems need only limited inputs of feed and fertilizers. As the density increases, more feed has to be supplied and at the higher end of the range, so-called 'intensive' farming, machines have to be put in the ponds to mix and aerate the Water.

17.7 Mariculture

Mariculture is the cultivation of marine organisms in seawater, usually in sheltered coastal waters. Examples of mariculture are; farming of marine fish, marine crustaceans (such as shrimps), molluscs (such as oysters) and seaweed.

17.8 Pen and Cage Culture

Pen and cage culture involve the rearing of fish within fixed or floating net enclosures supported by frameworks made of bamboo, wood, or metal, and set in sheltered, shallow portions of lakes, bays, rivers, and estuaries. Yields from pen and cage culture are generally high, with or without supplemental feeding depending on the natural productivity of the water body.

Both fish pens and fish cages are built around the same basic design concept: a net enclosure supported by a rigid framework. They differ, however, in a number of respects. Firstly, a pen does not have a net bottom; the edges of its net walls/fences are anchored to the lake bottom/substrate by means of bamboo pegs and the lake bottom is the pen bottom. In comparison, a cage is like an inverted mosquito net with the cage bottom made of the same netting material used for its four sides.

Secondly, fish pens theoretically have no limit to their size/area while cages cannot exceed 1 000 m² in area for reasons of the quantity of material required for cage construction (due to the need for a flooring) and manageability of operation (cages have to be lifted and the fish scooped out and not harvested using nets as in pens).

Thirdly, design of the structures and methods of construction are different. Fish pens are fixed structures; fish cages may either be fixed or floating. Fish pens for milkfish culture in Laguna de Bay, Philippines consist of a nursery pen within the grow-out pen/enclosure. Cages are individual units for either seed production or grow-out; they are, however, usually installed in clusters or modules with a common framework.

Cages are usually used to farm fin fish in fresh and sea water. The cages are placed in either a sheltered bay in the sea or in a dam off a river for freshwater fish species. These cages are usually anchored to the sea/dam floor and have walking platforms or boat access for feeding and harvesting.

The development and adoption of pen culture as a popular technology has not been widespread, though, perhaps because of its site-specific requirements like its suitability mainly in shallow environments. The wider popularity of cage culture as compared to pen culture may be due to its greater flexibility in terms of setting the structures. For example, cages may be installed in bays, lagoons, straits, and open coasts as long as they are protected from strong monsoonal winds and rough seas. Floating cages can also be set up in deep lakes and reservoirs, and in rivers and canal systems, and even in deep mining pools which could not be used otherwise for culture due to harvesting difficulties.

17.8.1 Factors to be Considered in the Selection of Cage/Pen Sites

Two major factors that should be considered for the selection of cage/pen sites are:

1. Biological

It is the requirement of the species to be cultured such as:

a. Temperature: The survival and growth rate of the cultured species will be better at the optimum temperature of the cultured species.

b. Salinity (brackish waters)

c. Oxygen: Oxygen content is related to temperature and salinity. The abundance of other organisms (plankton, benthos) within the pen would deplete oxygen.

d. Water polution: Pollution (usually due to feed loss and excreta) of the farm (enclosure) area would depend on the species of fish, their size, feed given, *etc*. Faeces and food particles sink to the bottom. If the water flow is low and enclosure bottom is bounded by reefs or other thresholds, the wastes will accumulate and would contaminate the farm. Artificial pumping of water to have a good circulation in the enclosure is also suggested.

e. Feeding rate and pattern: Feeding rate, food quality and quantity depends on the species and the stocking density. The excess nutrient problem (which causes fouling) would be less if there is a good circulation of water through the enclosure, but the nutrients can also be utilized economically through polyculture.

f. Crowding (stocking density): The stocking density depends on the availability of oxygen and the rate of oxygen demand of feed remains and excreta.

g. Predator control: To protect the pens against predators, the enclosures have to be completely sealed by nets and wires; safety nets are to be used against predatory fish and seals, and birds are to be scared away.

   i. Some predators in tropical countries are:

     1. Rats: Controlled by dogs or rat guards/traps.

     2. Turtles: Controlled by perimeter nets.

3. Monitor lizards: Controlled by perimeter net or nets with feet buried in the earth.

4. Birds: Controlled by monofilament net laced cover on the cage top.

5. Puffer fish: Controlled by stretched tight netting.

ii. Some predators in temperate countries are:

1. Dog fish: Controlled by ensuring that bottom and sides of net are stretched tight.

2. Mink: Controlled by dogs.

h. Fouling by algae and mussels: The water around the enclosure would be rich in nutrients, therefore algae and mussels also grow attached to the framework and nets of the enclosure. Fouling makes the net heavier and prevents the exchange of water in the pen. Fouling can be reduced by cleaning the nets in chemicals.

2. Physical

It is the structural requirements of the cage/pen. The site should be considered from the aspect of exposure to winds and waves, and it would be necessary to design the structures above and below water, adequate to withstand hurricanes and typhoons as well. For structures below water level, wave force induced by wind blowing over a period of time is to be considered.

17.9 Floating Rafts at Sea (or Shellfish Culture)

Rafts are positioned in sheltered bays at sea with ropes or bags hanging from them in the water column. These are usually used to farm mussels and oysters. In the case of mussels they attach themselves to the ropes and feed on plankton in water column until they are fully grown and ready for harvesting. Oysters on the other hand are reared in hatcheries and then placed in bags and also feed on plankton until they grow to harvesting size. This type of system is known as an extensive system because there is very little human input into the system.

17.10 Onland Flow Through System

These systems are usually a set of tanks which are situated on land and are not directly open to surrounding water bodies. A flow through system is a system which requires the high volumes of water from the sea or freshwater depending on the species being farmed. The water is drawn from the sea or dam and pumped through the system, only being used once. This method is globally being used for farming trout, sturgeon (for caviar) and tilapia.

17.11 Onland Re-circulating System

Recirculating systems are made up of a series of tanks based on land which are not open to the surrounding water body. The water is re-circulated through the system and filtered, cleaned, purified and re-oxygenated on a continuous basis.

17.12 Raceways

Raceways come directly off a river diverting the flow of water down a series of channels in which fish are farmed. The water is usually then treated before it is diverted back into the river. This method is commonly used to farm trout in Himachal Pradesh, India.

17.13 Integrated

Integrated Multi-Trophic Aquaculture (IMTA) is a practice in which the by-products (wastes) from one species are recycled to become inputs (fertilizers, food) for another. Fed aquaculture (for example, fish, shrimp) is combined with inorganic extractive (for example, seaweed) and organic extractive (for example, shellfish) aquaculture to create balanced systems for environmental sustainability (biomitigation), economic stability (product diversification and risk reduction) and social acceptability (better management practices).

In a number of countries in Asia (*e.g.*, China, Nepal, Indonesia and also to some extent in Mizoram and north eastern hill (NEH) region of India), freshwater fish culture is integrated with the farming of crops, mainly rice, vegetables and animals (usually pigs, ducks, and chickens). This leads to greater overall efficiency of the farming system as wastes/by-products of one component are used as inputs in another. For example, poultry of pig manure can be used to fertilize the fish pond and the vegetable garden and the waste vegetables can be fed to the fish and the pigs.

Multi-Trophic refers to the incorporation of species from different trophic or nutritional levels in the same system. This is one potential distinction from the age-old practice of aquatic polyculture, which could simply be the co-culture of different fish species from the same trophic level. In this case, these organisms may all share the same biological and chemical processes, with few synergistic benefits, which could potentially lead to significant shifts in the ecosystem. Some traditional polyculture systems may, in fact, incorporate a greater diversity of species, occupying several niches, as extensive cultures (low intensity, low management) within the same pond. The "Integrated" in IMTA refers to the more intensive cultivation of the different species in proximity of each other, connected by nutrient and energy transfer through water.

Ideally, the biological and chemical processes in an IMTA system should balance. This is achieved through the appropriate selection and proportions of different species providing different ecosystem functions. The co-cultured species are typically more than just biofilters; they are harvestable crops of commercial value. A working IMTA system can result in greater total production based on mutual benefits to the co-cultured species and improved ecosystem health, even if the production of individual species is lower than in a monoculture over a short term period.

Sometimes the term "Integrated Aquaculture" is used to describe the integration of monocultures through water transfer. For all intents and purposes however, the terms "IMTA" and "integrated aquaculture" differ only in their degree of descriptiveness. Aquaponics, fractionated aquaculture, IAAS (integrated agriculture-aquaculture systems), IPUAS (integrated peri-urban-aquaculture systems), and IFAS (integrated fisheries-aquaculture systems) are other variations of the IMTA concept.

17.14 Temperature Problems in Deeper Ponds

Water has a high heat capacity and unique density qualities with maximum density at 4°C. If water temperatures in a pond are nearly equal at all pond depth, the nutrients, dissolved gases and fish wastes too will be almost equally distributed through the pond. This usually happens during spring. As the days become warmer, the water at the surface become warmer and lighter while the water underneath forms a cooler and denser layer resulting in **stratification** of the water. These different densities between the upper and lower layers prevent circulation of the colder bottom water. Due to reduced photosynthesis and air contact of bottom layer water, its dissolved oxygen levels get significantly reduced. The oxygen levels gets further reduced through decomposition of waste products which settle to the pond bottom.

The stratification may last for several weeks and the main loss occurs when there is a sudden rain. This will decrease the temperature of the upper warmer water layer allowing it to mix with the oxygen-poor-layer below. Decomposing materials in the oxygen-poor layer are again mixed evenly throughout the pond, resulting in an overall reduction in the dissolved oxygen level. Fish previously able to avoid the oxygen depleted layer are now susceptible to low-dissolved oxygen syndrome and possibly death. Mechanical aeration, pond shelters and water circulation may give some relief to the farmers especially during summer.

17.15 Ice Cover Problem

The fish may suffer from low dissolved oxygen in the area where ponds get covered with ice in winter. Although ice cover does not obstruct photosynthesis, and fishes consume less oxygen at colder temperature, fishes may suffer from low dissolved oxygen as under extended ice cover other gases (carbon dioxide, hydrogen sulphide, methane, *etc.*) may build up to dangerously high levels. Mechanical aeration is probably the most reliable way of preventing an ice buildup by keeping large areas of the pond free of ice.

17.16 Importance of Water Quality in Aquaculture

The water quality greatly affects fish culture, especially in re-circulating aquaculture system. It is necessary to understand the physical and chemical qualities of water because fishes are totally dependent upon water to breathe, feed and grow, excrete wastes, maintain a salt balance, and reproduce.

17.16.1 Physical Characteristics of Water

Water can hold large amounts of heat with a relatively small change in temperature thus allowing a body of water to act as a buffer against wide fluctuations in temperature. The larger the body of water, the slower the rate of temperature change. Furthermore, aquatic organisms take on the temperature of their environment and cannot tolerate rapid changes in temperature. Water has very unique density qualities.

Natural water contains several chemical elements that directly or indirectly affect the growth of aquaculture organisms.

17.16.2 Water Balance in Fish

In contrast, fish rely heavily on their gills for elimination of most nitrogen waste products, excreting primarily ammonia. The salinity of water in the ocean is more concentrated than that of the fish's body fluids. In this environment water is drawn out of fish gills, but salts tend to diffuse inward. Hence, marine fishes drink large amounts of sea water and excrete small amounts of highly salt-concentrated urine.

In fresh-water fish, the salt is constantly being lost through the gills, and large amounts of water enter through the fish's skin and gills. This is because the salt concentration in a fish (approximately 0.5 per cent) is higher than the salt concentration of the water in which it lives. Because the fish's body is constantly struggling to prevent the **diffusion** of water into its body, it excretes a large amount of water. As a result, the salt concentration of the urine is very low. By understanding the need to maintain a water balance in freshwater fish, one can understand why using salt during transport is beneficial to fish.

17.16.3 Sources of Water

Many of the negative chemical and environmental factors associated with most operations have their origins in the source of water selected. Final site selection has to be made based on both the quality and quantity of water available. The most common sources of water used for aquaculture are wells, springs, rivers and lakes, groundwater, and municipal water. Of the sources mentioned, wells and springs are considered to be consistently of high quality.

17.16.4 Water Quantity

The beginning aquaculturist usually underestimates the quantity of water required for commercial production. It is generally accepted that a minimum rate of 32 gallons per minute (gpm) is required for each surface hectare of ponds. With this in mind, a 40-hectare fish farm will need to have wells capable of producing 1,300 gpm of water. Such large volumes are required to replace water lost due to evaporation and seepage. In addition, the farmer may have several ponds to fill quickly during the spawning season. In **raceway culture,** it is advisable to have a minimum flow rate of 500 gpm. Even water re-circulating systems that recycle water, require large quantities of water. If a 100,000 gallon capacity water recirculating operation exchanges 10 per cent of the water daily, it will require 10,000 gallons of water per day.

17.16.5 Water's Physical Factors

17.16.5.1 Temperature

After oxygen, water temperature may be the single most important factor affecting the welfare of fish. Fish are cold-blooded organisms and assume approximately the same temperature as their surroundings. The temperature of the water affects the activity, behaviour, feeding, growth, and reproduction of all fishes. Metabolic rates in fish double for each 18°F rise in temperature.

Fish are generally categorized into warmwater, coolwater, and coldwater species based on optimal growth temperatures.

Ideally, species selection should be based in part on the temperature of the water supply. Any attempt to match a fish with less than ideal temperatures will involve energy expenditures for heating or cooling. This added expense will subsequently increase production costs.

Temperature also determines the amount of dissolved gases (oxygen, carbon dioxide, nitrogen, *etc.*) in the water. The gas is more soluble in cooler water.

See above, **Temperature problems in deeper ponds**

17.16.5.2 Suspended Solids

Suspended solids is a term usually associated with plankton, fish wastes, uneaten fish feeds, or clay particles suspended in the water. Suspended solids are large particles which usually settle out of standing water through time.

17.16.5.3 Plankton

Turbidity caused by phytoplankton (microscopic plants) and zooplankton (microscopic animals) is not directly harmful to fish. Phytoplankton (green algae) not only produces oxygen, but also provides a food source for zooplankton and filter feeding fish/shellfish. Phytoplankton also uses ammonia produced by fish as a nutrient source. Zooplankton is a very important food source for fry and fingerlings such as hybrid striped bass and yellow perch. However, excessive amounts of algae can lead to increased rates of respiration during the night thereby consuming extra oxygen. Excessive phytoplankton buildups or "blooms" which subsequently die will also consume extra oxygen. Any wide swings between day and night oxygen levels can lead to dangerously low oxygen concentrations.

17.16.5.4 Fish Wastes

Suspended fish wastes are a serious concern for water re-circulating culture systems. Large amounts of suspended and settled solids are produced during fish production. Fish waste particles can be a major source of poor water quality since they may contain up to 70 per cent of the nitrogen load in the system. These wastes not only irritate the fish's gills, but can cause several problems to the biological filter. The particulate waste can clog the biological filter, causing the nitrifying bacteria to die from lack of oxygen. Particulate waste can also promote the growth of bacteria that produces—rather than consumes ammonia.

17.16.6 Water's Chemical Factors

17.16.6.1 Photosynthesis

Photosynthesis is one of the most important biological activities in standing pond aquaculture. Many water quality parameters such as dissolved oxygen, carbon dioxide, pH cycles, and nitrogenous waste products are regulated by the photosynthetic reaction in phytoplankton. Simply stated, photosynthesis is the process by which phytoplankton uses sunlight to convert carbon dioxide into a food source and to release oxygen as a by-product. In addition to supplying oxygen in fish ponds, photosynthesis also removes several forms of nitrogenous wastes, such as ammonia, nitrates, and urea.

17.16.6.2 Dissolved Gases

17.16.6.2.1 Oxygen

Dissolved oxygen (DO) is by far the most important chemical parameter in aquaculture. Low-dissolved oxygen levels are responsible for more fish kills, either directly or indirectly, than all other problems combined. The amount of oxygen consumed by the fish is a function of its size, feeding rate, activity level, and temperature. Small fish consume more oxygen as compared to large fish because of their higher metabolic rate. The amount of oxygen that can be dissolved in water decreases at higher temperatures, *vis-a-vis* with increase in altitudes and salinities

To maximize fish production, stock greater amounts of fish in a given body of water than found in nature. At times during summer it may be necessary to supply supplemental aeration to maintain adequate levels of dissolved oxygen.

Fish are not the only consumers of oxygen in aquaculture systems; bacteria, phytoplankton, and zooplankton consume large quantities of oxygen as well. Decomposition of organic materials (algae, bacteria, and fish wastes) is the single greatest consumer of oxygen in aquaculture systems. Problems encountered from water recirculating systems usually stem from excessive ammonia production in fish wastes. Consumption of oxygen by nitrifying bacteria that break down toxic ammonia to non-toxic forms depends on the amount of ammonia entering the system.

Oxygen enters the water primarily through direct diffusion at the air-water interface and through plant photosynthesis. Direct diffusion is relatively insignificant unless there is considerable wind and wave action. Several forms of mechanical aeration are available to the fish farmer. The general categories are:

1. Paddlewheels
2. Agitators
3. Vertical sprayers
4. Impellers
5. Airlift pumps
6. Venturia pumps
7. Liquid oxygen injection
8. Air diffusers

Mechanical aeration can also increase dissolved oxygen levels. Because of the lack of photosynthesis in indoor water re-circulating systems, mechanical means of aeration is the only alternative for supplying oxygen to animals cultured in these systems. Oxygen depletions can be calculated, but predictions can be misleading and should never be substituted for actual measurements.

17.16.6.2.2 Carbon Dioxide

Carbon dioxide (CO_2) is commonly found in water from photosynthesis or water sources originating from limestone bearing rock. Fish can tolerate concentrations of 10 ppm provided dissolved oxygen concentrations are high. Water supporting good fish populations normally contain less than 5 ppm of free carbon dioxide. In water used for intensive pond fish culture, carbon dioxide levels may fluctuate from 0 ppm in the afternoon to 5-15 ppm at daybreak. While in recirculating systems carbon dioxide levels may regularly exceed 20 ppm. Excessively high levels of carbon dioxide (greater than 20 ppm) may interfere with the oxygen utilization by the fish.

There are two common ways to remove free carbon dioxide. First, with well or spring water from limestone bearing rocks, aeration can "blow" off the excess gas. The second option is to add some type of carbonate buffering material such as calcium carbonate ($CaCO_3$) or sodium bicarbonate (Na_2CO_3). Such additions will initially remove all free carbon dioxide and store it in reserve as bicarbonate and carbonate buffers. This concept is discussed in further detail under alkalinity.

17.16.6.2.3 Nitrogen

Dissolved gases, especially nitrogen, are usually measured in terms of " per cent saturation". Any value greater than the amount of gas the water normally holds at a given temperature constitutes supersaturation. A gas supersaturation level above 110 per cent is usually considered problematic.

Gas bubble disease is a symptom of gas supersaturation. The signs of gas bubble disease can vary. Bubbles may reach the heart or brain, and fish die without any visible external signs. Other symptoms may be bubbles just under the surface of the skin, in the eyes, or between the fin rays. Treatment of gas bubble disease involves sufficient aeration to decrease the gas concentration to saturation or below.

17.16.6.3 Ammonia

Fish excrete ammonia and lesser amounts of urea into the water as wastes. Two forms of ammonia occur in aquaculture systems, ionized and un-ionized. The un-ionized form of ammonia (NH_3) is extremely toxic while the ionized form (NH_4^+) is not. Both forms are grouped together as "total ammonia". Through biological processes, toxic ammonia can be degraded to harmless nitrates. Ammonia (NH_3) is first converted to nitrites (NO_2^-) and then to nitrates (NO_3^-).

In natural waters, such as lakes, ammonia may never reach dangerous high levels because of the low densities of fish, but the fish farmer must maintain high densities of fish and, therefore, runs the risk of ammonia toxicity. Un-ionized ammonia levels rise as temperature and pH increase.

Percentage of Total Ammonia that is Un-ionized at Various Temperatures and pH

pH	54°F (12.2°C)	62°F (16.7°C)	68°F (20°C)	75°F (23.9°C)	82°F (27.8°C)	90°F (32.2°C)
7.0	0.2	0.3	0.4	0.5	0.7	1.0
7.4	0.5	0.7	1.0	1.3	1.7	2.4
7.8	1.4	1.8	2.5	3.2	4.2	5.7
8.2	3.3	4.5	5.9	7.7	11.0	13.2
8.6	7.9	10.6	13.7	17.3	21.8	27.7
9.0	17.8	22.9	28.5	34.4	41.2	49.0
9.2	35.2	42.7	50.0	56.9	63.8	70.8
9.6	57.7	65.2	71.5	76.8	81.6	85.9
10.0	68.4	74.8	79.9	84.0	87.5	90.6

To determine un-ionized ammonia concentration, multiply total ammonia concentration by the percentage which is closest to the observed temperature and pH of the water sample. For example, a total ammonia concentration of 5 ppm at pH 9 and 20° C would be: 5 ppm total ammonia 28.5 per cent = 1.43 ppm.

Toxicity levels for un-ionized ammonia depend on the individual species; however, levels below 0.02 ppm are considered safe. Dangerously high ammonia concentrations are usually limited to water recirculation system or hauling tanks where water is continually recycled and in pond culture after phytoplankton die-off. However, the intermediate form of ammonia–nitrite has been known to occur at toxic levels (brown-blood disease) in fish ponds.

17.16.6.4 Buffering Systems

A buffering system to avoid wide swings in pH is essential in aquaculture. Without some means of storing carbon dioxide released from plant and animal respiration, pH levels may fluctuate in ponds from approximately 4-5 to over 10 during the day. In recirculating systems constant fish respiration can raise carbon dioxide levels high enough to interfere with oxygen intake by fish, in addition to lowering the pH of the water.

17.16.6.5 pH

The quantity of hydrogen ions (H+) in water will determine whether it is acidic or basic. The scale for measuring the degree of acidity is called the pH scale, which ranges from 1 to 14. A value of 7 is considered neutral, neither acidic nor basic; values below 7 are considered acidic; above 7, basic. The acceptable range for fish culture is normally between pH 6.5-9.0.

17.16.6.6 Alkalinity

Alkalinity is the capacity of water to neutralize acids without an increase in pH. This parameter is a measure of the bases, bicarbonates (HCO_3-), carbonates (CO_3—) and, in rare instances, hydroxide (OH-). Total alkalinity is the sum of the carbonate and bicarbonate alkalinities. Some waters may contain only bicarbonate alkalinity and no carbonate alkalinity.

The carbonate buffering system is important to the fish farmer regardless of the production method used. In pond production, where photosynthesis is the primary natural source of oxygen, carbonates and bicarbonates are storage area for surplus carbon dioxide. By storing carbon dioxide in the buffering system, it is never a limiting factor that could reduce photosynthesis, and in turn, reduce oxygen production. Also, by storing carbon dioxide, the buffering system prevents wide daily pH fluctuations.

Without a buffering system, free carbon dioxide will form large amounts of a weak acid (carbonic acid) that may potentially decrease the night-time pH level to 4.5. During peak periods of photosynthesis, most of the free carbon dioxide will be consumed by the phytoplankton and, as a result, drive the pH levels above 10. As discussed, fish grow within a narrow range of pH values and either of the above extremes will be lethal to them.

In recirculating systems where photosynthesis is practically non-existent, a good buffering capacity can prevent excessive buildup of carbon dioxide and lethal decreases in pH. It is recommended that the fish farmer maintains total alkalinity values of **at least** 20 ppm for catfish production. Higher alkalinities of at least 80-100 ppm are suggested for hybrid striped bass. For water supplies that have naturally low alkalinities, agriculture lime can be added to increase the buffering capacity of the water.

17.16.6.7 Hardness

Water hardness is similar to alkalinity but represents different measurements. Hardness is chiefly a measure of calcium and magnesium, but other ions such as aluminum, iron, manganese, strontium, zinc, and hydrogen ions are also included. When the hardness level is equal to the combined carbonate and bicarbonate alkalinity, it is referred to as carbonate hardness. Hardness values greater than the sum of the carbonate and bicarbonate alkalinity are referred to as non-carbonated hardness. Hardness values of **at least** 20 ppm should be maintained for optimum growth of aquatic organisms. Low- hardness levels can be increased with the addition of ground agriculture lime.

17.16.6.8 Other Metals and Gases

Other metals such as iron and sodium, and gases, such as hydrogen sulphide, may sometimes present special problems to the fish farmer. Most complications arising from these can be prevented by properly pre-treating the water prior to adding it to ponds or tanks. The range of treatments may be as simple as aeration, which removes hydrogen sulphide gas, to the expensive use of iron removal units. Normally iron will precipitate out of solution upon exposure to adequate concentrations of oxygen at a pH greater than 7.0.

Suggested water-quality criteria for aquaculture hatcheries or production facilities. Salmonid quality standards with modification for warmwater situations. Concentrations are in ppm (mg/l). (Source: Modification from Wedemeyer, 1977; Piper, *et al.* (Larsen), 1982)

Chemical	Upper Limits for Continuous Exposure and/or Tolerance Ranges
Ammonia (NH₃)	0.0125 ppm (un-ionized form)
Cadmium [a]	0.004 ppm (soft water < 100 ppm alkalinity)
Cadmium [b]	0.003 ppm (hard water > 100 ppm alkalinity)
Calcium	4.0 to 160 ppm (10.0-160.00 ppm [d])
Carbon dioxide	0.0 to 10 ppm (0.0-15.0 ppm [d])
Cholorine	0.03 ppm
Copper [c]	0.006 in soft water
Hydrogen sulphide	0.002 ppm (Larsen - 0.0 ppm)
Iron (total)	0.0 to 0.15 ppm (0.0-0.5 ppm [d])
Ferrous ion	0.00 ppm
Ferric ion	0.5 ppm (0.0-0.5 ppm [d])
Lead	0.03 ppm
Magnesium	(needed for buffer system)
Manganese	0.0 to 0.01 ppm
Mercury (organic or inorganic)	0.002 ppm maximum, 0.00005 ppm average
Nitrate (NO₃-)	0.0 to 3.0 ppm
Nitrite (NO₂-)	0.1 ppm in soft water, 0.2 ppm in hard water
	0.03 and 0.06 ppm nitrite-nitrogen
Nitrogen	Maximum total gas pressure 110 per cent of saturation
Oxygen	5.0 ppm to saturation; 7.0 to saturation for eggs or broodstock
Ozone	0.005 ppm
pH	6.5 to 8.0 (6.6-9.0 [d])
Phosphorus	0.01 to 3.0 ppm
Polychlorinated biphenyls (PCBs)	0.002
Total suspended and settleable solids	80.0 ppm or less
Total alkalinity (as CaCO3)	10.0 to 400 ppm (50.0-4.00.0 ppm [d])
per cent as phenolphthalein	0.0 to 25 ppm (0.40 ppm [d])
per cent as methyl orange	75 to 100 ppm (60.0-100.0 ppm [d])
per cent as ppm hydroxide	0.0 ppm
per cent as ppm carbonate	0.0 to 25 ppm (0.0-40.0 ppm [d])
per cent as ppm bicarbonate	75 to 100 ppm
Total hardness (as CaCO₃)	10 to 400 ppm (50.0-400.0 ppm [d])
Zinc	0.03-0.05 ppm

a. To protect salmonid eggs and fry. for non-salmonids 0.004 ppm is acceptable

b. To protect salmonid eggs and fry, for non-salmonids 0.03 ppm is acceptable

c. Copper at 0.005 ppm may suppress gill adensione triphosphatase and compromise smoltification in anadromous salmonids.

d. Warm water situations

17.17 Aquaculture Production Systems and Practices

Region	Major Culture Species	Major Culture Systems	Major Culture Practices	Scope for Future Development/ Needs for Further Expansion
ASIA	At least 75 species; diverse fresh-water and marine species, including high-value shrimps, molluscs, seaweeds, with carps and seaweeds dominating production	Traditional extensive to intensive	Fish ponds Fish pens and fish cages Floating rafts, lines, and stakes for molluscs and seaweeds	Development of culture-based fisheries in inland lakes, rivers, floodplains, and permanent and temporary reservoirs and barrages Resource enhancement programmes integrated with environmental management
PACIFIC	Mussels and oysters, red seaweeds	Intensive/semi-intensive to extensive	Hanging lines for mussels and pearl oysters Offshore cages for salmon Pond culture for shrimps, tilapia, catfish, milkfish Freshwater pens for crayfish	Production of high-value species for select markets Small-scale aquaculture for local markets Improved management of fishery resources, particularly reef fisheries
LATIN AMERICA	50 species of fish, crustaceans, and molluscs, including freshwater fish and marine shrimps in South America and molluscs in Central America	Extensive to semi-intensive and intensive	Offshore cage farming of Pacific and Atlantic salmon, ocean ranching in southern ocean, semi-intensive farming of marine shrimp in coastal ponds and extensive farming of freshwater fish in ponds	Production of species for export and marine shrimp and salmon
AFRICA	>26 freshwater fish; the most important being tilapia and common carp, molluscs and oysters also	Mainly extensive, rural-based, integrated with poultry and animal husbandry, rice-fish farming; some intensive in raceways and floating cages	Fish pond culture for fresh-water fish, raceways and floating cages for marine species	Increased emphasis on higher value catfishes for urban markets, on marine species of fish and crustaceans for select national market and export Culture-based fisheries in lakes and reservoirs Development of coastal lagoons which are almost totally unexploited

Contd...

Contd...

Region	Major Culture Species	Major Culture Systems	Major Culture Practices	Scope for Future Development/ Needs for Further Expansion
MEDITERRA-NEAN	>50 individual species, mostly freshwater and brackishwater fishes - most important being salmonids and carps; oysters and mussels	Well-diversified modern practices, with highly technical and intensive systems in developing countries and semi-intensive and extensive elsewhere	Fish ponds Fish cages Ocean ranching	Production of high-value species of tourism and export Integrated coastal zone management
CARIBBEAN	About 16 species of tilapias, carps, marine shrimp and, freshwater prawns, oysters and seaweeds		Floating cages in reservoirs Fish pond farming in freshwater Culture-based fisheries in reservoirs Rope production of molluscs	Priority is for aquaculture production for local markets

Source: ADCP Aquaculture Regional Profiles, 1989b

(18)

Aquaponics

Aquaponics (or pisciponics) is a method of food production that integrates aquaculture (growing aquatic animals) with hydroponics (growing plants) where inorganic mineral nutrients and pesticides are almost completely eliminated. This symbiotic relationship facilitates a sustainable system with little input necessary. Good bacteria build up, which then convert the toxins produced from fish waste into nutrients used by plants. By absorbing these nutrients, the plants filter the water, giving the fish a livable environment. This cycle helps keep the tank in good shape for both fish and plants.

Producing food aquaponically is almost completely organic because there is no need for fertilizer; the fish waste is all that is necessary for the plants to grow. Herbicides also are not needed because there is no soil used to grow the plants, and could even be harmful to the fish. Aquaponics is a great way to sustainably grow fresh fish and vegetables for a family, to feed a village, or generate a profit in a commercial farming volume.

Aquaponics systems range from those designed for hobby or backyard food production through to those designed for commercial scale production of fish and plants for sale. Aquaponics systems are closed loop facilities that retain and treat the water within the system. In aquaponics, fish, microbes (beneficial bacteria) and plants all thrive together in a symbiotic relationship and help each other to grow naturally and organically. The water containing fish excretion and uneaten fish food is circulated (either continuously or at specified time interval using a timer controlled water pump) from the fish tank to the grow beds, which are filled with expanded clay or some other growing media. Nitrifying bacteria in the system convert fish wastes into plant-usable nutrients which are used by plants as their main nutrient supply. The water is filtered by the plants and medium (and/or mechanical filter if attached) and is returned by gravity or pump back to the fish tank, giving the fish clean water to live in. Both the plants and the fish contribute to the cycling process of aquaponics; the fish provide the nutrients for the plants and the plants filter the water so that the fish are able to live.

Natural chemicals and the fish food are the only additives to the aquaponics system. Anything added to the system to boost plant growth will most probably harm the fish and possibly the bacteria colony. There are a few exceptions to this, including the use of liquid seaweed, small amounts of chelated iron, and a few minerals to adjust pH.

The micro-organisms that inhabit an aquaponic system can be sub-classified into a further array of organisms such as bacteria, fungi, phytoplankton, zooplankton, *etc.* The bacteria in aquaponics systems are critical to the operation and ecology of the system. These bacteria break down the fish waste and uneaten fish feed into plant nutrients. The bacteria perform critical processes that lead to, assist and drive the biological balance of the system. It is these bacteria that perform chemical processes in the water that make the water livable and usable by the fish and the plants that we wish to grow. Bacteria are so small that they cannot be seen by the naked human eye. However, they can also make up a large proportion of the overall system biomass (biomass is the weight of all the living things in a system). Bacteria do utilise some of the nutrients contained in the aquaponic system. They use elements like carbon, oxygen, nitrogen, phosphorous, potassium and calcium, along with many others, to build their own cells, just like we do, fish do and plants do. However, bacteria gain access to the energy they require by performing and assisting chemical reactions that release energy. For example, nitrification, the conversion of ammonia to nitrate is mediated (performed) by several species of nitrifying bacteria in aquatic systems. The reason the bacteria convert ammonia to nitrate is not because these bacteria like or prefer nitrogen in the form of nitrate more than nitrogen in the form of ammonia (unlike our fish, to which ammonia can be toxic). The reason is that when ammonia (NH_3) is converted to nitrate (NO_3) the exchange of hydrogen ions (H) for oxygen atoms (O_2) releases energy, and it is from this energy source that bacteria derive the energy to run their own metabolism. It is a positive by-product of this ability of the bacteria in our aquaponic systems that they also perform chemical conversions that are critical for fish and plant health. Like any aquatic organism, the bacteria in aquaponic systems are dependent on the water as the medium within which they live their lives. Therefore, bacteria are also dependent upon the water being in the best chemical condition it can be so they live in an environment that is amenable to their continued health and well being.

18.1 Aquaponics vs. Hydroponics

See Table 1.

Table 1: Comparison of Hydroponics and Aquaponics

Hydroponics	Aquaponics
Plants are grown under highly optimized conditions. Beneficial bacteria can be added for better plant growth and to combat pests and diseases.	It is an ecosystem where plants, fish, bacteria, and worms all live together in a beautifully balanced symbiotic relationship. It is a multi-faceted system where one component affects one or more other components.
Food	
Only Vegetarian food. Fish are not grown or harvested.	Vegetarian and fish (or other aquatic animals) food.
Nutrients	
Nutrient management in hydroponics is quite complex and hectic task as compared to aquaponics.	In aquaponics this is managed by simply achieving a state of balance within its ecosystem
Hydroponic nutrients are generally delivered in mineral salt form.	Fed by the organic waste from the fish, which has very little salts.
The nutrient solution in hydroponics must be regularly replaced to balance the nutrient solution.	Only top up of the fish tank with water is required. The water is replaced only if there is a severe, unexpected problem.
With time, the nutrient concentration in the solution will get imbalanced.	A healthy aquaponics system just keeps getting better and better the longer it operates.
Different composition of nutrients for each crop.	Fish waste converted to nitrates for all crops.
Over fertilizing will immediately burn the plants.	No over fertilization in a fully cycled and well established system.
Optimal pH	
5.5 to 6.0	Compromised pH levels between the plants, fish and bacteria. It is usually 6.4 to 7.0
Water temperature	
Temperature is maintained according to the requirements of plant variety and stage.	The selection of those fish and plants should be made that survive under same optimum temperatures.

Contd...

Table 1–Contd...

	Hydroponics	Aquaponics
EC	EC should be regularly monitored and adjusted.	Only nitrates are to be measured. If nitrates are low (close to zero), more fish should be added to the system. If nitrates are high (above 50) more grow beds and/or plants should be added.
Use of pesticides	Toxic chemicals can be used to control pests.	Pesticides would kill the fish and beneficial bacteria in the system.
Diseases and pests	Hydroponics revolves around a sterile environment	Aquaponics embraces all micro-organism as they each play an important part in the growing process. Aquaponics tends to have less diseases and pest problem.
System start time	System can be started soon after it gets completely installed and plants and nutrients are added.	It takes few days to about a month to start the system because it needs nitrifying bacteria to be developed through a process called 'cycling'. These nitrifying bacteria convert ammonia from the fish waste into the nitrates for the plants.
Flood and drain cycle	Once every 3 to 8 hours depending on the water holding capacity of media, concentration of nutrients in the nutrient solution, plants and seasons.	Flooding for 10-20 minutes every 45-60 minutes. This is because in aquaponics the grow bed has additional role of being the filter for the fish waste.
Depth of grow beds	Enough to hold plant roots. Also depends on the type of system.	Depth is almost double than that of hydroponics as aquaponics grow bed serves a dual role of both home for the plants and bio-filter for the fish waste. It should be deep enough to not only filter the liquid waste, but also to provide an excellent home for composting red worms and the heterotrophic bacteria needed to break down the solid waste from the fish. Most media based aquaponics gardeners use 30 cm deep grow beds filled with an inert media.
Change interval of water solution	Should be changed regularly because the nutrient solution builds up salts and chemicals in the water.	Only needs top-up with freshwater.

18.2 Advantages and Disadvantages

18.2.1 Advantages

☆ The system is self-circulatory and self-sustaining: Once the system is installed and established, there is no need to water or fertilize the crops.

☆ Minimum effort with no gardening problems when using sealed greenhouse

☆ Low water requirement: Less than hydroponics and up to one tenth relative to aquaculture and soil grown crops.

☆ Low power requirement when planned for gravity flow

☆ Environment friendly. No solution waste dumping or runoff which is common to both hydroponics and aquaculture.

☆ Almost free from chemicals, pesticides, or fertilizers.

☆ Tasty and healthy fish and vegetables.

☆ Worms, bacteria, hardy fish, duckweed, simple filters create amazing harvests at low cost.

18.2.2 Disadvantages

☆ Regular monitoring and control of water quality (pH, oxygen, nitrates, temperature) especially in new system.

☆ Failing to properly cycle new system (up to a month in warm temperature, longer when cold) results in poor water quality, fish shock, and dead fish

☆ Hard to maintain the optimal condition for all the three life forms (fish, plants and bacteria).

☆ Over feeding of fish and too few plants in the system would kill the fish.

☆ Power outages without back-up system: high fish mortality (lack of oxygen, too much ammonia).

18.3 Aquaponics Operation

The working of an aquaponics system generally revolves around the **input-process-output** cycle. Inputs are given from outside, which is processed by the various components and life forms in the system and finally the output is in the form of edible or ornamental plants and aquatic animals.

Input

☆ The main inputs to the system are water, fish feed for aquatic animals and oxygen.

☆ In addition, selected minerals or nutrients such as iron can be added for plants.

☆ Dissolved oxygen.

❖ pH up or pH down.

Fish
harvest

Vegetable
harvest

Worms eat
food scraps

Fish excrete
Ammonia

Vegetables
absorb Nitrates

Nitrosomonas sp.
coverts Ammonia
to Nitrites

Nitrobacter sp.
coverts Nitrites
to Nitrates

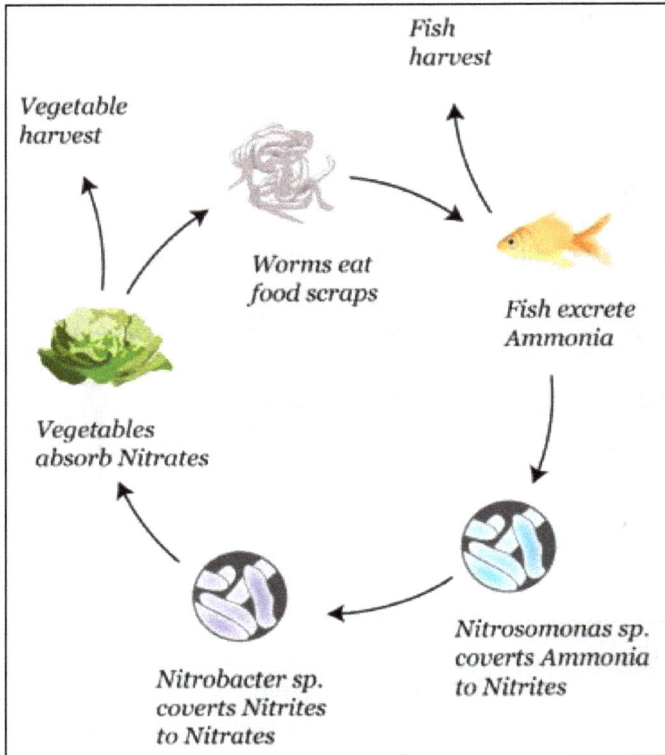

Aquaponics Cycle

Process

The processing units are:

- ☆ Fish: Continuously release wastes that are the main source of nutrients for the plants.
- ☆ Filters: Used to filter uneaten feed and fish wastes. It is also a home for beneficial bacteria.
- ☆ Aeration/Oxygenation
- ☆ Media: Home for beneficial bacteria and worms. Also provides support for plants.
- ☆ Aquarium heater: Maintains the water temperature in the fish tank.
- ☆ Beneficial bacteria: Responsible for converting fish waste to plants nutrients (ammonia to nitrite and nitrite to nitrate).
- ☆ Plant roots. Absorb nutrients and cleans the water for fish.

Output

The main outputs are vegetables or fruits and edible aquatic species.

18.4 Main Components of an Aquaponics System

Aquaponics system can be constructed in number of different ways. Despite this diversity there are three sub systems in any aquaponics set-up: rearing tank or fish tank, bacteria colony in the grow bed or filters or in a separate mineralization unit, and hydroponics sub system. Each sub system contributes to the functioning of the entire aquaponics system. The input to rearing tank is given from outside in the form of fish feed, water and oxygen. The output of this sub system, *i.e.* fish waste and uneaten feed are the main sources of input for bacteria colony where bacteria converts ammonia released from fish waste to nitrates. Plants in the hydroponics sub system absorb the nitrates and clean the water for the fish. This water is then sent to rearing tank for the fish. Several other components or subsystems too are responsible for the effective working of aquaponics system.

Typically an aquaponics system contains some or all of the following components:

☆ **Rearing tank or Fish tank:** These are tanks for raising and feeding the fish. The fish tank is the major component of the entire aquaponics system and can be of a glass or fibre aquarium or any other clean container (for example, a plastic tub, bucket or barrel) that can hold enough water for the fish to thrive. The fish tank should be made from the materials that ensure minimum exposure of the fish to potential toxicity. Two safest materials for making fish tanks are fibreglass and polyethylene (PE). Other materials that can be used include stainless steel or PE as a liner. Fish tank may be a single tank configuration (as is often used in hobby-scale aquaponics systems) or a multi-tank configuration (as is often used in commercial scale aquaponics systems). Water temperature control would be easier and cheaper if fish tank is buried in ground.

　　☐ **Water movement through the fish tank:** Water movement through the fish tank should be optimized to provide the best possible water quality. Optimized water movement through fish tanks is related to several key parameters:

　　　　✦ The flow rate through the fish tank, *e.g.* litres/hour.

　　　　✦ The water turnover rate (*e.g.* fish tank volumes exchanged per hour).

　　　　✦ The optimization of the fluid dynamics of flow that ensures the complete volume of water enters and leaves the tank in the required way.

　　☐ **Size of fish tanks:**

　　　　✦ A larger tank can contain greater number of fish.

　　　　✦ The greater number of fish will produce greater amount of waste.

　　　　✦ The increased amount of waste means more nutrients for plants.

　　　　✦ More plant nutrients mean more plants can be grown.

- ✦ Some examples of large fish tanks are:
  - ▼ Open ponds
  - ▼ Large stock tanks
  - ▼ Swimming pools
  - ▼ Fibreglass tanks
- ☐ **Fish tank shape and design**: The shape and design of the fish tank depends on many factors such as:
  - ✦ Water flow and mixing
  - ✦ Gas exchange
  - ✦ Fish density
  - ✦ Water quality requirements
  - ✦ Solids removal
  - ✦ Treatment of dissolved fish waste
- ☐ **Fish tank Surface to volume ratio**: Majority of the gas exchange (*e.g.* carbon dioxide release and oxygen uptake) of the water body in the tank occurs at the water surface. In the deep and narrow tanks where the surface area is too low as compared to water volume, enough oxygen cannot enter the water body and enough carbon dioxide is not released from the water body. Therefore the surface area to volume ratios should be as high as possible
- ☐ **The rearing techniques:** The rearing techniques can be sequential, stock splitting and multiple rearing units.
  - ✦ **Sequential rearing**: It is quite simple to implement. Fish with different age groups are added in the same tank. The fish that become large enough to be harvested are taken out and small new fish are introduced in the tank. The drawback of this technique is that it can induce stress in the fish that are not fully-grown for market when others are taken; it also makes it difficult to keep track of stock records.
  - ✦ **Stock splitting**: In this technique, the fish are split into two different tanks randomly when the first tank reaches carrying capacity. Again the drawback of this technique is the stress induced by transferring the fish. This stress can be detrimental to their overall growth.
  - ✦ **Multiple rearing units**: In this system, populations start at different ages and are transferred to larger tanks when the fish are big enough.
- ☐ **Stocking Density**: Stocking density is the weight of fish related to volume of water. The accepted approach for stating fish stocking density is the weight of fish (kg) per litres (volume) of water. This is usually stated as kg/m^3 (1 m^3 of water=1000 litres). Most commercial aquaponics growers keep fish at densities up to nearly 50kg/m^3 in

aerated systems and up to approximately 150kg/m³ or higher in direct oxygen injected systems. Hobby growers usually keep fish at densities below 20kg/m³.

☆ **Gravel for tank bottom:** The gravel in the fish tank serves as a home to the nitrifying bacteria that convert ammonia to nitrite and then to nitrate, which can be used by the plants. These gravels can be of different colours and provide an attractive look to the aquarium. These gravels should be washed thoroughly before adding in the tank as these are often dusty and unwashed gravel will cloud the tank water.

☆ **Grow beds (hydroponics subsystem):** The portion of the system where plants are grown by absorbing excess nutrients from the water. It is also a home for beneficial bacteria. There are many types of media that can be used in the media based grow beds. Each media has its pros and cons.

☆ **Growing medium (for media based systems):** It is a porous inert material used to hold the plant roots and maintains moisture. Some examples are: perlite, expanded clay pebbles, peat moss, pea gravel and coconut coir. Media of gravel or non-calcium based rock is cheapest; baked clay is light but expensive. Small size increases surface area for beneficial bacteria growth.

☆ **Sump:** The lowest point in the system where the water flows to and from which it is pumped back to the rearing tanks. The sump is the one place where water is pumped in the system. This is a good place to add water if the system has lost any.

☆ **Auto-siphons:** Auto-siphons can be used in the system to make sure that the water levels will never get too low for the fish and the plants to survive and thrive. Auto-siphons work by taking run-off water from the plants and bringing it back to the tank.

☆ **Red worms:** Red worms not only deal with the solid buildups in the media based systems but also help in managing plant disease and harmful insects and nematodes.

☆ **Water pump and tubing:** Water pump is used to circulate water in the system, usually from the fish tank to the grow bed. The pump can be simple that runs continuously or timer operated. In a simple aquaponics, after the water is pumped into the grow bed and is allowed to drain back to the fish tank through gravity. Enough tubing may be needed to circulate the water from the outlet on the pump to the top of the grow bed and to allow it to drain back from the outlet of the grow bed to the fish tank.

☆ **Aeration pump, air stones and tubing (**used to aerate water in the fish tank): It blows air into the tank water for both the fish and the plants. Tubing connects the air pump to an air stone at the bottom of the tank. The air stone breaks the stream of bubbles coming from the air pump into micro-bubbles and increases the oxygenation in the water.

✰ **Filters:**

☐ Biofilter: It is a place where the nitrification bacteria can grow and convert ammonia into nitrates.

☐ Mechanical filtration (for removal of solid waste): A solid removal filter should be used if there is more fish waste than can be controlled by biofilter and plants. The type of solid removal system depends on how much organic waste is produced in the system that may form a buildup in the grow beds or settle down in the tank. Balanced ratio of fish and plants helps to minimize the solid waste in the system.

✰ **Mineralization tank:** If solids are removed from the system using sophisticated solid removal filters, mineralization is done separately in separate units. In such systems, there is no or very low competition for the dissolved oxygen in fish tanks or in the grow beds, and the development of anaerobic zones is very rare.

✰ **Aquarium heater and aquarium shade cloth cover** (used to maintain the water temperatures in the fish tank): Although for small aquaponics systems aquarium heater and aquarium shade cloth cover can be used to prevent sudden temperatures shifts from occurring, greenhouse will do the same for larger systems. Also a larger volume of water of 1,000 litres and more will offer stability and moderate sudden temperature swings in the system.

✰ **Grow lights:** Light is necessary for the overall development of the plants because plants use carbon dioxide, water and light along with many minerals to produce food for themselves. There is no need to install artificial grow light for an outdoor hydroponic system or even for indoor hydroponic system if it is kept at a place where sufficient sunlight is available. If the hydroponic system does not get enough sunlight, the grow lights can be used. The choice of grow lights for the hydroponic system depends on the plant type (variety), plant size, plant stage (flowering and fruiting), type and size of hydroponic system being used, place (indoor or outdoor), and season of the year (such as summer or winter). There are many types of grow lights that are specifically designed for indoor hydroponics system. Some of them are fluorescent hydroponics grow lights, metal halide hydroponics grow lights, high pressure sodium grow lights and special LED grow lights.

✰ **Light for fish tank:** It is generally used in indoor systems (aquariums or aquaponics fish tank with ornamental fish). Most aquariums have a florescent light to see the fish and monitor their health. These lights also add extra beauty to the aquariums.

✰ **Fish and plants (or seeds):** Fish produce waste (source of plant nutrients) and plants purify the water by consuming the nutrients.

✰ **Water:** A good quality clean, de-chlorinated and free from pests and diseases should be used.

18.5 Sump

In an aquaponics system, sump is the lowest point where the water flows to, and from where it is pumped back to the rearing tanks. The sump is the one place from where water is pumped in the system. Its main purpose is to enable the fish tank water height to remain constant regardless of how full or empty are the grow beds. This is a good place to add water if the system has lost any and to provide more system stability with regard to pH and temperature. Small fish can also be put grown in the sump. Sump is usually placed at a lower point than the grow beds and is where the grow beds drain to. A sump tank can be placed at a lower or constant height with the fish tank.

A sump tank should be large enough to handle the situation when all of the grow beds are full at once. The grower need to add the total water volume of the grow beds, then subtract the displacement effect of the grow media (for hydroton it has been recorded that 55 per cent of the water is displaced, and 62 per cent with 1.8" gravel). The grower also needs to calculate the minimum amount of water that must remain in the sump tank to cover the pump. This volume should be added to the grow bed water volume calculation and this total is the minimum volume required for the sump tank. If required, two sump tanks can be easily connected using bridge siphon.

There can be many different variations in which the sumps can be utilized. Some of these variations are given below.

☆ The water is pumped from the sump tank to the fish tank using a water pump. From the fish tank, water is drained to grow beds and then from there it is drained back to sump tank.

☆ The water from the sump is pumped to both the grow beds and the fish tank and both the fish tanks and grow beds are designed to drain it back to the sump.

☆ The water is pumped from the sump tank to grow beds. From the grow beds, water is drained to fish tank and then from there it is drained back to sump tank.

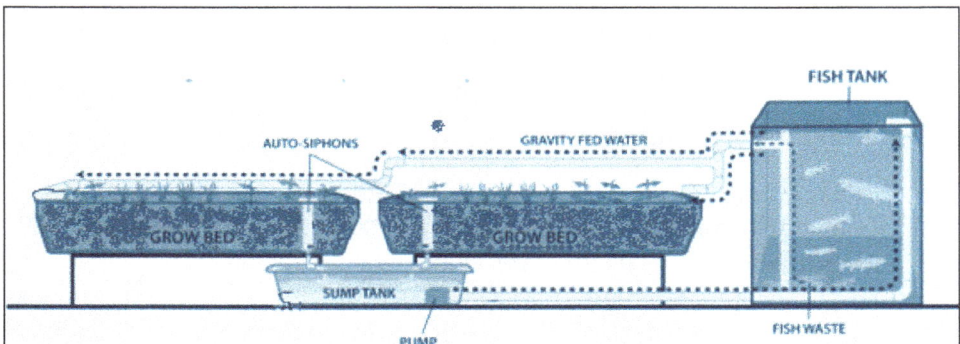

Aquaponics with One Sump, One Fish Tank and Two Grow Beds.

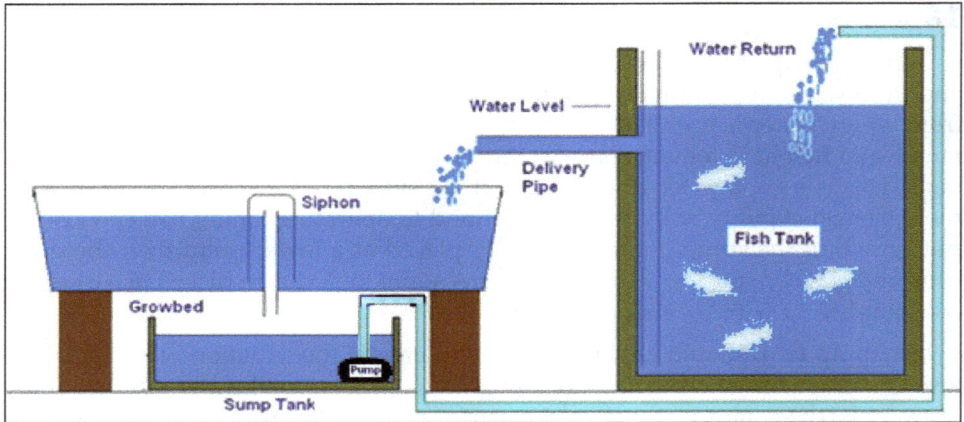

Aquaponics with One Sump, One Fish Tank and One Grow Bed.

☆ Water is pumped from sump to fish tank. Another pump in the fish tank pumps water in the fish tank to grow beds. Water from the grow beds is drained by gravity to sump.

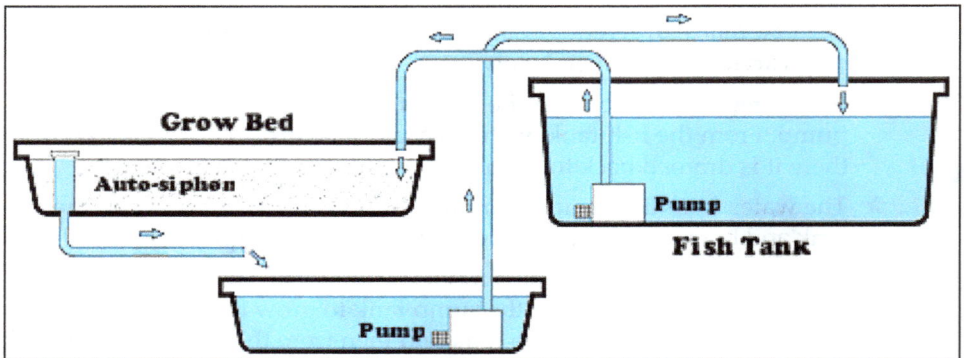

Sump Tank

18.6 Types of Aquaponics Systems

Aquaponics system can be designed as solution culture (without media) or medium culture.

The three main widely used aquaponics systems are:

1. Raft system (also called float, deep channel, and deep flow): Plants are grown in floating styrofoam boards in a tank separate from the fish tank. Raft systems are limited to certain plants to be grown.

2. Nutrient film technique (NFT): Plants are grown in long, narrow channels with a thin film of water containing nutrients which flows through them to bring nutrients to the plants' roots.

3. Media filled beds: Media filled beds are simply containers filled with a growing medium, like gravel, perlite, or hydroton, in which the plants roots are held, then they go through a flood and drain sequence to bring nutrients to the roots. If the top 5 to 6 cm layer of the media is kept dry, it helps in preventing weeds, fungus and harmful bacteria from growing, and also save worms from being drowned. The ideal depth of media beds is 30 to 45 cm or more depending on the type of plants to be grown in them. The grow bed shell should extend 5 cm above the media bed with an overflow pipe to return water to the sump tank if stoppage occurs in return plumbing (if sump tank is not being used, the overflow pipe should return water to the fish tank).

The first two methods are more common in commercial-size operations, while the last method is most commonly used in backyard operations. Other methods include 'static solution culture', 'wick system', 'drip system', *etc.*

18.7 Electricity Cost in Aquaponics System

Although a grower has to spend a great amount on electricity bill in an aquaponics system (for running aerators, pumps, artificial light and for maintaining temperature), this cost on electricity can be saved to a great extent by proper design methodologies and proper selection of fish and plant species.

For example:

☆ Instead of using aerators, the grow beds can be arranged so that the water from it falls from enough height to splash back into the fish tank, mixing air into the water.

☆ The most basic form of mechanical filtration is to use a foam block underneath the irrigation pipe entering the grow beds. This is cleaned on a regular basis, and stops large amounts of sludge and uneaten food, it is a cheap and very easy way of creating mechanical filtration.

☆ The electricity cost on maintaining proper temperatures can be saved to a great extent by intelligently selecting the fish and plant species.

☆ Short tube lengths: Unnecessary long tubes to transfer water should be avoided because water pump needs more power to transfer water to the beds or fish tanks through long tubes. The costs on electricity bills can be completely eliminated for small systems by doing manual labour to irrigate plants in the system. In such systems, the following items will be needed.

☐ A tank to hold the fish (like a large plastic tub)

☐ A container for the plants means to elevate plants above fish tank. It should be placed at enough height so that the water from it falls from enough height to splash back into the fish tank, mixing air into the water

☐ A watering device.

❑ To start up this system, it is important to put the fish in the fish container at least a week before. Also, before watering the plants, the fish-rearing tank should be swirled and then using a watering can, this water containing fish waste can be manually transferred to plants grow bed. In this system, the fish-rearing tank will need to be cleaned periodically and it is important to flood containers at least three times a day.

18.8 Backup System

No matter how well designed and controlled the system is, the aquaponics gardeners must keep a backup power supply ready in case of power failure. A well stocked system filled with fish will not last long enough after power failure. The fish may die due to low or no oxygen in the water. The plants will also not survive due to unavailability of water and oxygen for a long time. Therefore the gardeners need to have an alternate option for power supply in the system to run water pumps and air pumps. Although generators can be used as an alternate power supply, it would require atleast one person to be permanently available at the aquaponics site. The best backup system would be inverter system that will switch automatically over to internal batteries in case of power failure and when the power comes back, the internal battery gets recharged and the system is automatically switched to the main supply. Other better option would be to install a solar system to recharge the battery and run pumps in the system.

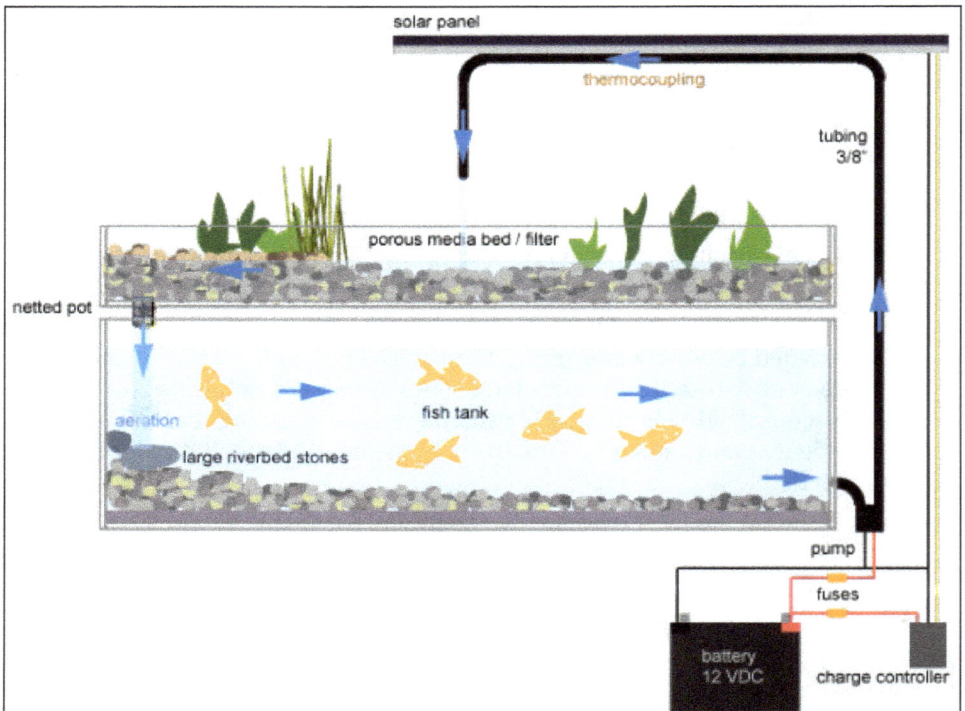

Aquaponics System with Backup Power Supply.

Some growers use air pumps with internal rechargeable batteries in them. Normally they run while plugged into the power, pumping air through air lines and air stones in the fish tank. When the power goes out they switch automatically over to their internal batteries and continue to pump air into the water keeping the fish alive for the life of the battery within the unit, often up to 10 hours. When the power comes back, the internal battery is recharged.

18.9 Mineralization Tank

A mineralization tank may be required in raft or NFT system, especially when solid removal filters are being used in the system. In its simplest form it contains some type of porous media, water and something to provide aeration (usually air stone). The solids may then be directed either manually or can be configured so that they may be directed automatically to the solid mineralization device for mineralization and breakdown and then the mineralized nutrients may easily be added back into the aquaponics system either manually or automatically. Aerated mineralization tanks may be configured in a number of ways; some contain complex baffle systems and screens to separate the water containing the dissolved nutrients from the remaining fish waste solids, some have automatic water and solids addition and automatic nutrient-rich water removal systems, *etc.*

Mineralization tank may not be needed in a media-filled bed where the solids remain in the system trapped in the media. If media beds are sized correctly, and designed to encourage aerobic conditions, then they should operate well as solid's filtration and mineralization devices.

18.10 Plumbing

There are many factors that must be considered carefully while plumbing an aquaponics system. These factors depend on the design, space and location of an aquaponics system being constructed.

18.10.1 Pipes

Usually PVC or UPVC pipes are used for plumbing in almost all the aquaponics systems. This is because they are durable, cheap, easily available and above all they come in standard sizes worldwide with wide range of adapters and connectors. Other pipes such as agricultural pipe, flexi-pipe, bamboo, hosepipe, *etc.* can also be used but before using any pipe it should be made clear that they are not toxic for fish or plants. Iron pipe is not recommended as it gets easily corroded and copper pipes are toxic to the fish.

18.10.2 Connectors

Two common types of connectors that can be used in aquaponics are threaded connectors and slip connectors.

☆ Threaded connectors are ones that screw into one another and are designated as male and female.

☆ Slip connectors as the name suggests, just slip into one another. In order to preserve pipe diameter you will find that most connectors are female, and the male pipe simply fits directly into it.

☆ Examples of connectors are: 90 elbows, 45 elbows, 90 tee fittings, ball valves, bulkheads, reducers, couplings.

18.10.3 Fitting Pipes through the Growbeds

When connecting a pipe through a growbed or a fish tank, generally bulkhead fittings or something like a uniseal is used.

☆ Bulkheads come in various sizes and shapes and are easily available in any homestores. These can either be purposely-designed adapters or could be standard male and female adapters with some rubber washers. The bulkhead is a good option for plumbing through the bottom of the growbed when using a standpipe.

Bulkhead

☆ Uniseals are rubber rings that fit into the holes that have been drilled into the tank. They clamp around the hole making a watertight connection and then the PVC pipe can be slotted into the seal. The seals usually allow the pipe to be installed in only one direction, thus providing a watertight seal between the pipe and the connector. Uniseals are a cheap and easy method of putting a pipe through a tank, and they can also be used with rounded surfaces thus making them particularly useful for plumbing into barrels and other such rounded containers.

Plumbing with Uniseal

18.11 Aquaponics Greenhouse

Although a greenhouse is not an essential part of an aquaponics system, it provides enormous protection for the system compared to being exposed to the rain, hail, wind, *etc.* The light, temperature and humidity for a small indoor aquaponics system can be maintained using artificial light, ventilation, aquarium heater and aquarium cover. During winter season insulation, solar panels and aqua heaters can be used to maintain the fish tank temperature. The best suited environment for all living creatures (fish, plants and bacteria) in a large commercial aquaponics system is best maintained in a greenhouse or growroom. See Greenhouse (Chapter 4–"Environmental Conditions for Plants")

18.12 Success of Aquaponics System

The success of an aquaponics system depends on the following factors:

☆ Nitrification (a proper nitrogen cycle)/Mineralization
    ❑ Beneficial bacteria
    ❑ Denitrification

☆ Good water quality
- ❏ Water hardness
- ❏ Ammonia, nitrite and nitrate levels
- ❏ Water temperature
- ❏ pH
- ❏ Alkalinity
- ❏ Carbon dioxide
- ❏ Dissolved oxygen
- ❏ Water testing

☆ Filtration
- ❏ Biofilters (colonies of nitrifying bacteria)
- ❏ Solid removal system or mechanical filtration

☆ Size of aquaponics system and fish to plant ratios
- ❏ Fish stocking densities

☆ Flow rate of water
- ❏ Proper design and size of pipes

☆ Ratio of water volume to surface area of the fish tank

☆ Water change

☆ Nutrient supplement (calcium, potassium and iron)

☆ Feeding of fish

☆ Fish stress

☆ Using salt

☆ Pests and diseases

☆ Fish species for aquaponics

☆ Plant species

☆ Use of biological pest control

18.13 Nitrification (System Cycling)

One of the most significant functions in an aquaponics system is nitrification (aerobic conversion of ammonia into nitrates). Ammonia is gradually released into the water through the excreta and gills of fish, and its higher concentration will be toxic for the fish. Nitrogen found in ammonia is not readily taken up by plants, so no matter how high the ammonia levels get in the fish tank, the plants will not be getting much nutrition from it. Nitrification reduces the toxicity of the water for fish, and allows the resulting nitrate compounds to be removed by the plants for nourishment. Therefore, it becomes necessary to establish the **nitrogen cycle** quickly and with minimal stress on the aquarium's inhabitants. Aquarium filtration has advanced from the old box filters filled with charcoal and glass wool to under-gravel filters, then trickle filters, and most recently - fluidized bed filters. Every advance has been made to improve upon the effectiveness of biological filtration which in turn increases the efficiency of the nitrogen cycle.

18.13.1 Nitrifying Bacteria

There are several genera of bacteria that will oxidize ammonia to nitrite and nitrite to nitrate. Bacteria take the unusable aquaponics fish waste and create a near perfect plant fertilizer. The most common genus of a family of autotrophic bacteria called **nitrifying bacteria** are *Nitrosomonas* (that convert ammonia into nitrites) and *Nitrobacter* (convert nitrites into nitrates), respectively. Ammonia can also be converted into other nitrogenous compounds through healthy populations of these bacteria. Beneficial bacteria in aquaponics not only aid in breakdown of organic material, but also prevent harmful bacteria from taking hold. Since the nitrification process acidifies the water, non-sodium bases such as potassium hydroxide or calcium hydroxide can be added for neutralizing the water's pH. In addition, selected minerals or nutrients such as iron can be added in addition to the fish waste that serves as the main source of nutrients to plants.

The bacteria responsible for nitrification process will form a biofilm on all solid surfaces of the system which are in constant contact with the water. In aquaponics systems where no media are being used (*e.g.* floating raft systems); the submerged plant roots and aquarium filters can accumulate enough bacteria for nitrification. Most aquaponics systems include a biofiltering unit, which helps facilitate growth of these microorganisms.

Nitrification is optimal at high dissolved oxygen and low levels of organic matter. The nitrification process also destroys alkalinity, produces alkalinity (H) and lowers pH.

When an aquaponics system gets constructed and becomes ready to be used for the first time, there will be virtually no beneficial bacteria present. A buildup of the natural colony of bacteria can take 20-30 days, sometimes up to 8 weeks. It will be a good idea to establish the colonies of *Nitrosomanas* and *Nitrobacter*, before adding the fish and plants to the aquaponics system. Although these beneficial bacteria will establish themselves naturally (it will just take a little longer to run the system) as these are naturally found in the air and water, these can be brought from other already running systems such as aquariums and ponds. Some methods to establish beneficial bacteria build-up are:

☆ Allow a buildup of bacteria colony naturally. It will take a little longer.

☆ From water, gravels and filters of already running disease free aquaponics systems.

☆ **It can be manually added by purchasing from aquaponics stores.** Recommended if the aquaponics is to be started immediately.

☆ **Fish feed:** Before adding plants and fish in the system, start cycling the system using fish feed. The feed starts to break down on the bottom of the fish tank, and in the growbeds it will release ammonia for the bacteria to feed on.

☆ Some other things that can be added for establishing bacteria colony are mentioned below.

　❑ Cow dung or cow pee: At least 15-20 day old manure. Fresh dung is not recommended.

　❑ Urea fertilizers: This is a fairly straightforward method of cycling a system.

　❑ Household ammonia: Use of only food grade ammonia is recommended.

　❑ The dead prawn or fish: Place some rotting fish or crustacean in the system to induce a source of ammonia for the bacteria to feed on.

Note: *Anything added in the system should not be overdosed.*

The amount of time for establishing bacteria colony depends on variable factors such as:

☆ **Water temperature**: Optimum is 25-30°C. Nitrifying bacteria will most probably die at temperature below 1.1°C or above 40.5°C. *Nitrobacter* is less tolerant to low temperatures than *Nitrosomonas*.

☆ **pH range**: If the pH drops to 6.0, all nitrification is inhibited.

　❑ Optimum pH range for *Nitrosomonas* is 7.8–8.0.

　❑ Optimum pH range for *Nitrobacter* is 7.3-7.5.

　❑ Compromise point between the needs of the fish and the needs of the plants is 6.8-7.0.

　❑ Bacteria will stop ammonia to nitrate conversion if there is much pH fluctuation and prolonged levels outside of the system's acceptable range.

☆ **Dissolved oxygen**: Maximum nitrification will occur if dissolved oxygen levels exceed 80 per cent saturation. If dissolved oxygen concentrations drop to 2.0 mg/l (ppm) or less, nitrification will be inhibited. *Nitrobacter* is affected more by low dissolved oxygen than *Nitrosomonas*.

☆ **Depth of grow bed**: 30-35 cm will be better.

☆ **Light:** Nitrifying bacteria are sensitive to light, especially to blue and ultraviolet light. Once their colony is developed, there is no problem.

☆ **Chlorine**: It will surely kill beneficial bacteria and fish. To neutralize chlorine water conditioner or chlorine neutralizer that is suitable for use with fish can be used.

☆ **Other additives**: Anything that may harm beneficial bacteria or fish should not be added or applied over the aquaponics system. For example pesticides and growth enhancers.

Note: *If a fishless cycling technique is being used, the fish should be added once nitrates are present and the ammonia and nitrite levels have peaked and declined below 1.0 ppm.*

Monitoring of ammonia, nitrite, nitrate, pH and water temperature should be done regularly (daily during cycling). Aquarium water test kits can be used to measure the amount of ammonia, nitrite and nitrate levels in ppm (parts per million) as well as pH and temperature. It is recommended to test everyday once the source of ammonia is added. When the presence of nitrates is confirmed, and the ammonia and nitrite concentrations have both dropped to 0.5 or lower, the system will be fully cycled. The pH and temperature affects both the cycling rate and the health of fish and plants in the system. A pH test kit and a submersible thermometer can be used to measure the pH and temperature, respectively.

18.14 Denitrification

Denitrification usually occurs under anaerobic conditions (no oxygen) when anaerobic bacteria gets developed and may convert back nitrate to nitrite which is toxic to fish. During denitrification (under anaerobic conditions), the nitrogen in the aquaponics system may be converted to many different forms and eventually to nitrogen gas (N2) which bleeds out of the system. The denitrification process raises the pH of the water in the aquaponics. If too much denitrification occurs due to too many anaerobic zones, the pH can rise to neutral (7) or even above, which is not good, as the buffers cannot be added to adjust pH to supplement potassium and calcium nutrients which are essential for plant growth. A well designed and managed aquaponics system should operate under aerobic conditions where the pH is constantly falling.

Note: *Decreasing nitrate levels is beneficial when raising fruiting plants such as tomatoes as high nitrate levels promote vegetative growth and reduce fruit set. High nitrate levels stimulate the growth of leafy green plants.*

18.15 Earthworm (Red Worm)

A good way to deal with solids buildup in media based aquaponics is the use of worms, which break down the solid organic matter (waste from the fish, leftover fish feed and other biodegradable materials) and make them more bio-available to the plants through their waste called worm casting or vermicompost. Apart from converting waste into nutrients, red worms help in managing plant disease and harmful insects and nematodes.

A handful of composting red worms can be added to each grow bed once the system is fully cycled and fish have been added.

18.16 Good Water Quality

Only good quality de-chlorinated water which is free from pests and diseases should be added in an aquaponics system. Good water quality in the system insures the growth of healthy fish and plants. The water quality depends on various factors such as its source, pH, ammonia, nitrite/nitrate, bacteria, temperature, oxygen, and alkalinity. Water quality in the system can be maintained by:

☆ Using de-chlorinated and disease free water (best would be RO water).

☆ Proper monitoring and maintenance of water temperature and pH levels.

☆ Using good filtration techniques.

☆ Proper control of concentration of ammonia, nitrite, nitrate and carbon dioxide.

☆ Proper aeration.

☆ Proper feed

18.16.1 Water Hardness

Water to be used in aquaponics may have many dissolved compounds. Therefore it is necessary to first identify the dissolved elements in the water and must be rectified accordingly.

18.16.2 Ammonia, Nitrite and Nitrate Levels

High concentration of ammonia and nitrite is toxic for fish.

18.16.2.1 Ammonia

Toxicity levels for un-ionized ammonia depend on the individual fish species; however, levels below 0.02 ppm are considered safe. Normally ammonia concentration should be maintained below 1 ppm, but anything over 0.25 ppm can be harmful and may cause stress in fish, which can leave the fish more susceptible to disease. Majority of ammonia is non-toxic at water pH slightly below neutral, *i.e.* 7.0.

If ammonia levels exceed the tolerance levels of fish, 1/3 of the tank's water should be replaced with fresh, clean and de-chlorinated water, preferably with water having nearly same temperature and pH levels.

There may be a sudden rise in ammonia levels in a new or not fully cycled system or if the fish are overfed, especially when temperature conditions are warmer than normal. Ammonia poisoning is known to cause extensive tissue damage, in particular the gills and kidney, physiological imbalances, decreased resistance to disease, impaired growth and even death. It can be seen physically by reddening of the eyes. Therefore, a regular test for ammonia is necessary, especially if the system is new or the fish are overfed.

18.16.2.2 Nitrite

Nitrite poisoning inhibits the uptake of oxygen by red blood cells, also known as brown blood disease. The higher concentration of nitrite (above 10 ppm) in the fish tank will result in the death of fish due to **brown blood disease**. In such case, 1/3 of the water in the fish tank should be replaced with fresh, de-chlorinated water and sodium chloride salt should be added to the system. This can be done by adding about 1 part per thousand (1 kg of salt per 1000 litres of water) using non-iodized salt. Pool salt or water softener salt can be added but table salt should be avoided as it contains potentially harmful anti-caking agents. It will be better to use the salt which is specifically formulated for fish. The salt should be first dissolved completely in some container and then this salt solution should be added in the fish tank. Salt helps mitigate nitrites because the chloride ions bind with the nitrites and thereby help keep some of the nitrite out of fish. The fish feeding must be stopped and the aeration should be increased to maximum until the nitrite levels decline to below 1.0.

18.16.2.3 Nitrate

Although high nitrate levels won't kill the fish, they will feel the impact of nitrates by the time the levels reach 100 ppm, particularly if levels remain there for some time. The resulting stress leaves the fish more susceptible to disease. A low to no nitrate reading in the fish tank means that the aquaponics system is properly balanced and confirms that the vegetation in the system is flourishing.

For a high nitrate reading in the fish tank, either the number of fish should be reduced or number of plants and/or grow-beds should be increased to absorb up the excess nutrients. The growth of plants will give a good indication of the health of the system. Algae growth in the tank is also a good indicator of high nitrates in the system.

18.16.3 Water Temperature

Water temperature is a cross-affecting component. It affects the oxygen levels, the amount of unionized ammonia (ammonia not yet converted to nitrite ions) and the water salinity. Warm water has less oxygen, a greater proportion of unionized ammonia and more salinity as compared to cold water.

Maintaining proper temperature of water in fish tank will enhance the growth and quality of both fish as well as plants. The selection of those fish and plants should be made that survive under same optimum temperatures and thrive well under the native environment. Fish species are temperature dependent. Each species of fish has a desirable temperature range, and depending on the climate, heating or cooling of the water may be needed to keep fish happy. There must not be a large fluctuation of water temperatures and conditions. Plants too do not like rapid temperature changes; especially in the root zone, but as compared to fish they are more tolerant to temperature changes. It is recommended that the change in temperatures should not be more than 3°F per day. Therefore, before adding water to the fish tank, it should be allowed to come to the same temperature as the water in the tank.

Different species of fish require slightly different water temperatures and conditions. The water temperature for most plants can be maintained between 17-27 C. The grower must remember that aquaponics is a compromise between the needs of the fish, the plants and the bacteria.

Although for small aquaponics systems aquarium heater and aquarium shade cloth cover can be used to prevent sudden temperature shift from occurring, greenhouse will do the same for larger systems. Also a larger volume of water of 1,000 litres and more will offer stability and moderate sudden temperature swings in the system. Water temperature can be easily controlled if fish tank is buried in ground.

Note: Fish get stressed and won't eat if the temperature is too cold or too hot.

If plants and fish with entirely different temperature requirements are to be grown in the same system, the provisions should be made so that the water from the fish tank should be pumped to some other container where its temperature is maintained for use in plants and then transferred to grow bed. The purified water from the grow bed is again first collected in a container that adjusts its temperature as per the need of fish and then this water is transferred to fish tank. This will either require two extra containers and extra drainage systems or temperature adjustment can be performed in the drainage pipes. The investment would be quite high. Therefore it would be better to have the fish and plants that will survive under similar conditions.

The following figure shows temperature adjustment for fish and plants.

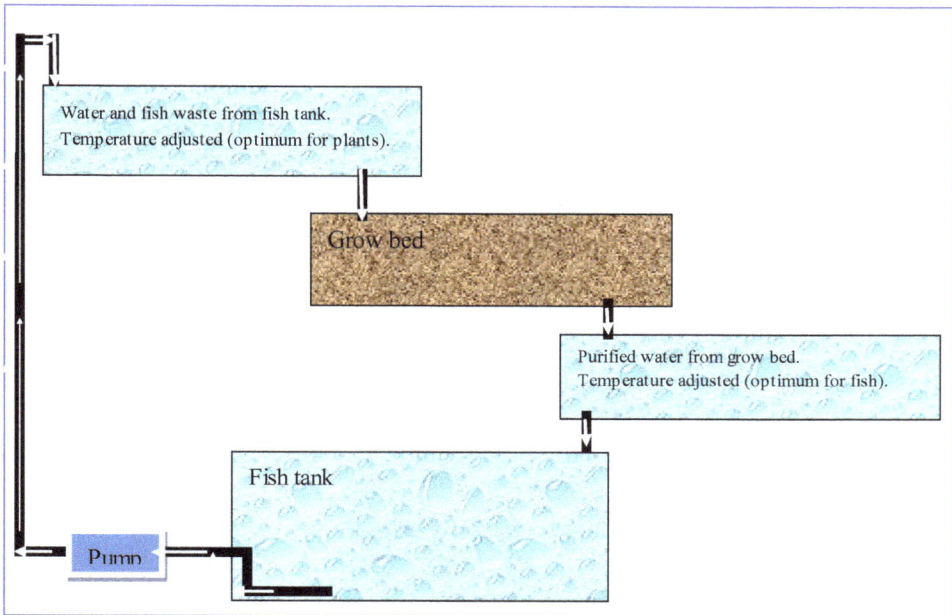

Fish are generally categorized into warmwater, coolwater, and coldwater species based on optimal growth temperatures as shown below.

	Fish		Bacteria	Plants
Coldwater Optimum 55-65ºF (12.8-18.3ºC)	Coolwater Optimum 65-75ºF (18.3-23.9ºC)	Warmwater Optimum 75-90ºF (23.9-32.2ºC)	Optimum 77-86ºF (25-30ºC)	Optimum for most plants 62º - 80°F (17-27°C)
Example: Rainbow trout	Example: Walleye and yellow perch	Example: Tilapia and catfish	Will most probably die at tempe-rature below 34ºF (1.1ºC) or above 115ºF (46.1ºC). *Nitrobacter* is less tolerant to low temperatures than *Nitrosomonas*	

The temperatures in the fish tank are maintained as compromised optimum levels for fish, bacteria and plants. For example:

☆ For cold water fish: 60-65°F (15.6-18.3°C)

☆ For cool water fish: 70-75°F (21.1-23.9°C)

☆ For warm water fish: 75-80°F (23.9-27°C)

This example is just a basic idea, actual compromised optimum temperatures depends on the fish and plant species being grown in the system.

Prolonged water temperatures outside of the system's acceptable range will cause the following issues:

☆ Fish: Poor food consumption and/or stress leading to disease vulnerability

☆ Plant:

❒ Root disease

❒ Slow growth leading to foliage disease and pest vulnerability

☆ Bacteria: Ammonia conversion is slowed or completely stopped leading loss of fish and plants.

18.16.4 pH

The pH influences many water quality parameters such as percentage of NH_3 vs NH_4 and solubility of plant nutrients.

In aquaponics, fish, microbes (beneficial bacteria) and plants all thrive together in a symbiotic relationship and help each other to grow naturally and organically. Therefore the pH and temperature of the water should be maintained at compromised levels between the plants, fish and bacteria. The pH is usually 6.4 to 7.3 (good results obtained by us at 6.8-7.0).

Optimum pH Levels for Fish, Bacteria and Plants

Life Forms in Aquaponics										
Fish (Freshwater)		Bacteria		Plants (for most plants)					Compromised Ideal Level for Mature Aquaponics	
Ideal for most Fish	Adjust themselves	Nitrosomonas	Nitrobacter	Usual Range	Perfect Range	Can be Allowed to Float Between	Optimum for Vegetative Growth	Optimum for Flowering and Fruiting	Usual Range	Showed Good Results
6.5-7.5	6.0-8.5	7.8–8.0	7.3-7.5	5.5-6.5	5.8-6.3	5.5 - 7.0	6.0-6.3	5.7-5.9	6.4-7.3	6.8-7.0

Although some fish may be happy at pH below 6.0, most of the fish prefer slightly alkaline water and the range should not go below 6.0. Also if the pH drops to 6.0, all nitrification is inhibited. At pH above 7.0, ammonia toxicity gets increased (higher pH readings suggest higher ammonia concentrations) which will kill the fish in the tank. Most freshwater fish will do well in a pH range of 6.0–8.5. Fish will adjust to different pH ranges, but only if done very slowly.

The plants prefer pH between 5.5 and 6.5 and nitrifying bacteria do well at pH 7.3-8.0. The preferred pH of all life forms in an aquaponics system has a different pH range. Therefore in an aquaponics system pH should be maintained at compromised levels somewhere between 6.4 and 7.3. Most aquaponics farmers maintain the pH slightly below 7 *i.e.* 6.8. There are two reasons behind this:

1. If ammonia is present in the system, majority of it is non-toxic at water pH slightly below neutral 7.0.

2. Many plants like a pH of around 6.0.

In aquaponics the plants will produce some oxygen through photosynthesis during the day but they will produce carbon dioxide at night. The carbon dioxide will lower the pH hence pH may be lower in the morning than in the afternoon. Fish too release carbon dioxide during respiration. In aquaponics systems, constant fish respiration can raise carbon dioxide levels high enough to interfere with oxygen intake by fish, in addition to lowering the pH of the water.

If pH gets too low, anaerobic bacteria may get developed in some or many parts of the media bed, which produce acids. In such case, plants with very large systems should be removed because these create pockets where air cannot access. If the pH is too high, it is generally a sign that the system biofilters are not keeping up with the fish's production of ammonia. More plants can be planted to sort out the problem.

18.16.4.1 Adjusting pH

If the pH is out of range, it should be changed slowly because a sudden and large pH swing is very stressful on fish. The pH reading should be shifted towards the target pH slowly, with no more than 0.2 to 0.3 per day. It will be better if the pH is tested and buffered (if required) every day because in a fully established system the pH will drop only a little bit in a day and it will require only a small shift to bring it to optimum.

There are several products that can be used to quickly and easily adjust the pH up or down as needed. The pH in aquaponics water can be easily checked and monitored with a simple pH meter, paper test strips or liquid test kits.

The safest way to adjust the pH in an aquaponics system is to use a tiny bit of the pH up or pH down adjusters. The instruction labels of these pH adjusters must be carefully read before adding them in water. The pH in an aquaponics system can be raised using calcium hydroxide, potassium carbonate (or bicarbonate) or potassium hydroxide.

The carbonates buffer the pH making it hard to adjust to the desired level accurately. Citric acid should never be used as it is anti-bacterial.

Note: If vegetables with very different pH requirements are to be grown in the same grow bed, the heavy feeder plants should be planted where the water flows into the grow tray, favouring their pH requirement and the light feeder plants at the end where the water flows out. It will be best to grow blooming plants (tomatoes, peppers, cucumbers, etc.) and non-bloom plants (lettuces and other green leafy plants such as basil, herbs, etc.) in separate grow trays so that the grower can closely monitor and adjust the pH levels for each variety.

18.16.4.2 Stable pH

Stable pH is a condition that arises due to de-nitrification in anaerobic zones producing alkalinity. Such periods when pH does not decline is considered detrimental because calcium and potassium cannot be supplemented. If such condition arises, the filter tanks should be cleaned regularly (*e.g.* twice a week) and the deposits of organic matter from the grow-beds should be removed.

18.16.4.3 Nitrification and pH

Nitrification is at its peak when the pH is at 6.8-7.0 or a little above. Coincidentally, this is the compromise point between the needs of the fish and the needs of the plants. When the pH drops below 7.0, nitrification slows down, and at a pH less than 6.0 nitrification nearly stops. The nitrification process generates nitric acid, which will lower the pH of the water. The variance of pH may also be due to other factors such as alkalinity and buffering capacity of the water, chemistry of the water and temperature.

pH fluctuation and prolonged levels outside of the system's acceptable range will cause the following issues:

- ☆ Fish: Stress leading to disease vulnerability
- ☆ Plant:
    - ❑ Nutrient lockout *i.e.* unable to take up available nutrients
    - ❑ Slow growth leading to foliage disease and pest vulnerability
- ☆ Bacteria: Stop converting ammonia to nitrate leading to fish and plant loss.

18.16.5 Alkalinity

Alkalinity is a buffer that neutralizes acid and its concentration is expressed as the equivalent concentration of calcium carbonate ($CaCO_3$). Base can be added to increase alkalinity. The acceptable level of alkalinity in aquaponics has a broad range between 59 mg and 300 mg/litre.

18.16.6 Carbon Dioxide

The CO_2 level should not exceed 20 mg/litre because at higher levels the fish cannot absorb enough oxygen through their gills. In systems with diffused aeration, CO_2 buildup is not a problem as it is vented off to the atmosphere. In aquaponics use of pure oxygen is not recommended, therefore CO_2 buildup is not a problem.

18.16.7 Dissolved Oxygen

The dissolved oxygen is important for the optimal growth of all life forms in an aquaponics system (fish, plants and beneficial bacteria). It is advisable to use diffused aeration (air stones and porous hose) in an aquaponics system and measure dissolved oxygen frequently in a new system and seldom in a mature system. This can be done by test kits and DO meters. The recommended levels of the dissolved oxygen in an aquaponics system is 5 milligrammes per litre or higher. The best levels of dissolved oxygen are close to 80 per cent saturation, which translates as 6-7 milligrammes per litre. Levels below 3 milligrammes per litre are a hardship on aquatic life and below 1 milligramme per litre are fatal. The amount of oxygen the water can hold depends on the properties of the water, particularly temperature. The oxygen holding capacity of water declines at higher temperature.

Oxygen in the fish tank reduces to a great extent soon after feeding. The amount of the dissolved oxygen in the system is affected by certain factors as shown below:

☆ Time of the day
  ❑ Highest during late afternoon
  ❑ Lowest during early morning
☆ Weather: Goes down during cloudy and/or rainy days
☆ Stocking density of fish (more fish, less oxygen)
☆ Temperature (higher temperature, less oxygen)
☆ Salinity of water (high concentration of dissolved salts, less oxygen)
☆ Use of air diffusers (smaller bubbles, more oxygen)
☆ Fish waste (more waste, less oxygen)

The behaviour of fish will change when the dissolved oxygen levels in the fish tank lowers to the fish desired levels. They may stop eating, gasp at the water surface, gather at the inflow of water and become susceptible to diseases.

If the water in the fish tank is not properly aerated, fish may die within 45 minutes. Even if death is not immediate, gill damage can be permanent and slowly, and the fish population will fall. Therefore it is necessary to have a backup power system that can be used in case of power failure for aerators and for pumps that circulates water. Water aerators can be purchased at an aquarium supply store but it is not the only way to add oxygen to the fish tank. The water flowing out of the grow beds can be arranged so that it falls from enough of a height to splash back into the fish tank, mixing air into the water.

18.16.8 Water Testing

It is recommended that the water in the system should be regularly tested for ammonia, nitrite, nitrate and pH levels. Ideal ranges for these tests are as follows:

☆ Ammonia less than 1
☆ Nitrite less than 1
☆ Nitrate optimal 2.5
☆ pH between 6.8 - 7.0

18.17 Filtration

Good filtration will ensure good water quality. Inadequate filtration (through small pump, through the grow-bed or external bio-filter) will result in water clarity issues especially if the fish load is increased. A bad water smell means there is trouble in aquaponics. If the levels of water impurities become too high, the fish feeding should be stopped and a partial water change is required. Also uneaten food should be removed from the bottom of the tank as it will start to decompose and result in a buildup of ammonia in the system. Two basic requirements for aquaponics filtration are:

1. Biological filtration (biofiltration):
2. Mechanical filtration:

18.17.1 Biofiltration

Biofiltration in aquaponics can be defined as treatment of wastes using biological agents (aerobic bacteria) which remove ammonia and nitrite. Biofilters are a habitat for bacteria (and ideally composting red worms) that processes wastes such as fish excreta and leftover fish feed to a form that the plants can uptake. In a media based system, the medium (either gravel or expanded clay) acts as the biofilter, allowing bacteria to proliferate on its surface area. In raft system, the combination of plants and raft surface area is the biofilter.

Additional biofilters can be incorporated in the system. These are light weight and easy to handle and usually contains biomedia; a plastic media designed for maximum surface area to allow colonization of bacteria. This biofilter container can be located between the fish tank and the grow beds, usually after simple mechanical filtration. It is not necessary in raft and media-filled systems because there is enough surface area for the bacteria to colonize to a healthy level. However, in a NFT system, extra colonization space must be provided for a healthy colony to stabilize. This extension is called a biofilter.

Establishing the bio-filter begins when ammonia is first added to the system, either by fish or synthetically, in a process referred to as 'cycling'. Cycling is complete when enough bacteria have established themselves to convert all the ammonia, and all the subsequent nitrites to nitrates and the levels of ammonia and nitrites are zero. The bio-filter matures over time, and becomes increasingly more efficient at converting fish waste into plant food. It requires 2-4 weeks or even more for sufficient bacteria populations to develop. During this period fish must be fed at a very slow rate and ammonia and nitrite levels have to be measured daily. Usually ammonia levels rise and then decline followed by a rise and decline in nitrite levels. When TAN (total ammonia nitrogen - which is the sum of the gaseous and ionic forms) and nitrite levels get declined and nitrate levels are increased, a biofilter gets established. It will be better if the biofilter is established with ammonia compounds, prior to adding fish in the system.

An additional natural help keeping the grow bed media clean is to put earthworms in the grow beds. Worms thrive on fish waste.

Note: Fish can tolerate higher TAN, if the pH is less than 7.0.

18.17.2 Solid Removal System

A solid removal filter should be used if there is more fish waste than can be controlled by biofilter and plants. The type of solid removal system depends on how much organic waste is produced in the system that may form a buildup in the grow beds or settle down in the tank. Balanced ratio of fish and plants helps to minimize the solid waste in the system.

Mechanical filtration is used to stop solids (fish excreta, uneaten feed, biological growth) from entering the grow beds. These intermediate mechanical filters help collect the solid and facilitate conversion of ammonia and other waste products prior to delivery to hydroponic vegetables. Also these solids are large enough to settle to the bottom of the tank in a short time period. Filtering the nutrient water to remove particles reduces the waste build up in the grow beds and helps increase availability of oxygen to the plant roots. The filtered waste can be used for worm beds, fertilizer for fruit trees, or other plants in the ground.

In a smaller system, a submersible aquarium filter can be used to collect most of the settled solids from the tank. Larger commercial systems usually use clarifiers and netting. Netting is set up after the clarifier to catch excess organic waste that has escaped the clarifier. This netting needs to be cleaned once to twice a week. When solids are allowed to enter grow beds, over time buildup occurs and can cause anaerobic areas within the grow beds to develop, which in turn causes toxic substances to develop.

The most basic form of mechanical filtration is to use a foam block underneath the irrigation pipe entering the grow beds. This is cleaned on a regular basis, and stops large amounts of sludge and uneaten food, it is a cheap and very easy way of creating mechanical filtration. The suspended solids that do not settle can be removed in the filter tanks. These suspended solids adhere to orchard netting. Filter tanks should be cleaned once or twice a week.

More professional methods of mechanical filtration involve using swirl separators/clarifiers, which take either all or part of the fish tank water, passing it through and collecting the solids. These solids are removed manually, and are a great fertilizer for soil based plants around the house.

18.17.2.1 Clarifier

Most growers use clarifiers to remove solids from the water column. This can be done in multiple ways. Conical clarifiers and settling basins facilitate the solids settling out of the water column. Clarifier should be large enough to store the required water volume to meet the retention time required. Generally solids wastes sink and can be captured at the bottom of a clarifier (settling basin or a conical clarifier). Another way to remove the solids is a micro screen drum filter that removes organic matter in a backwashing process.

Removing solids is usually required in the raft and NFT systems because in a media-filled bed, the solids are caught in the media, where they can then biodegrade without interfering in the function of any other system components. Clarifiers can

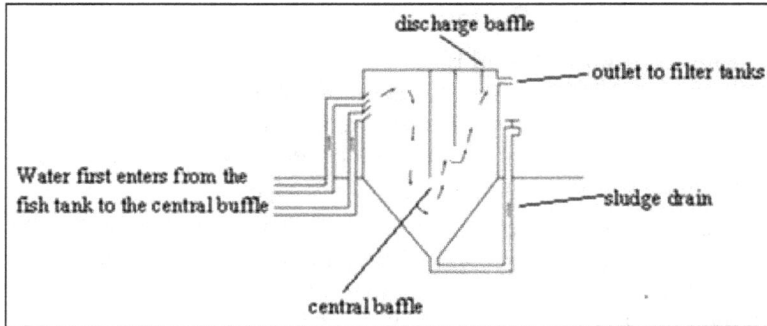

Clarifier

also be used in a media-filled system and is helpful when lots of solid waste is present.

Water first enters from the fish tank to the central baffle. Then it moves to leave either through the discharge baffle or into the outlet to filter tanks or out through the sludge drain. A series of baffle obstacles would make suspended particles easier to trap as the water flows down, up and out through the tank. The settled solids in the clarifier can be removed either by manually opening the valve fitted at the end of sludge drain pipe or using a manual siphon something similar to toilet siphon.

18.17.2.2 Swirl Filter

Swirl filter rotates the water very slowly in a sweeping arc. Remember the water needs to move very slowly with the outlet pipe located in the centre of the tank, just at the surface to gently capture water and send it on its way. The solids have plenty of time to sink as they gently "swirl" in a grand almost motionless spiral arc that traps matter and gently allows it to fall to the bottom of the tank.

Swirl filter consists of a round tank of the required water volume to meet the retention time required. It is slightly sloped inward and the water enters the tank in a way that is tangential to the tank wall (meaning that the water enters the tank and is directed along the side wall of the tank). This makes the water in the device to spin or swirl around in the tank in a circular motion allowing solids to settle (due to gravity) in the centre of the swirl tank. The outlet located at the bottom centre of the tank is used to easily clean the settled solids *via* simple hydraulic pressure (a valve is opened and if the outlet is lower than the water height in the swirl tank, then the solids are simply sucked out of the bottom centre of the swirl tank and removed with the water flow).

Clarifiers and swirl filters use the sedimentation techniques of utilizing the force of gravity upon the solid particles to settle them at the bottom of the tank. The most basic sedimentation device is a rectangular tank where the water enters one end, flows slowly though the tank to allow the solids to settle and exits the other end. The solids are then regularly removed from the base of the sedimentation tank in some way.

The rate at which a solid particle falls is directly related to the mass of the particle. Also, if the water flows speedily through the sedimentation tank then majority of solid particles may not get enough time to fall to the bottom and can exit the sedimentation tank with the water flow. Therefore, the key requirements in designing and sizing sedimentation devices are:

1. The flow rate of the water through the sedimentation device, The flow rate through the sedimentation device is set by the flow rate of water exiting the fish tank and by the volume of the sedimentation device

2. The retention time of the water in the device (which is related to the flow rate and device volume). The longer the retention time, more and smaller solid particles will fall to the bottom of the tank and be separated.

Clarifiers and swirl filters cannot remove all the solid matters and a further series of tanks stuffed with netting are used to trap solids before the water flows on its way to the raft system growing the vegetables.

The particles would eventually coat the pipes and fittings of the tanks with a thick sludge; therefore it needs periodic and laborious cleaning.

A mineralization tank filled with porous media can be used before the clarifier, especially in raft or NFT systems. In this tank nitrifying bacteria convert the waste into elements that are readily used by the plants. This process also creates gases such as hydrogen sulphide, methane, and nitrogen. Therefore, a degassing tank is needed to help release these into the air. This is not needed in a media-filled bed because the solids remain in the system trapped in the media.

18.17.2.3 Screen Filters

Screen filters use some sort of material that restricts the solids but allows water to pass through it. Screening out solids depends on the pore size of the screen material. There are many different materials that may be used as screens, including mesh screens (similar to the screen material used for "screen printing"), plastic woven screens and filter mats. Even cotton or foam can be used in screen filters.

Static screen filter: The screen does not move and the water passes through it. It is often used in small systems and can be easily self-build using the screen material being located in some type of housing to support it. Static screen filters do need active management as screens need to be regularly removed, washed and re-inserted. The main disadvantage is that it may clog and need to be cleaned frequently when fish numbers or biomass gets too high.

Moving screen filter: The screen moves so that its entire surface may be exposed to the water to be filtered.

18.17.2.4 Drum Screen Filter

The most popular style of screen filter for large commercial systems is the drum screen filter which is an automatic, self-cleaning filter. In this approach, the screen is attached to a drum and the unfiltered water enters the inside of the drum and flows through the screen to the outside. Inside the drum there is a switch that

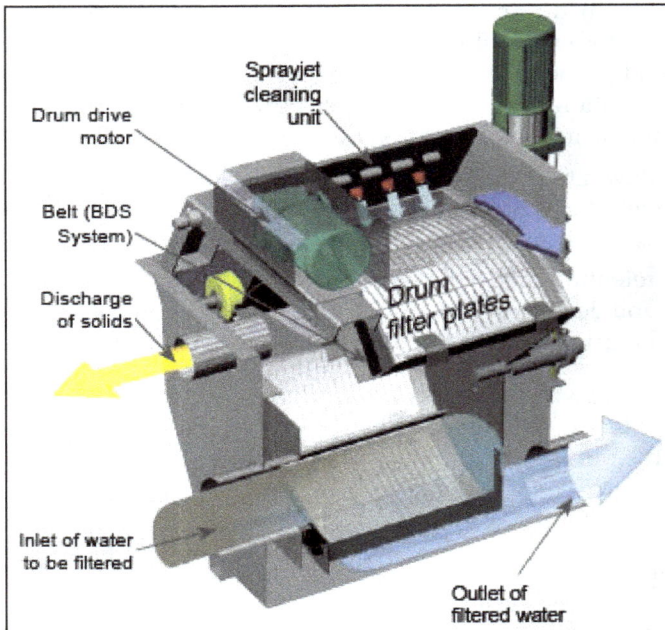

senses the water height. When the screen starts to clog with solids, the water doesn't pass through the screen as easily and the water height inside the drum begins to rise. If it raises enough, it trips the switch and the filter enters what is known as the "**back flush**" mode. In this back flush mode a series of high pressure sprays fire on the outside of the drum and wash the collected solids off the screen and into a channel within the drum that is directly opposite the high pressure spray nozzles. This cleans the solids off the screen and collects them in the channel which then directs the solids to an outlet point. Similar to the drum screen filter is the belt screen filter, which operates in the same way except that instead of using a drum, the filter utilizes a belt that rotates.

18.17.2.5 Media Bed Filtration

Media beds filtration is similar to screen filtration. Media beds provide pores (the interstitial spaces between the media particles) for the water to flow through. If the solid fish waste particle is larger than the pore between the media particles, it is blocked and separated from the water column. If the media beds are not sized sufficiently to meet the solids load, they can quickly clog and therefore, the water may flow out of the top of the media bed and not through the media. The constant cleaning out media bed filters require lots of labour and needs to stop the system for some time.

Important Notes:

☆ A solid removal filter may not be needed in a fully cycled media based aquaponics system containing proper fish and plant ratios, and enough established bacteria colonies and earthworms. These beneficial life forms would completely decompose the fish waste into plant nutrients.

☆ An aerated mineralization tank filled with porous media can be used before the clarifier or other solid removal filter, especially in raft or NFT systems. In this tank, nitrifying bacteria convert the waste into elements that are readily used by the plants. This process also creates gases such as hydrogen sulphide, methane, and nitrogen. Therefore, a degassing tank is needed to help release these into the air. Mineralization tank may not be needed in a media-filled bed where the solids remain in the system trapped in the media.

☆ Using a mineralization tank and solid removal filters, provides clean water with the nutrients trapped within water.

18.18 Importance of Mineralization/Nitrification Unit and Solid Removal Filters

If solids are allowed to stay in the main aquaponics system for mineralization, aerobic bacteria that perform the major role in mineralization will compete for dissolved oxygen in the grow beds as well as in fish tank. Firstly, the solids are broken down to organic macromolecules (large molecules of an organic nature), which consumes system-based dissolved oxygen. These macromolecules may be released from the media bed and may roam around in the main aquaponics system and further mineralization may occur in those areas where we do not want it to compete for dissolved oxygen (*e.g.* the fish tanks or deep flow beds).

Another point to be noted is that if the solids get accumulated in some parts of the system, then anaerobic zones may get developed there resulting in anaerobic mineralization. These anaerobic mineralization may release some harmful elements that may harm the fish and plants in the system.

In aquaponics systems where solids are removed using solid removal filters and mineralization is done separately in separate units, there is no or very low competition for the dissolved oxygen in fish tanks or in the grow beds, and the development of anaerobic zones is very rare.

18.19 Size of an Aquaponics System and Fish to Plant Ratios

The fish to plant ratio depends on the amount of waste the fish produce and the number of plants required to consume the nutrients available from the fish waste. The amount of waste the fish will produce depends on the number of fish, their species and feeding habit, and of course the quality of the feed given to them. Therefore it is necessary for the grower to first decide the species of plants and number, and species of fish to be grown in the system and should be aware of the approximate amount of waste that particular fish species will produce.

The number of fish to be grown depends on the size and weight of the fish and fish tank. For example, approximately 1 pound of fish per 5–7 gallons of tank water (0.5 kg per 20-26 litres). The number of plants to be grown depends on the waste produced by these fish (generally 1 pound of fish for every 1 sq ft(.1 sq m) of grow bed surface area, assuming the beds are at least 30 cm deep). The size of the grow bed depends on the number, type and density of plants to be grown. For example, 30 lettuce plants can be grown per square metre.

The size of an aquaponics system and fish to plants ratio depends on the following factor:

☆ Number, type and size of fish to be grown.

☆ Volume of water these fish needs to live happily. For example, 1 pound of fish per 5–7 gallons of tank water (0.5 kg per 20-26 litres).

☆ Amount and quality of fish feed required to feed these fish.

☆ Amount of waste these fish will produce.

☆ Number and type of plants required to consume the nutrients available from the fish waste produced by these fish.

☆ Amount of space required by these plants.

☆ The grow bed must be at least 30 cm deep to allow for growing the widest variety of plants and to provide complete filtration and a good place to nitrifying bacteria.

Note: For young and small fish, the number of plants should be reduced in proportion to the size of the fish and their corresponding feed rate/waste production.

18.19.1 Recommended Fish Stocking Densities

Stocking density is the weight of fish related to volume of water. The accepted approach for stating fish stocking density is the weight of fish (kg) per litres (volume) of water. This is usually stated as kg/m^3 (1 m^3 of water=1000 litres). Most commercial aquaponics growers keep fish at densities up to nearly $50kg/m^3$ in aerated systems and up to approximately $150kg/m^3$ or higher in direct oxygen injected systems. Hobby growers usually keep fish at densities below $20kg/m^3$.

18.19.2 Flow Rate of Water

The flow rate of water through the fish tank is dependent on the density of fish being kept in the tank.

For example,

☆ For low to medium densities (up to 15 kg/m^3): An exchange rate of one half of the fish tank volume per hour is the minimum recommended.

☆ For densities above 15 kg/m^3 a minimum exchange rate of 1 fish tank volume per hour is recommended.

Some factors that help in deciding and maintaining the desired flow rate are:

☆ Volume of water to move: Calculating the total volume of water to be moved per hour in order to maintain good water quality for the fish in the system. This can be calculated by summing up the following:

❑ Number and volume of grow beds in the system.

☐ Volume of fish tank.

☐ Volume of the sump tank.

The water volume of the fish tank must be exchanged at a rate that is adequate to meet the water quality requirements of the fish. The water velocity must not be too high or too low; if it is too high, then the fish will be wasting more energy to constantly swim against the flow; if it is too low, then the fish may not swim enough and suffer alternative negative effects to their health; the water should be made to flow at a rate that allows the solid wastes to be concentrated in a region of the tank that allows their removal; it should be exchanged often enough to ensure that the dissolved ammonia does not reach at levels that may harm the fish; it should be directed into the tank to ensure adequate mixing and to ensure the required water velocity is achieved but not exceeded.

☆ Tank Shape and size

☐ Round or Rectangular: Although round shaped fish tanks are usually recommended for commercial large-scale aquaponics systems, hobby-scale aquaponics practitioners can use rectangular or other fish tanks because they usually keep fish at low stocking densities and so a proportion of the requirements and performance attributed to round fish tanks may be flexible. The round tanks have no corners in the vertical dimension; hence the water is free to flow into and out of the tank uninhibited. Corners in fish tanks have the ability to slow or fasten water flows in particular tank regions. Corners in the tank may also cause areas of low to no water exchange and as a result, the water in these areas may be retained in the fish tank for extended periods of time and this may affect the water quality in the tank. The solids in a circular tank will usually settle at the centre of the tank, hence the outlet or submersible pump can be placed at its centre. In a non-circular tank, it is very difficult to determine that where the solids will actually settle and where the outlet or submersible pump should be placed.

☆ How to move water

☐ The height up to where the water is to be pumped.

☐ Using the advantage of gravity to move water around with the minimal of mechanical intervention. For example, if the grow beds are higher than the fish tank then water should be pumped from the fish tank to the grow beds and should be allowed to flow back from the grow beds to the fish tank by gravity.

☆ Type of pump: The pump must be of good quality. It should be decided on whether to continuously run the pump or to use the timer operated pump to have a timed flood and drain system. The power required to achieve the flow rate must be efficient and economical.

☐ Its flow rate and its head pressure. The flow rate tells how much water the pump can move in litres per minute or per hour. The head pressure then adjusts this measurement based on how high the water is being pumped.

☐ The pump chart should be checked carefully before its installation.

☐ A slightly bigger pump will also allow you to increase the size of your system if you want to expand it at a later date.

☆ Pipe design

☐ The water flows more quickly through the straight pipe and slowly through the pipe having series of bends.

☐ The pipes need to be big enough to also handle desired volume of water and to have various controls and safety measures in place just in case there is any problem.

☐ Flow rate also decreases as the length of pipe increases.

☐ Joints, and bends in the plumbing will decrease the flow rate of the water.

☆ Pipes

☐ The diameters of pipes delivering water to the growbeds may be smaller than those removing the water from them. This will ensure that the beds can drain properly and not overflow, or at least able to adequately handle the volume of water being delivered to the growbeds.

☐ It is recommended that the diameter of the pipes should be kept slightly larger than necessary. This will reduce the chance of blockages in the pipes.

☐ Over time, debris may build up on the inside of the pipes and this will affect the rate of flow in the pipes. Pipes may need to be cleaned every few months in order to ensure a constant and unobstructed flow of water.

☆ The Connectors: Two common types of connectors that can be used in aquaponics are threaded connectors and slip connectors.

If the water in the fish tank is not properly mixed and if water does not properly flows through the tank (as sometimes occurs in non-round fish tanks), then some of the total water body in the system may develop areas of high solids and high dissolved ammonia load, and may also cause zones of non-optimized water chemistry such as low oxygen, high or low pH, EC, dissolved oxygen or water temperature.

Hobby aquaponics grower often maintain low fish stocking densities; therefore these factors are usually of very less concern because the water quality may be maintained adequately based on the fact of the low fish stocking rate.

The correct water flow through fish tanks ensures:

☆ Timely removal of fish wastes and uneaten feed.

☆ Timely and adequate biofiltration of the dissolved solids.

☆ Adequate dissolved oxygen levels in the entire water volume.

☆ Nearly uniform temperature, pH and EC in the entire area of the water volume.

18.19.3 Ratio of Water Volume to Surface Area of the Fish Tank

Majority of the gas exchange (*e.g.* carbon dioxide release and oxygen uptake) of the water body in the tank occurs at the water surface. In the deep and narrow tanks where the surface area is too low as compared to water volume, enough oxygen cannot enter the water body and enough carbon dioxide is not released from the water body. Therefore the surface area to volume ratios should be as high as possible.

18.20 Water Change

Until not very necessary, any large water change should not be done in an aquaponics system as it will add to the fish stress. There will be a large and sudden pH swing which is not good for the fish in the tank. A self regulation ball valve that tops water into the fish tank as the water level drops due to evaporation/ transpiration is a gentle way to keep the water in the fish tanks constantly replenished. Also not more than 1/3 part of water should be changed.

18.21 Nutritional Supplements for Plants and pH Control

Most of the nutrients required by plants in adequate quantities are supplied by fish waste and fish feed. There are some deficiencies however that might appear in plants and can be addressed by adding supplements. Most systems require calcium, potassium and iron supplements and sometimes a little magnesium is also required. The nutrient supplements should be carefully used as some can harm the fish in the system.

In a mature well functioning system, nitrification of ammonia causes pH to drop over time and the best way to adjust the pH and add nutrients (calcium and potassium) at the same time is to alternate between potassium hydroxide and calcium hydroxide. Some people prefer potassium carbonate and calcium carbonate, but the carbonates buffer the pH making it hard to adjust to the desired level accurately. Chelated iron can be used to add iron to the system.

Phosphoric acid can be used to lower the pH because the plants can also make use of the phosphorus which the grower likes to supplement for fruiting and flowering nutrient hogs such as tomatoes or cucumbers. Rock phosphate can be used to add phosphorous if pH is not to be adjusted.

An important aspect in the system that is needed to ensure a steady supply of plant nutrients is pH. If pH is not optimal, an accumulation of toxic materials will build up, and nutrient generation by the bacteria will be halted. In addition, pH outside the range will make it harder for the plant roots to take up nutrients and can harm the fish. The optimum pH levels in an aquaponics system are around 6.8 to 7.

Nutrient supplements are usually not needed in a properly managed matured system with:

☆ Good fish feed

☆ pH at 6.8 to 7

☆ Well oxygenated water

☆ Proper temperatures

Under these controlled conditions, the bacteria will provide more than enough nutrients for the plants.

18.22 Fish Feed

One of the primary inputs to an aquaponics system is the fish feed, and its quality not only determines the health of the fish in an aquaponics system, but also considerably affects the health and yield of plants in the system. Appropriate feeding is especially important in aquaponics as it is the main determinate of quality and quantity of nutrients available to plants. Proper observation and maintaining records of eating habits of the fish will be very helpful at later stages of the system.

The easiest way to feed the fish is to buy commercial fish feed from the pets or aquarium shops. There are varieties of commercial fish feed for different fish species that are easily available with the pets or aquarium shop. These fish food are available as food for specific fish species and generalized food that suits the need for most of the fish. Apart from commercial feed, homemade or live fish food such as worms, snails, larvae of black soldier fly, red worms and duckweed (type of plant that grows very quickly and has a lot of protein) can be used. A combination of these can also be used to feed the fish.

Note: To keep the fish healthy in an aquaponics system, they should be fed with high quality feed and their environment should be kept stress-free. The stress that affects fish health can be physical, chemical, and biological. If the fish are stressed, they will stop eating.

18.22.1 What to Feed?

Although hobby growers can select the fish food as per their suitability and availability, for a commercial aquaponics system, professionally formulated feed is recommended. The carnivorous fish such as perch and trout require a feed that has high levels of proteins. The omnivorous fish such as tilapia, koi and catfish require a feed that has low levels of proteins. Also, younger fish need more protein as compared to mature fish.

18.22.2 When and How Much to Feed?

Most aquarium fish such as goldfish, carps, silver shark, *etc.* can survive without food for weeks but may die due to over eating. Generally it is better to feed little and often, at least twice a day, ideally 3-4 times. As a guideline, mature tilapia typically consumes about 1 per cent feed of their body weight in a day, while fingerlings can consume as much as 7 per cent. Any uneaten food should be removed from the system before it sinks and rots consuming oxygen from the water while increasing ammonia levels. Therefore it is necessary to feed the fish only as much as they eat within few minutes (5-6 minutes) and after that the uneaten feed should be removed. It will be better to feed a floating pellet as it will give better idea of what's being eaten.

18.23 Fish Stress

Fish thrive best when all their survival needs are met and are not stressed. Fish in a natural environment have a healthy slime coating that prevents bacterial infections and parasites from attacking them. Fish are more susceptible to stress than plants. Fish stress can be physical, chemical and biological.

☆ **Physical and psychological stress**: Physical stress includes:

☐ Environmental conditions

✦ Temperature: All fish have a temperature range within which they will thrive, and a wider range within which they will survive. Since fish are cold-blooded animals, therefore they do not have the ability to spend energy to maintain a constant internal body temperature like human. If the water temperature goes outside of their optimal (or, thriving) range, fish will eat less, or stop eating all together, and they become more susceptible to disease.

✦ Sudden exposure to light: Fish usually get frightened upon sudden exposure to light. They may reflexively run in random direction and may even bang against the walls of their tank to escape the light.

☐ Sudden vibration: Fish can feel or hear vibrations with their entire bodies. Tapping or knocking against the wall of a tank will also cause them undue stress.

☐ Water velocity: Fish originating from still lake waters do not like much movement in their tank water.

☐ Water change: A large water change is also one of the causes for stress in fish. Especially if there is a sudden pH swings because the source water is usually different from the water in the main fish tank.

☐ Transportation: Traveling and transporting large numbers of fish can have a drastic effect on their slime coat, leading to weakening their immune system. When introducing new fish into your tank, take special care to adapt them to the water temperature and also the pH conditions gradually with minimal disruption to their daily routine.

☐ Unnecessarily touching or netting: This will again create unnecessary stress for fish. Until and unless not required netting of fish should be avoided. Fish should be left undisturbed so that they can feel safe at the back of the tank. The proper weave of net should be used to prevent the fish from being snared and gilled.

☐ Harassment from other fish: Mixing of incompatible fish species in the same tank, especially decent with unsuitable fish species that are predatory in nature and territorial to the other fish.

☆ **Chemical stress**: The poor quality of water in the tank is the main cause of chemical stress for fish.

    ❐ High ammonia or nitrite levels: A high level of ammonia and nitrite are toxic to fish and is the main cause of their stress, especially at the beginning of an aquaponics system when the fish are introduced to the system before the nitrifying bacteria have been fully established.

    ❐ Anaerobic zone in grow beds: Anaerobic zones get buildup in the grow beds and may release some toxic substances.

    ❐ Low pH: Maintaining a very low pH (below 6.0) can also be stressful.

    ❐ Insufficient filtration and low dissolved oxygen are also other forms of chemical stress.

    ❐ Overstocking of tank: Overstocking the tank with fish decreases the water quality.

☆ **Biological stress**: Stress caused by diseases (viruses, bacteria, fungi and parasites) are biological stresses. Although a small amount of sea salt can be added in the water to get rid of many parasites in the fish tank, it should be done carefully as it may harm all the life forms (plants, fish and beneficial bacteria) in the system. It is recommended that the grower must consult the experts before adding anything else in the tank to control the disease.

The best defence for fish is their own immune system. The less they are stressed the better will be their immune system.

18.24 Using Salt

It is common to use salt in aquaponics to get rid of fish parasites and to de-stress the fish. Fish release minerals into the water when they are stressed and if this condition persists for a long time, the salt loss can be fatal for them. Adding salt increases the resistance of the fish to disease and infections, because it has a positive effect on the mucous layer of the fish. The normal recommended salinity level to dose the fish is in the range of 0.5-3 ppt [parts per thousand (kg/1000 litres of water)]. Many plants such as strawberries will not tolerate salt levels beyond 3 ppt. A higher dosage of salts may harm all the life forms (plants, fish and beneficial bacteria) in the system.

Table salt contains potentially harmful anti-caking agents; therefore sea salt would be a better choice for the aquaponics systems. It will be better to use the salt which is specifically formulated for fish. The salt should be first dissolved completely in some container and then this solution should be added in the fish tank.

18.25 Plant's Pests and Diseases

See Plant Enemies (Chapter 8)

    a) The best way to deal with plant's pests and diseases in an aquaponics system is by using the physical methods already discussed in Plant Enemies (Chapter 8).

1. A strong blast of water from a hose may kill and wash away the pests.
2. Pests like caterpillars can be hand-picked and dropped into soapy water.
3. Potato beetles can be removed by shaking off the plants onto a sheet.
4. Bright sticky traps are effective for insects.
5. Beneficial insects such as Ladybugs beetles, Praying mantis, Trichogramma Wasp, *Encarsia formosa, etc.* can be used to control various insects and their larvae. See Plant Enemies (Chapter 8).

b) Homemade or commercial organic pesticides can be used. The grow bed media and fish tank should be covered with plastic sheet or other things to keep too much of the spray from getting into the system. This may harm beneficial bacteria and fish, and may disturb the pH levels in the system.

Pesticides or anything else that will harm the fish and beneficial bacteria should never be used. Any diseased plant should be immediately removed from the system. The main point is that anything done to treat plants should never stress the fish in the system.

18.26 Fish Species for Aquaponics

Fish are one of the primary outputs that the growers get from an aquaponics system. They not only provide an extra source of food to the grower, but their waste too is utilized as the primary source of nutrients for plants. Fish are the most delicate members among all the three life forms (fish, beneficial bacteria and plants) and therefore need to be dealt with carefully and sensitively. Although aquaponics fish are generally grown for human consumption, the ornamental fish such as gold fish, carps or other ornamental species too can be grown by the aquaponics gardeners who don't like killing fish for food.

Any freshwater fish including edible species such as tilapia, barramundi, silver perch, *etc.* and ornamental (non-edible) fish species such as koi and goldfish that would be happy in the grower's aquaponics environment, can be grown. Although freshwater fish are the most common aquatic animals raised using aquaponics, freshwater crayfish, mussles and prawns are also sometimes used. Each species of fish will have an ideal temperature and pH range, and an optimum temperature and pH within this. This optimum condition may not always be met as it can conflict with the temperature and pH range of the plants and/or the bacteria. Remember that the environment of an aquaponics system is a compromise between the plants, fish and bacteria (in general, at 21-23 C temperature and 6.4-7.0 pH range).

Although cooling and heating equipments can be used to maintain the temperature of the fish tank, this expenditure can be saved by making proper selection of fish and plant species and using insulation and aquarium cover cloth to stabilize the temperature. It will be better to grow the fish which are native to grower's area and are more tolerant to changes and can survive fluctuations in water quality or temperature.

Different fish species having similar living environment (same requirement for temperature, oxygen, pH and above all same food) can be grown together in the same system. Aggressive carnivores should never be mixed with omnivores and care should be taken that fish of nearly same size are being grown together. Fast species should not be mixed with slow species as they will eat up the feed quickly and nothing or very less food will be left for the slower fish. Larger and aggressive fish usually attacks smaller ones. Also some smaller fast and aggressive fish will attack the fins and gill of larger decent fish. A very large aquaponics system such as aquaponics ponds, which will have enough space, can successfully accommodate carnivores, omnivores, fast and slow species.

Some examples of fish that can be grown in an aquaponics system are:

Tilapia (Freshwater Edible Species)

They make great eating and are one of the most important commercial aquaponics fish species. Tilapias are of many types, with wildly different characteristics. The female tilapia grows slower than male tilapia and scientists came up with a method of treating all of the tilapia fingerlings with male hormones.

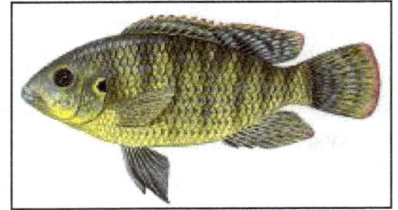

Feeding habit

☆ Omnivorous

☆ Mature tilapia typically consumes about 1 per cent of their body weight in a day.

☆ Fingerlings can consume as much as 7 per cent.

☆ **Feed**: Pellet fish food, duckweed, algae, worm, *etc.*

Preferable pH Range

7-8

Note: The pH range of 7-8 is on the alkaline side of the scale and the plants prefer a more acidic environment. Most aquaponics farmers keep their pH levels between 6.8-7.0, which is a compromise for the fish and the plants. Remember that the fish will adjust to different pH ranges, but only if done very slowly.

Salinity

They grow well up to salinities of 16 to 20 parts per thousand (sea water is 35 ppt). Salt can be added to discourage pests and to create a warmer layer in a deep pond during the winter.

Temperature

Tilapia prefers warm water. Every breed of tilapia is different.

☆ Temperature range for some telapia: 12 C–36 C

★ Preferable temperature range for some tilapia: 25° - 30° C and for some others: between 31 C and 36 C

★ Some may tolerate temperatures as low as 8° or 9°C for short periods (overnight).

DO Levels

Low oxygen requirements. 3- 10 mg/l (can survive at as low as 1 but not advised)

Tolerance

These are hardy fish and are tolerant to changes.

★ It can survive fluctuations in water quality or temperature. Tolerates pH shifts, temperature changes, high ammonia and low dissolved oxygen.

Rainbow Trout

Freshwater edible species that require clean and cold water

Feeding Habit

★ Carnivorous

★ Will eat smaller fish in the system

★ **Feed**: Pellet fish food, smaller fish, insects, *etc.*

pH

6.5- 8.0

Temperature

★ Able to survive within a temperature range of 0-25 C

★ Grows at the water temperature range of 10-20 C.

★ Optimum temperature of 15-18 C

★ Water temperature should never exceed 23 C

★ Stops feeding above 21 C and below 10 C.

★ 18 C is regarded as the optimum for metabolism in rainbow trout. Higher temperature would assist higher level of metabolism and growth as well.

DO Levels

★ Should not be less than 5.5 mg/l

★ Above 8 mg/l is considered suitable for trout culture

Tolerance

Sensitive to changes in pH, water quality and water temperature.

Silver Perch

Freshwater edible species.

Feeding Habit

☆ Omnivorous

☆ **Feed**: readily accept pellets and do well on a low protein, high fibre diet. Eats insects, small crustaceans and vegetation

pH

6.5- 8.0

Temperature

☆ They can survive in water temperatures ranging from 2 C to 35 C.

☆ Comfort level (within range) is 10°–30°C

☆ The optimum temperature range for commercial production is 23-28 C.

☆ Temperature below 5°C and above 32°C is stressful for them.

DO Levels

☆ Above 3mg/l and free ammonia levels of less than 0.1mg/l should be maintained.

Tolerance

They are quite hardy and can tolerate a range of water quality parameters.

Goldfish and Koi

Freshwater ornamental species of carp

Goldfish

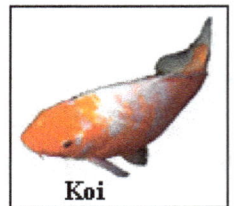
Koi

Feeding habit

☆ Omnivorous

pH

7.0-8.6

Temperature

☆ Surviving temperature: 8° to 32°C

☆ Thriving temperature: 18° to 24°C

DO Levels

☆ Low oxygen needs

Tolerance

They are quite hardy and can tolerate a range of water quality parameters.

Jade Perch

Warm water edible fish.

Feeding Habit

☆ Omnivorous

☆ Jades readily accept pellets and do well on a low protein, high fibre diet.

☆ They are aggressive feeders, and care should be taken to ensure that all the fish in the tank get fed, not just the dominant fish in the population.

pH

☆ Tolerates between: 6-9

☆ Desired range: 6.5-8.5

Temperature

☆ Grows best in water temperature above 24°C.

☆ Aeration needed to increase when temperature exceeds 30°C.

☆ Growth rate declines rapidly when temperature falls below 20°C.

☆ Most farmers cease feeding at temperature below 16°C.

DO Levels

☆ Require 4 mg/l or greater for maximum growth and survival.

Salinity

Up to 5g/l of sodium chloride are acceptable for long term exposure.

Tolerance

They are quite hardy. The biggest issue with jade perch is temperature. Jades stop metabolizing food at water temperatures below 18°C, and will start to show stress at 16°C, and death at prolonged periods below 14°C.

18.27 Plants

Plants need to deal with two main environments; air (at the leaf and stem surfaces) and water (at the root surfaces). The roots too deal with these two environments of air and water; sometimes they are dry and exposed to the air within the medium that they are growing in, and sometimes they are wet due to the water that surrounds them in the medium. Therefore, plants need good air quality along with water conditions that suit them and the processes that occur at the root surface. In aquaponic systems, the fish produce waste, that waste is processed by the bacteria and the processed waste is then taken up by the plants as nutrients. Plants do not filter out nutrients and leave the water behind; they gain access to the nutrients by taking water in across their root surfaces. Therefore, plants are constantly taking up and internalizing water, hence the quality of water in the system is critical. The nutrients that the plant is taking up across its roots can also be affected by the chemistry of the water and if the water chemistry is incorrect, this can manipulate the chemical form that the nutrient is in and this can mean that the nutrient is either in a form that the plant does not prefer, or is in a form that the plant cannot even access. Hydroponic farmers well know that if the pH of their water is incorrect, then this can stop the plant from gaining access to the nutrients and this is seen as nutrient deficiency. So, plants, like the fish and bacteria, are also dependent upon the water being in the best chemical condition.

Almost any plant that can be grown hydroponically can also be grown in aquaponics system. Although plants can be added to the system as soon as the cycling process gets started, it is suggested that only leafy vegetables and herbs such as lettuce and basil or the crops that have low nutrient requirements should be grown during first few months of the system. When the system becomes mature enough to provide most of the nutrients, *i.e.* when the nitrogen cycle becomes fully established, other plants such as tomatoes and cucumber can be added to the system. Plants can take up nitrogen at all stages of the cycling process to varying degrees; from ammonia, nitrites and nitrates, but they will grow better and give high yield when cycling is complete and the bacteria are fully established.

In an aquaponics system, several different varieties of crops can be grown in the same grow bed as long as their needs for air and water are within a comparable range as well as their pH, nitrate and ammonia level requirements. The major source of nutrients for all crops is fish feed and fish waste. This is in contrast to the hydroponics where calculated nutrients for the grown crop are mixed in the water to form nutrient solution.

It is not recommended to grow those plants in aquaponics systems that have fast growing roots, such as mint. An aggressive root system will grow into the piping and overtake the system. Also, if possible, the blooming plants (such as tomatoes, peppers, zucchini, green beans, cucumbers, *etc.*) and the non-bloom plants (such as lettuces and green leafy plants) should be grown in separate grow beds.

The growers should always remember that the environment of an aquaponics system is a compromise between the plants, fish and bacteria (in general, at 21-23 C temperature and 6.4-7.0 pH range). Therefore the pH, dissolved oxygen and the temperature of the water in the fish tank must be maintained considering the suitability of all life forms that are going to thrive in the same system. Also any major and sudden changes in temperature and pH can have adverse effects on fish health and of course on plant health too.

18.28 Future of Hydroponics and Aquaponics

Due to drastic increase in global population and scarcity of arable land, the future of hydroponics and aquaponics are certainly looking up. The world is going to need vertical farms because conventional agriculture can't handle the food requirement of all people. High-rise buildings can be transformed into vertical greenhouses to produce fruits, vegetables, and flowers in an urban setting, close to millions of hungry consumers.

Indoor agriculture is more efficient and is approximately four to six times more productive as compared to outdoor agriculture because of the year-round growing season.

⑲
Folkewall (Living Wall)

Folkewall (also known as living walls, green walls, biowalls, ecowalls, or vertical gardens) is a specially designed hydroponics growing system with the dual functions of growing plants vertically and purifying wastewater. Vertical farming helps in efficient use of space and purification of percolating water which may be greywater.

In its simplest design, folkewall is a wall of hollow concrete slabs with compartments that have proper openings on one or both sides of the wall. The hollows are filled with inert material like LECA-pebbles (light expanded clay aggregate), gravel, vermiculite or perlite. It is designed to allow the water to travel over the longest possible treatment path through the wall along with the pebbles.

The water is brought in at the top, and percolates following a zig-zag pattern inside the wall. The water is forced to percolate in a zigzag pattern by using plastic sheets. A film of beneficial bacteria grows over the pebbles, releasing the nutrients in the percolating greywater.The plant roots grow between the inert material and absorbs nutrients from the water. At the bottom of the wall a container collects the purified water which can then be used for household use other than drinking, for watering the garden, or it can be returned to the top of the wall. The water feeding the plants in the folkewall must be free of heavy metals and/or unsafe pollutants, notably human waste and this may require using source-separating toilets.

The best suitable crops in this type of system are fast growing herbaceous crops. Perennials that take time to fully grow such as trees or shrubs are not recommended for the Folkewall.

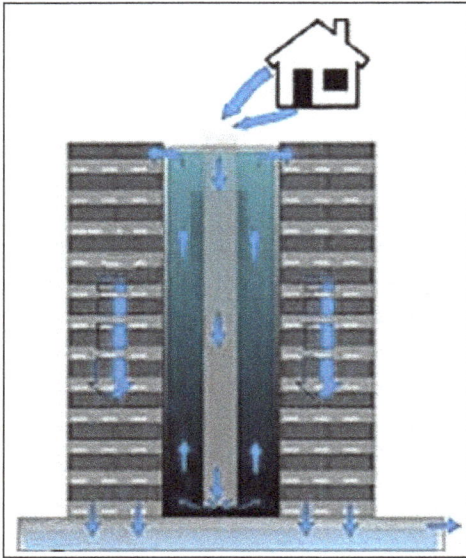

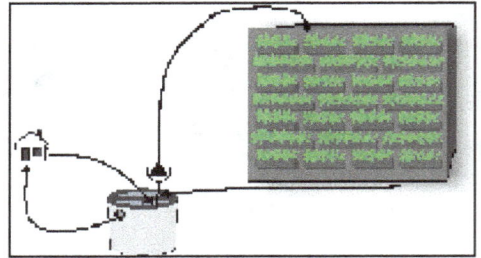

Designs of Folkewall

19.1 Advantages

☆ Better utilisation of the irrigation water: Most of the evaporation goes through the leaves of the plants, which is good in a dry climate.

☆ Space is efficiently utilized and can be practiced in a greenhouse or in open in frost free climate.

☆ It works as a heat exchanger and temperature buffer in a greenhouse where the wall is combined with greywater purification.

☆ Purification of the percolating water, so greywater can be used as irrigation water. With a long enough passage through the living wall, the water is sufficiently clean for re-use as utility water.

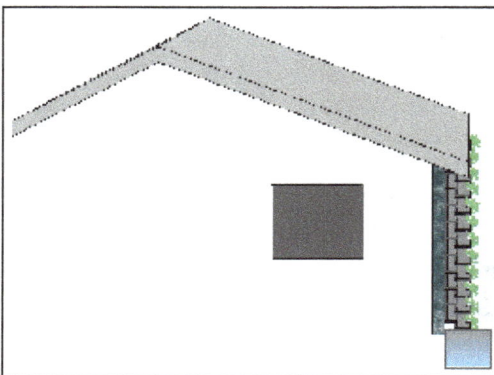

Folkewall Images

☆ In hot climate, the folkewall gives a cooling effect to the building.

☆ Low-cost housing: the combined use of folkewalls and source-separating toilets would reduce the infrastructure cost.

☆ In warm climates, the food production can go on all the year around, simultaneous with the water purification.

Glossary

Abscisic acid (ABA): A plant growth-inhibiting hormone.

Abscission layer: Specialized cells, usually at the base of a leaf stalk or fruit stem, that trigger both the separation of the leaf or fruit and the development of scar tissue to protect the plant.

Abscission: The dropping of leaves, flowers, or fruit by a plant. This can result from natural growth processes (*e.g.*, fruit ripening), from growth inhibiting harmone or from external factors such as temperature or chemicals.

Absorption: The intake of water, nutrients and/or other materials through root or leaf cells.

Acid - Refers to medium or nutrient solution with a low pH; an acidic solution has a pH below 7. The lower the pH, the more acidic is the medium or nutrient solution. See pH.

Adjuvant: Also called surfactant. A substance that, when added to a pesticide or foliar spray nutrient solutiuon, reduces the surface tension between two unlike materials (*e.g.*, spray droplets and a plant surface), thus improving adherence.

Adventitious root: A root in an unusual place, often where a branch contacts soil or damp material. A plant cannot be reproduced from cuttings or layering unless adventitious roots develop.

Adventitious: Growth not ordinarily expected, usually due to the result of stress or injury. A plant's normal growth comes from meristematic tissue, but adventitious growth comes from nonmeristematic tissue.

Aeration: Supplying air or oxygen to the root zone (nutrient solution and medium). In some hydroponic systems, a nutrient solution is aerated using an aquarium air pump and air stone. Mechanically loosening or puncturing soil or other media to increase permeability to water and air. ·

Aerobic bacteria: Aerobic bacteria get developed under aerobic conditions. See anaerobic condition.

Aerobic condition: Presence of sufficient oxygen. A well designed and managed aquaponics system should operate under aerobic conditions.

Aeroponics: A variation of hydroponics in which the roots of a plant are suspended in air and are consistently or intermittently misted with fine droplets of nutrient solution. No medium is needed with this method and usually only small plants that need no support are grown this way.

Aggregate: Medium usually grow rocks, gavel, or lava rocks that are roughly the same size, and used as an inert hydroponic medium.

Alkaline: Refers to medium or nutrient solution with a high pH; any pH over 7 is considered alkaline. The higher the reading, the more alkaline the medium or nutrient solution. See pH.

Ammonia in aquaculture: Two forms of ammonia occur in aquaculture systems, ionized and un-ionized. The un-ionized form of ammonia (NH_3) is extremely toxic while the ionized form ($NH4+$) is not. Both forms are grouped together as "total ammonia". Through biological processes, toxic ammonia can be degraded to harmless nitrates. Ammonia (NH_3) is first converted to nitrites (NO_2-) and then to nitrates (NO_3-).

Ammonia-type nitrogen: Gardeners recognize nitrogen as being available in two forms; ammonia-type nitrogen and nitrate-type nitrogen. Most plants prefer to obtain their nitrogen in the nitrate form. Urea and sulphate of ammonia are unsuitable for hydroponics as they need to undergo conversion from ammonia-type to nitrate-type by soil bacteria before plants can use them.

Ammonium in hydroponics: In Hydroponics nutrient solution, ammonium concentration should represent no more than ~20 per cent of the required total nitrogen concentration. An excess can cause damage to roots and stem bases, particularly in younger plants, which results in poor growth and yield. It is generally accepted that plants will not uptake nitrogen in the ammonium ($NH4+$) form. It must first be oxidized into nitrate (NO_3-). Most plants prefer to obtain their nitrogen in the nitrate form. Urea and sulphate of ammonia are unsuitable for hydroponics as they need to undergo conversion from ammonia-type to nitrate-type by soil bacteria before plants can use them.

Ampere (amp): The unit used to measure the strength of an electric current. For example, a 20 amp circuit describes the strength of the electrical current it can handle. A 20-amp circuit is overloaded when drawing more than 17amps.

Anaerobic bacteria: Anaerobic bacteria gets developed under anaerobic conditions (no oxygen). See anaerobic condition.

Anaerobic condition: Absence of free oxygen. When solids are allowed to enter grow beds (especially in aquaponics system), over time buildup occurs and can cause anaerobic areas within the grow beds to develop, which in turn causes toxic substances to develop.

Anion: A negatively charged ion. Plant nutrient examples include nitrate (NO3-), phosphate (H$_2$PO$_4$-), and sulfate (SO$_4$$^{2-}$). See cation.

Annual: A plant that normally completes its entire life cycle in one year or less: Marigolds and tomatoes are examples of annual plants.

Anthocyanin: A blue, violet, or red flavonoid pigment found in plants.

Apex: The tip of a stem or root.

Apical bud: A bud at the tip of a stem.

Apical dominance: The inhibition of lateral bud growth by the presence of the hormone auxin in a plant's terminal bud. Removing the growing tip removes auxin and promotes lateral bud break and subsequent branching, usually directly below the cut.

Apical meristem: A region of actively dividing cells at the tip of a growing stem or root.

Aquaponics: The integration of aquaculture (the raising of marine animals, such as fish) with hydroponics; the waste products from the fish are treated and then used to fertilize hydroponically growing plants.

Asexual reproduction: See vegetative propagation.

Atomic weight: The relative weight of an atom.

Autotrophic nutrition: A form of nutrition in which complex food molecules are produced by photosynthesis from carbon dioxide, water, and minerals.

Auxin: Classification of plant hormones; Auxins are responsible for foliage and root elongation.

Axil: The upper angle formed by a leaf's stalk (petiole) and the internodes above it on a stem.

Axillary bud primordium: An immature axillary bud.

Axillary bud: A bud that forms on an axil.

Bacillus thuringiensis (Bt): A bacterium used as a biological control agent for many insect pests; primarily mosquitoes, fungus gnats, and caterpillars.

Bacteria: Single cell organisms with no chlorophyll.

Bacterial soft rot: See botrytis.

Bark: All the tissues, collectively, formed outside the vascular cambium of a woody stem or root.

Beneficial insect: An insect that helps gardening efforts. May pollinate flowers, eat harmful insects (*e.g.* ladybug) or parasitize them, or break down plant material in the soil, thereby releasing nutrients. Some insects are both harmful and beneficial. For example, butterflies can be pollinators in their adult form, but destructive in their larval (caterpillar) form.

Biodegradable: Able to decompose or break down through natural bacterial action; Substances made of organic matter are biodegradable.

Biosolids: A by-product of wastewater treatment sometimes used as a fertilizer.

Bleach: A cleansing solution. Generally one part laundry bleach mixed with ten parts water can be used as a fungicide. This solution can be used to clean the equipments.

Blight: Rapid, extensive discoloration, wilting, and death of plant tissue.

Bloom booster/blossom booster - Fertilizer high in phosphorus (P) that increases flower yield.

Blossom-end-rot (BER): A physiological and nutritional disorder on fruit creating a black, leathery, sunken appearance on the blossom end of the fruit–often associated with poor watering, root death, and calcium deficiency.

Blotch: A blot or spot (usually superficial and irregular in shape) on leaves, shoots, or fruit.

Boron (B) - Micronutrient for plants.

Botrytis: A fungal disease promoted by cool, moist weather. Also known as gray mold or fruit rot.

Branch: A subsidiary stem arising from a plant's main stem or from another branch.

Breathe: While stomata "breathe" CO_2, the roots "breathe" oxygen.

British Thermal Unit (BTU): Amount of heat required to raise the temperature of 1 pound of water 1 F.

Bud: A small protuberance on a stem or branch, sometimes enclosed in protective scales, containing an undeveloped shoot, leaf, or flower.

Budding: The grafting of a bud onto stock of a different plant. The bud is the scion.

Budstick: A shoot or twig used as a source of buds for budding.

buffering: The ability of a nutrient solution or raw water to resist changes in pH

bulb: 1. An underground storage organ consisting of a thin, flattened stem surrounded by layers of fleshy, dried leaf bases. Roots are attached to the bottom. 2. The outer glass envelope or jacket that protects the arc tube of an HID lamp.

BURN - Leaf tips that turn dark from excess fertilizer and salt burn.

C:N ratio: The ratio of carbon (C) to nitrogen (N) in organic materials. Materials with a high C:N ratio (high in carbon) are good bulking agents in compost piles, while those with a low C:N ratio (high in nitrogen) are good energy sources.

CALCIUM (Ca) - Calcium is vital in all parts of plants to promote the translocation of carbohydrates, healthy cell wall structure, strong stems, membrane maintenance and root structure development. Calcium is a macronutrient.

Calcium carbonate (CaCO$_3$): A compound found in limestone, ashes, bones, and shells; the primary component of lime.

Canker: A localized lesion on a limb or trunk, usually due to disease or injury. Part of the bark or wood appears to be eaten away or is sunken.

Canopy: (1) The top branches and foliage of a plant. (2) The shape-producing structure of a tree or shrub.

Capillary action: The action by which water molecules bind to the surfaces of soil particles and to each other, thus holding water in fine pores against the force of gravity.

Capillary water: Water held in the tiny spaces between soil particles or between plant cells.

Carbohydrate: The cellulose, starches and sugars in plants are carbohydrates. It is a neutral compound of carbon, hydrogen and oxygen.

Carbon dioxide (CO$_2$) - A colorless, odorless, tasteless gas in the air necessary for plant life. Occurs naturally in the atmosphere at .03 per cent.

Caterpillar: See larva.

Cation: A positively charged ion. Plant nutrient examples include calcium (Ca++) and potassium (K+). See anion.

Cell wall: The outer covering of a plant cell.

Cell: This is the basic structural unit of both plants and animals. A cell contains the nucleus, chloroplasts and membrane.

Cellular respiration: The chemical breakdown of food substances, resulting in the release of energy.

Cellulose: This is a complex carbohydrate that causes the plant to stiffen. A tough stem will contain cellulose.

Chelate: A complex organic substance that holds micronutrients, usually iron, in a form available for absorption by plants.

Chiller: A refrigeration unit used to reduce the temperature of water or nutrient solution.

Chinampas (floating gardens): Layers of mud and vegetation are used to suspend crops over freshwater lakes.

Chlorine (Cl): This micronutrient is essential for photosynthesis, where it acts as an enzyme activator during the production of oxygen from water. It is a chemical that purifies water.

Chlorophyll: The green pigment in plants. Responsible for trapping light energy for photosynthesis.

Chloroplast: A specialized component of certain cells. Contains chlorophyll and is responsible for photosynthesis.

Chlorosis: The condition of a sick plant with yellowing leaves due to inadequate formation of chlorophyll. Chlorosis is caused by a nutrient deficiency, usually iron or nitrogen; nutrient deficiencies are themselves often caused by a pH that is out of the acceptable range.

Clay: Soil that is created from very tiny mineral and organic particles (less than 0.002 mm in diameter). Clay is generally not suitable for container gardening.

Climber: A plant that climbs on its own by twining or using gripping pads, tendrils, or some other method to attach itself to a structure or another plant. Plants that must be trained to a support are properly called trailing plants, not climbers.

Clone: A plant produced through asexual reproduction including, but not limited to, cuttings, layering and tissue culture.

Closed system: In closed hydroponic system, the same nutrient solution is re-circulated and the nutrient concentrations are monitored and adjusted according to the requirements. Closed hydroponics can be both simple hydroponic systems (such as 'deep water culture') as well as sophisticated hydroponics (such as NFT).

Color Spectrum: The band of colors (measured in nm) emitted by a light source.

Compaction: Pressure that squeezes soil into layers that resist root penetration and water movement. Often the result of foot or machine traffic.

Compaction: Soil condition that results from lightly packed soil: Compacted soil allows for only marginal aeration and root penetration

Companion planting: The practice of growing two or more plants together in the hope that the combination will discourage disease and insect pests.

Compost: The product created by the breakdown of organic waste under conditions manipulated by humans. It is high in nutrient content.

Conductivity: It is the dissolved salts in most water that allows it to conduct electricity and is usually measured in S/cm.

Controlled environmental agriculture (CEA): The growing of plants in structures as greenhouses that permit the regulation of optimum environmental conditions for the crop year-round regardless of ambient weather conditions.

Copper (Cu): This micronutrient is an internal catalyst and acts as an electron carrier; it is also believed to play a role in nitrogen fixation.

Corm: An underground storage organ consisting of the swollen base of a stem with roots attached to the underside. Crocus and gladiolus are examples of plants that form corms.

Cormel: A small, underdeveloped corm, usually attached to a larger corm.

Critical photoperiod: The maximum day length a short-day plant, and the minimum day length a long-day plant, require initiating flowering.

Cross-pollination: The fertilization of a ovary on one plant with pollen from another plant, producing an offspring with a genetic makeup distinct from that of either parent.

Crown: (1) Collectively, the branches and foliage of a tree or shrub. (2) The thickened base of a plant's stem or trunk to which the roots are attached.

Cultivar: A specially cultivated variety of a plant that most often is reproduced vegetatively.

Cuticle: (1) A relatively impermeable surface layer on the spidermis of leaves and fruits. (2) The outer layer of an insect's body.

Cutin: (1) A waxy substance on plant surfaces that tends to make the surface waterproof and can protect leaves from dehydration and disease. (2) A waxy substance on an insect's cuticle that protects the insect from dehydration.

Cutting: A piece of leaf, stem, or root removed from a plant for asexual propagation. See asexual propagation. See vegetative propagation.

Cytokinin: A plant hormone primarily stimulating cell division. Visit our selection of plant hormones.

Damping-off: Fungus disease that attacks young seedlings and cuttings causing them to rot at the base of the stem and causes toppling over and death of the plants. It is caused by soil-borne fungi under too-moist planting media.

Decomposition: The breakdown of organic materials by microorganisms.

Deep flow Technique: See 1.5.6 Deep Flow Technique (DFT)–Pipe System

Deep water culture (DWC): See 3.4 Deep Water Culture or Direct Water Culture (DWC)

Denitrification: During denitrification (under anaerobic conditions), the nitrogen in the aquaponics system may be converted to many different forms and eventually to nitrogen gas (N2) which bleeds out of the system.

Detergent: A cleanser made from liquid soap. This is used in gardening as a pesticide.

Determinate: A plant growth habit in which the stems stop growing at a certain height and produce a flower cluster at the tip. Determinate tomatoes, for example, are short, early-fruiting, have concentrated fruit set, and do not require staking. See indeterminate.

Dieback: Progressive death of shoots, branches, or roots, generally starting at the tips.

Division: The breaking or cutting a part of a plant's crown for the purpose of producing additional plants, all genetically identical to the parent plant.

Dormancy: The annual period when a plant's growth processes greatly slows down.

Dormant bud: A bud formed during a growing season that remains at rest during the following winter or dry season. If it does not expand during the following growing season, it is termed a latent bud.

Dormant oil: A horticultural oil applied during the dormant season to control insect pests and diseases.

Drip aeration - A hydroponic method wherein air pressure from a small air pump is used to percolate nutrient solution out through a ring of feeder tubing which encircles the plant.

Drip irrigation: A type of irrigation system by which each plant is fed individually with a small drip tube and the flow is regulated by an emitter commonly used in most hydroponic systems.

Drip Line: The lines around the plant under the outermost blanch tips. The roots will seldom grow past the drip line.

Drip System: A system used in hydroponics that includes a hose with small emitters of water that drip out the nutrients one at a time.

Ebb-and-flow (or flood and drain) - A hydroponic system in which the medium, usually aggregate pebbles, is periodically flooded with nutrient solution and then drained again, feeding and aerating the medium and root system.

Ebb-and-Flow: See 3.3 'Ebb and Flow'.

Economic threshold: The level at which pest damage justifies the cost of control. In home gardening, the threshold may be aesthetic rather than economic.

Electrical conductivity (EC): A measure of the ability of a nutrient solution to conduct electricity, which is dependent upon the ion concentration and nature of the elements present. Visit our selection of solution testing equipment.

Elongate: Grow in length.

Enzyme: A biological catalyst that aids in a specific biochemical process, such as converting food from one form to another.

Epidermis: The outermost layer of cells covering a plant's leaves, roots, and young parts.

Ethylene: A gaseous plant hormone (C_2H_4) produced in abundance by ripening fruits and damaged tissues.

Exotic: Non-native.

Family: A broad group of plants with common characteristics.

Feed: The fertilizer

Feeder roots: Fine roots and root branches with a large absorbing area (root hairs). Responsible for taking up the majority of a plant's water and nutrients from the soil.

Fertilization: (1) The fusion of male and female germ cells following pollination. (2) The addition of plant nutrients to the environment around a plant.

Fertilizer analysis: The amount of nitrogen (N), phosphorus (as P_2O_5), and potassium (as K_2O) in a fertilizer, expressed as a percentage of total fertilizer weight. On the N-P-K fertilizer label, the percentage by weight of nitrogen (N) is always listed first, phosphorus (P) second, and potassium (K) third.

Fertilizer Burn: Over-fertilization: When too much fertilizer is used, the tips of the leaves curl and turn brown.

Fertilizer: A natural or synthetic product added to the soil or sprayed on plants to supply nutrients.

Fibrous root: A root system that branches in all directions, often directly from the plant's crown, rather than branching in a hierarchical fashion from a central root. See taproot.

Flower cluster/truss: A group of flowers that form from the stem of tomato plants which when pollinated produce the fruit.

Flower: The reproductive branch or structure of an angiosperm plant.

Fluorescent Lamp: An electrical lamp tube that is coated with fluorescent materials. It is low in heat and power consumption, generally used to grow root cuttings.

Foliage: The leaves or green part of the plant.

Foliar feeding: Misting plants with fertilizer solution, which is absorbed by the foliage.

Food: An organic substance that provides energy and body-building materials, especially carbohydrates, fats, and proteins.

Fruit: Botanically, a fruit is a ripened, mature ovary.

Fungicide: A product that destroys or inhibits fungus. Sulfur and copper sulfate are two common mineral fungicides.

Fungus: Any of a major group (Fungi) of saprophytic and parasitic spore-producing organisms usually classified as plants that lack chlorophyll and include molds, rusts, mildews, smuts, mushrooms, and yeasts. Common fungal diseases that attack plants are "damping-off," Botrytis, and powdery mildew.

Fusarium: Any of several fungal diseases that afflict plants; commonly called dry rot or wilt.

Gene: A unit of genetic inheritance. generative growth reproductive phase of a plant in which it produces flowers and fruit.

Gene: The part of the chromosome that influences the development of the plant itself. The genes are inherited properties of sexual propagation.

Genetic Make-Up: The totality of genes that are inherited from parent plants. The genetic make-up of a plant is the most important factor in its vigor, potency and vitality.

Genus: A group of related species, each of which is distinct and unlikely to cross with any other. A group of genera forms a family, and a group of families forms an order. See species.

Germination inhibitor: A chemical substance preventing seed germination.

Germination: The initial sprouting stage of a seed.

GPM: Gallons per minute

Grafting: The act of inserting a shoot or bud of one plant into the trunk, branch, or root of another, where it grows and becomes a permanent part of the plant.

Growing medium: Materials that are sometimes used in hydroponic growing to support the plant's roots and, sometimes, to hold nutrient.

Growing season: The period between the beginning of growth in the spring and the cessation of growth in the fall.

Growth regulator: A compound applied to a plant to alter its growth in a specific way. May be a natural or synthetic substance. See hormone.

Halide: The binary compound of halogen(s) that contain electro-positive elements.

Hardening off: To gradually acclimatize a plant to a more harsh environment. A seedling must be hardened-off before planting outdoors.

Hardy: Frost or freeze-tolerant. In horticulture, this term does not mean tough or resistant to insect pests or disease.

Herbaceous perennial: A herbaceous plant that dies back in the winter and regrows from the crown in the spring.

Herbaceous: A soft, pliable, usually barkless shoot or plant. Distinct from stiff, woody growth.

Herbicide: A chemical used to kill undesireable plants.

Hermaphrodite: A single plant with both male and female properties. The breeding of hermaphrodites is hard to control.

Hertz (Hz): A unit of a frequency that cycles one time each second: A home with a 60 hertz AC current cycles 60 times per second.

Heterotrophic nutrition: A form of nutrition in which the organism depends on organic matter for food, such as humans.

HID: High Intensity Discharge

Honeydew: A sticky substance excreted by aphids and some other insects such as aphids, scale and mealy bugs.

Hood: Reflective cover of a HID hydroponics lamp; A large, while hood is very reflective.

Hood: The reflective cover of HID hydroponic lamps.

Hormone: Chemical substance that controls the growth and development of a plant. Root-inducing hormones help cuttings root.

Host: A plant on which an insect or disease completes all or part of its life cycle.

Humidity (relative): Ratio between the amount of moisture in the air and the greatest amount of moisture the air could hold at the same temperature.

Hybrid: The offspring from two plants of different breeds, variety or genetic make-up.

Hydrated lime - Instantly soluble lime, used to raise or lower pH.

Hydrogen: The lightest of all gasses, hydrogen combines with oxygen to form water.

Hydroponics: A method of growing plants without soil. Plants usually are suspended in water or inert growing media, and plant nutrients are supplied in dilute solutions.

Immune: A plant that does not become diseased by a specific pathogen. See resistant, tolerant.

Indeterminate: A plant growth habit in which stems continue growing in length indefinitely. For example, indeterminate tomatoes are tall, late-fruiting, and require staking for improved yield. See determinate.

Indicator paper: A litmus type paper that changes color with specific levels of acid or base and is used to check ph indicator solution a solution that changes color with ph changes.

Inert: A chemically non-reactive material or medium used in hydroponics. It does not interfere with nutrient solutions. Inert growing mediums make it easy to control the chemistry of the nutrient solution.

Infection: The condition reached when a pathogen has invaded plant tissue and established a parasitic relationship between itself and its host.

Insecticidal soap: A specially formulated soap that is only minimally damaging to plants, but kills insects. These usually work by causing an insect's outer shell to crack, resulting in drying out of its internal organs.

Insecticide: Any material that kills insects. Includes numerous botanical products, both organic and synthetic.

Integrated pest management (IPM): A method of managing pests that combines cultural, biological, mechanical, and chemical controls, while taking into account the impact of control methods on the environment.

Intensity: The magnitude of light energy per unit: Intensity diminishes the farther away from the source.

Intensive gardening: The practice of maximizing use of garden space, for example by using trellises, intercropping, succession planting, and raised beds.

Internode: The portion of a stem between two nodes.

Ionized ammonia: see Ammonia in aquaculture.

Iron (Fe): This micronutrient acts as a catalyst in the photosynthesis/respiration process, and is essential for the formation of sugars and starches. Iron also activates certain other enzymes.

Juvenile stage: (1) The early or vegetative phase of plant growth characterized by carbohydrate utilization. (2) The first stage of an insect's life cycle, either as a larva or a nymph.

K: The chemical symbol for postassium.

Larva: The immature form of an insect that undergoes complete metamorphosis. Different from the adult in form. Also called a caterpillar.

Leach: To wash out soluble components of soil through heavy watering. This is used in hydroponics to flush out excesses of fertilizer salts.

Leaf curl: Leaf malformation due to overwatering, over fertilization, lack of magnesium, insect or fungus damage or negative tropism.

Leaflet: Small immature leaf

Leggy: Abnormally tall, with sparse foliage: Legginess of a plant is usually caused by lack of light.

Life Cycle: A series of growth stages through which plant must pass in Its natural lifetime: The stages for an annual plant arc seed, seedling, vegetative and floral.

Lignin: A tough, durable plant substance deposited in cell walls, especially in wood and coconut.

Lime: A rock powder consisting primarily of calcium carbonate. Used to raise soil pH (decrease acidity).

Litmus Paper: A chemically sensitive paper used for testing pH balance.

Lumen: Measurement of light output: One lumen is equal to the amount of light emitted by one candle that falls on one square foot of surface located one foot away from

Macronutrient: The major minerals that are used by plants in large amounts, consisting of nitrogen (N), phosphorus (P), potassium (K), sulfur (S), calcium (Ca), and magnesium (Mg).

Manganese (Mn) - This micronutrient activates one or more enzymes in fatty acid synthesis; it also activates the enzymes responsible for DNA and RNA production. Closely associated with copper and zinc, manganese also participates directly in the photosynthetic creation of oxygen from water.

Medium: The substrate or soilless material which supports the plant and absorbs and releases the nutrient solution in hydroponic horticulture.

Mesophyll: A leaf's inner tissue, located between the upper and lower epidermis, where raw materials (carbon dioxide and water vapor) are held for use in photosynthesis.

Metabolism: The sum of the biochemical processes of a living cell.

Micronutrient: A nutrient used by plants in small amounts, less than 1 part per million. Micronutrients include boron, chlorine, copper, iron, manganese, molybdenum, and zinc. Also called trace elements.

Mineral deficiency: When a plant is not receiving a required nutrient—at all or in an insufficient amount—a disorder will result.

Molecular weight: The relative weight of a molecule.

Molecule: A chemically bonded group of atoms.

Molybdenum (Mo) - This micronutrient is essential for nitrogen fixation and nitrate reduction.

Mottle: An irregular pattern of light and dark areas.

Mulch: A protective covering of organic compost, old leaves, etc.: Indoors, mulch keeps soil too moist, and possible fungus could result.

Mutation: A genetic change within an organism or its parts that changes its characteristics. Also called a bud sport or sport.

N: The chemical symbol for nitrogen.

Nanometer:.000 000 001 meter, nm is used as a scale to measure electromagnetic wave lengths of light: Color and light spectrums are expressed in nanometers (nm).

Necrosis: The dying of plant tissue, usually the result of serious nutrient deficiency or pest attack. Browning of leaf tissue due to a nutritional disorder.

Nectar: A gland secreting nectar.

Nectar: A sugary fluid secreted by some flowers.

Nematode: A microscopic roundworm, usually living in the soil. May feed on plant roots and can be disease pathogens or vectors. Others are beneficial parasites of insect pests. Visit our selection of beneficial insects.

NFT (nutrient film technique): A hydroponic method in which nutrient is fed into grow tubes or trays in a thin film where the roots draw it up. This "nutrient film" allows the roots to have constant contact with the nutrient and the air layer above at the same time. See **Chapter 1 (1.5.5 Continuous-Flow Solution Culture and NFT-Nutrient Film Technique).**

Nitrate (NO_3.): A plant-available form of nitrogen contained in many fertilizers and generated in the soil by the breakdown of organic matter. Excess nitrate in soil can leach into groundwater. See nitrogen cycle.

Nitrate-type nitrogen: See Ammonia-type nitrogen.

Nitrification: Aerobic conversion of ammonia into nitrates.

Nitrifier: A microbe that converts ammonium to nitrate.

Nitrifying bacteria: Bacteria that will oxidize ammonia to nitrite and nitrite to nitrate. The most common genus of a family of autotrophic bacteria (**nitrifying bacteria)** are Nitrosomonas (that convert ammonia into nitrites) and Nitrobacter (convert nitrites into nitrates), respectively.

Nitrobacter: See Nitrifying bacteria.

Nitrogen (N): A primary plant nutrient, especially important for foliage and stem growth. It is used in various forms to promote rapid vegetative growth, leaf, flower, fruit and seed development, and chlorophyll development; and to increase the protein content in all plants.

Nitrogen cycle: The sequence of biochemical changes undergone by nitrogen as it moves from living organisms, to decomposing organic matter, to inorganic forms, and back to living organisms.

Nitrogen fixation: The conversion of atmospheric nitrogen into plant-available forms by Rhizobia bacteria living on the roots of legumes.

Nitrosomonas: See Nitrifying bacteria.

Node: The point on a plant where a branch, bud, or leaf develops. On younger branches, it usually is marked by a slight swelling. The space on the stem between nodes is called an internode.

N-P-K: The acronym for the three primary nutrients contained in manure, compost, and fertilizers. The N stands for nitrogen, the P stands for phosphorus, and the K stands for potassium. On a fertilizer label, the N-P-K numbers refer to the percentage of the primary nutrients (by weight) in the fertilizer. For example, a 5-10-5 fertilizer contains 5 per cent nitrogen, 10 per cent phosphorous, and 5 per cent potassium.

Nutrient film technique (NFT): See **NFT**

Nutrient solution: The mixture of water and water-soluble nutrients which is provided to the plants for nourishment in a hydroponic systems.

Nutrient: Any substance that is essential for and promotes plant growth. These are food of the plant, such as N-P-K along with secondary and trace elements. See macronutrient, micronutrient.

Organic fertilizer: A natural fertilizer material that has undergone little or no processing. Can include plant, animal, or mineral materials.

Organic matter: Any material originating from a living organism (peat moss, plant residue, compost, ground bark, manure, etc.).

Organic: Made of, derived from or related to living organisms.

Organism: A living plant or animal.

Ornamental plant: A plant grown for beautification, screening, accent, specimen, color, or other aesthetic reasons.

Osmosis: The flow or diffusion that takes place through a semipermeable membrane typically separating a solvent and a solution that strives to bring about a condition of equilibrium.

Oxidation: The chemical process by which sugars and starches are converted into energy. In plants, this is also known as respiration.

Oxygen: Essential for respiration and formation of sugar, starch, and cellulose. Oxygen is about 88 per cent of the composition of water and plays a critical role in plant's growth. Plants obtain the oxygen they need through the stomata on the leaves, through the roots via the water and through the process of photosynthesis. It is involved in anion exchange between the roots and surrounding medium.

Oxygenation: The supplying of oxygen; usually refers to the needs of plants' roots oxygen deficit when oxygen is inadequate to support normal plant physiological processes. Visit our selection of air pumps for better oxygenation of your plants' roots.

P: The chemical symbol for phosphorus.

Parasite: An organism that lives off another host organism, such as fungus. It withdraws nutrients from its host.

Parthenocarpic: Development of fruit without fertilization.

Parts per million (PPM): A ratio figure that represents the amount of one substance that is in one million parts of another substance; commonly used to describe the relative concentrations of nutrient solutions.

Pathogen: Any organism that causes disease. Generally applied to bacteria, viruses, fungi, nematodes, and parasitic plants. Visit our selection of pesticides.

Peatlite mix: A soilless medium consisting of a mixture of peat, sand, vermiculite and/or perlite.

Pectin: A substance in cell walls binding cells together.

Perennial: A plant that completes its life cycle over a number of years, such as a tree or shrub.

Perlite: 1. Sand or volcanic glass which has been expanded by heat; perlite holds water and nutrients on its many irregular surfaces.

Petals: The usually showy structures around a flower's reproductive organs.

Petiole: The stalk of a leaf.

pH (potential hydrogen): The pH is a parameter that measures the acidity or alkalinity of a solution. The pH is usually measured on a scale of 1-14 and represents the concentration of hydrogen ions in solution. This value indicates the relationship between the concentration of free ions H^+ and OH^- present in a solution. Generally, it is used to determine whether a hydroponic solution is acidic or basic (alkaline). A solution is acidic if it has more positive hydrogen ions and is alkaline if it has more negative hydroxyl ions. See Chapter 5 (5.5.4 The pH (potential hydrogen)

pH Tester: Electronic instrument or chemical used to find where soil or water is on the pH scale.

Phloem: Photosynthate-conducting tissue. See xylem.

Phosphate: The form of phosphorus listed in most fertilizer analysis (P_2O_5).

Phosphor Coating: Internal bulb coating that diffuses light and is responsible for various color outputs.

Phosphorus (P): A primary plant nutrient, especially important for flower production. In fertilizer, usually expressed as phosphate (P_2O_5). It promotes and stimulates early growth and blooming and root growth.

Photoperiod: The relationship between the amount of light and dark in a 24 hour period.

Photosynthesis: The process by which plants use light energy, water and CO_2 to build chemical compounds (carbohydrates).

Phytotoxic: Toxic to a plant.

Pistil: The female sexual organ of a flowering plant, made up of the stigma, style, and ovary.

Pith: A region of parenchyma cells at the center of a stem.

Plant growth regulator: See growth regulator.

Pollen: A plant's male sex cells, which are held on the anther for transfer to a stigma by insects, wind, or some other mechanism.

Pollination: The transfer of pollen from a male anther to a female stigma, enabling fruits to set and develop.

Pollinator: An agent, such as an insect, that transfers pollen from a male anther to a female stigma.

Potash: The form of potassium listed in most fertilizer analysis (K_2O).

Potassium (K) - Potassium promotes disease resistance and good development of carbohydrates, starches and sugars, and it increases fruit production. Potassium is a macronutrient.

Pot-Bound: A plant that is bound or inhibited from normal growth by a container. A pot will contain the roots of the plant.

Ppen (non-recirculating) system: A hydroponic system in which the nutrient solution passes only once through the plant roots; the leachate is not collected and returned to a cistern for a repeated cycle.

Predator: An animal that eats another animal.

Primary nutrient: A nutrient required by plants in a relatively large amount. See macronutrient, N-P-K.

Propagate: To start new plants by seeding, budding, grafting, dividing, etc.

Prune: To remove plant parts to improve a plant's health, appearance, or productivity.

PVC Pipe: A plyvinyl chroide pipe that is commonly used, easy to reshape and readily available for gardens.

Quick-release fertilizer: A fertilizer that contains nutrients in plant-available forms such as ammonium and nitrate.

Relative humidity: The ratio of water vapor in the air to the amount of water vapor the air could hold at the current temperature and pressure.

Reservoir: The container in a hydroponic system which holds nutrient solution in reserve for use.

Resistant: A plant having qualities that make it retard the activities of a pathogen or insect pest. See immune, tolerant.

Respiration: The process within plants where sugars and starches are converted into energy.

Reverse osmosis: The process of removing minerals from water, which is forced by pressure through a differentially permeable membrane, filtering out the minerals; can happen when growers accidentally apply too strong of a nutrient to a plant's roots, leeching life out of the plant. Visit our selection of reverse osmosis systems.

Rhizome: A thickened underground stem that grows horizontally with bud eyes on top and roots below. Bearded iris is an example of a plant that produces rhizomes.

Rockwool: Inert, soilless growing medium consisting of woven, thin strand-like fibers made from molten volcanic rock and limestone, which is heated to over 1500°C, extruded, and formed into slabs, cubes and blocks.

Root cutting: A section of root prepared for the purpose of vegetative propagation.

Root hair: A delicate, elongated epidermal cell that occurs just behind a root's growing tip. Root hairs increase the root's surface area and absorptive capacity.

Root tuber: An enlarged, food-storage root bearing adventitious shoots.

Root: Generally, the underground portion of a plant. It anchors the plant and absorbs water and nutrients.

Rootstock: A plant which has an established healthy root system used for grafting, a cutting, or budding from another plant. Rootstocks are most commonly used with fruiting and flowering plants that are susceptible to root diseases. By grafting to a healthy vigorous rootstock, typical root diseases for a cultivar are prevented.

Rosette: A small cluster of leaves radially arranged in an overlapping pattern.

Rot: Decomposition and destruction of tissue.

Rotation: The practice of growing different plants in different locations each year to prevent the buildup of soil-borne diseases and insect pests, or the depletion of specific nutrients.

Salt: Crystalline compound that results from improper pH or toxic buildup of fertilizer. Salt will burn plants, preventing them from absorbing nutrients.

Sand: The coarsest type of soil particle (0.05 to 2.0 mm in diameter).

Sanitation: The process of removing sources of plant pathogens from a growing area, for example, by cleaning up plant debris and sterilizing tools and growing media.

Scale: (1) A modified leaf that protects a bud. (2) A type of insect pest.

Secondary nutrient: A nutrient needed by plants in a moderate amount: calcium, magnesium, and sulfur. See macronutrient, primary nutrient.

Seed: A reproductive structure formed from the maturation of an ovule and containing an embryo and stored food.

Seedling: A young plant, shortly after germination.

Selective pesticide: A pesticide that kills or controls only certain kinds of plants or animals.

Self-fruitful: A plant that bears fruit through self-pollination.

Self-pollination: The transfer of pollen from the anther to the stigma of the same flower.

Self-unfruitful: A plant that requires another variety for pollination.

Shear: To cut back a plant (as opposed to selective pruning or deadheading). Often used to regenerate plants with many small stems, where deadheading would be too time-consuming.

Shoot: One season's branch growth. The bud scale scars (ring of small ridges) on a branch mark the start of a season's growth.

Short-day plant: A plant requiring more than 12 hours of continuous darkness to stimulate a change in growth, *e.g.*, a shift from the vegetative to reproductive phase. See long-day plant, day-neutral plant.

Shrub: A woody plant that grows to a height of 3 to 12 feet. May have one or several stems with foliage extending nearly to the ground.

Soil: A natural, biologically active mixture of weathered rock fragments and organic material at the earth's surface.

Soilless mix: A sterile potting medium consisting of ingredients such as sphagnum peat moss and vermiculite.

Solubility: The ability of a fertilizer to dissolve in water to form a solution.

Soluble salts: A mineral residue often remaining in soil from irrigation water, fertilizer, compost, or manure applications.

Soluble: The ability to dissolve in water.

Species: The basic unit of plant or animal classification. Plants within an individual species have several characteristics in common. Most importantly, they can cross with one another, but normally not with members of other species. Classification of species is quite fluid, with periodic revision by botanists.

Spore: (1) The reproductive body of a fungus or other lower plant, containing one or more cells. (2) A bacterial cell modified to survive in an adverse environment.

Sprout: A recently germinated seed 2. small new growth of leaf or stem.

Stalk: On a male flower, the portion of the stamen that supports the anther.

Stamen: The male, pollen-producing part of a flower consisting of the anther and its supporting filament

Starch: The principal food-storage substance (a carbohydrate) of higher plants.

Stem cutting: A section of a stem prepared for vegetative propagation.

Stem: The leaf and flower bearing part of a plant.

Sterile: (1) Material that is free of disease organisms (pathogens), as in potting medium. (2) A plant that is unable to produce viable seeds.

Sterilization: The act of rendering something free from living cells. In hydroponics it is essential that all materials (especially any growing medium) used are sterile to avoid contaminating the hydroponic system. Steam and chemical agents are often used in this process.

Sterilize: To remove the dirt, bacteria and germs. Using a water solution with 10 per cent bleach can effectively sterilize hydroponic equipment.

Stomata: A small mouth-like opening or pore on the underside of leaves. The stomata are responsible for many of the life functions of the plant.

Stress: A physical or chemical factor that causes extra exertion by plants. A stressed plant will not grow as well, as a plant without stress.

Subspecies: A major division of a species, more general in classification than a cultivar or variety.

Succession planting: The practice of planting new crops in areas vacated by harvested crops.

Sugar: Food product of a plant.

Sump: Reservoir or receptacle that serves as a drain or holder for hydroponic nutrient solutions.

Surfactant: See adjuvant.

Symbiotic: Mutually beneficial.

Symptom : A change in plant growth or appearance in response to living or nonliving damaging factors.

Synthesis: Production of a substance, such as chlorophyll, by uniting light energy and elements or chemical compounds.

Tap Root: The main or primary root that grows from the seed: Lateral roots will branch off the tap root,

Terminal Bud: Bud at the growing end of the main stem.

Tissue culture: The process of generating new plants by placing small pieces of plant material onto a sterile medium. Also called embryo culture.

Tissue: A group of cells of the same type having a comon purpose. tissue analysis a laboratory analysis of plant tissue to determine levels of nutrients present.

Tolerant: A plant that will produce a normal yield even if infested by a disease or insect pest. See immune, resistant.

Total ammonia: see Ammonia in aquaculture.

Total dissolved salts - The amount of dissolved fertilizer salts, that are measured in water in parts per million.

Total dissolved solutes (TDS): The concentration of all the elemental ions present in a nutrient solution; electrical conductivity (EC) is a measure of TDS generally expressed as mS (milliSiemens) or mMho (millimhos).

Transpiration: The process of losing water in the form of vapor through stomata.

Transpire: Give off water vapor and by products via the stomata.

Trellis: A frame of PCV, lattice or small boards that supports the plants.

Tuber: An underground storage organ made up of stem tissue. Contains buds on the surface, from which shoots may arise. Potatoes are an example.

Tuberous root: An underground storage organ made up of root tissue. Sprouts only from the point where it was attached to the parent plant. Dahlias are an example.

Ultraviolet: Light with very short wave lengths, out of the visible spectrum.

Un-ionized ammonia: see Ammonia in aquaculture.

Vaporization: The evaporation of the active ingredient in a pesticide during or after application.

Variety: A strain of a plant having distinctive features that persist over successive generations in the absence of human intervention. Generally, variety applies to naturally-occuring strains, while cultivar applies to horticulturally-developed strains.

Vascular cambium: A narrow cylinder of cells that gives rise to secondary xylem and phloem. A lateral meristem.

Vascular plant: A plant which has water and food conducting tissues.

Vascular tissue: Water, nutrient, and photosynthate-conducting tissue. See xylem, phloem.

Vegetative propagation: The increase of plants by asexual means using vegetative parts. Normally results in a population of identical individuals. Can occur by either natural means (*e.g.*, bulblets, cormels, offsets, plantlets, or runners), or by artificial means (*e.g.*, cuttings, division, budding, grafting, or layering). Click here to learn more about how to clone and propagate plants.

Vein: A strand of xylem and phloem in a leaf blade.

Venation: The arrangement of veins in a leaf.

Vent: Opening such as a window or door that allows the circulation of fresh air.

Ventilation: Circulation of fresh air, fundamental to healthy indoor garden. An exhaust fan creates excellent ventilation.

Vermiculite: see chapter 6 (6.9 Vermiculite)

Vernalization: A low-temperature treatment promoting flowering.

Verticillium: Any of several fungal diseases that afflict plants; commonly called wilt. See fusarium.

Virus: An infectious agent too small to see with a compound microscope. Multipies only within a living host cell.

Water-holding capacity (WHC): The ability of a soil's or soilless media's micropores to hold water for plant use.

Wetting Agent: Compound that reduces the droplet size and lowers the surface tension of the water, making it wetter. Liquid concentrate dish soap is a good wetting agent if it is biodegradable,

Wick: Part of a passive hydroponic system using a wick suspended in the nutrient solution, the nutrients pass up the wick and are absorbed by the medium and roots.

Wilt: (1) Lack of freshness, turgor, and the drooping of leaves from a lack of water. (2) A vascular disease that interrupts a plant's normal uptake and distribution of water.

Wilting point (WP): The point at which water content within plant cells is low enough that cellular turgor is lost and the plant wilts.

Xylem: Water- and nutrient-conducting tissue. See phloem.

Zinc (Zn) - Like copper and manganese, zinc is linked to chlorophyll synthesis.

Zygote: A fertilized egg.

References

Aquaculture Development and Coordination Programme (ADCP), 1989a. Aquaculture Regional Profiles, Rome, FAO.

Aquaculture Development and Coordination Programme (ADCP), 1989b. Planning for aquaculture development. Report of an Expert Consultation held in Policoro, Italy, 26 July-2 August 1988. Rome. FAO, ADCP/REP/89/33, 68 p.

Brad Gabora and Wayne Wiebe, (eds) 1997. *Tomato Diseases: A Practical Guide for Seedsmen, Growers, and Agricultural Advisors.* Seminis Vegetable Seeds, Inc. Saticoy, California, USA.

Cooper, A. 1988. "1. The system. 2. Operation of the system". In: *The ABC of NFT. Nutrient Film Technique*, pp. 3-123. Grower Books (ed.), London, England.

Hoagland, D.R. and Arnon, D.I. 1950. The Water Cultural Method for Growing Plants without Soil. Circular 347. California Agricultural Experiment Station, University of California, Berkeley, CA.

Libia I. Trejo-Téllez and Fernando C. Gómez-Merino 2012. Nutrient Solutions for Hydroponic Systems. In:, Hydroponics - A Standard Methodology for Plant Biological Researches, Dr. Toshiki Asao (Ed.), InTech,

Lopes, M. and G. Walker. 1990. *You and Your Well: How to be responsible for your private water supply.* Issues in Water Quality. Fact Sheet, University of Massachusetts.

McMurtry, M.R., D.C. Sanders, and R.G. Hodson. 1997. Effects of biofilter/culture tank volume ratios on productivity of a recirculating fish/vegetable co-culture system. Journal of Applied Aquaculture. Vol. 7, No. 4. p. 33–51.

McMurtry, M.R., D.C. Sanders, J.D. Cure, R.G. Hodson, B.C. Haning, and P.C.S. Amand. 1997. Efficiency of water use of an integrated fish/vegetable co-culture system. Journal of the World Aquaculture Society. Vol. 28, No. 4. pp. 420–428.

Merle H. Jensen, Alan J. Malter 1995. Agricultura, p. 157. World Bank Publications, Washington, DC, USA. Available from: http://ag.arizona.edu/hydroponictomatoes/nutritio.htm

Merle H. Jensen, Alan J. Malter 1995. *Protected Agriculture: A Global Review*. World Bank Technical Paper Number 253. World Bank Publications, Washington, DC, USA.

Piper, R.G. *et al.* (editors). 1982. Fish hatchery management, pp. 517. USDI, U.S Fish Wildl. Ser., Washington, D.C.

Rakocy, James E. 1998. Integrating hydroponic plant production with recirculating system aquaculture: Some factors to consider. p.392–394. In: Proceedings of Second International Conference on Recirculating Aquaculture, Held July 16-19, Roanoke, VA.

Resh, H.M. 1995. *Hydroponic Food Production, A Difinitive Guidebook of Soilless Food-Growing Methods. 5th Ed*. Woodbridge Press Publishing Company, Santa Barbara, California, USA.

Resh, H.M. 2001. Hydroponic Food Production, 6th edition. New Concepts Press, Mahwah, NJ.

Salisbury, F. B. and Ross, C. W. 1992. *Plant Physiology*. Wadsworth Publishing Company, California, U. S. A.

Sneed, K. 1975. Fish farming and hydroponics. Aqua-culture and the Fish Farmer. Vol.2, No. 1. p. 11, pp. 18–20.

Steiner, A. A. 1984. The Universal Nutrient Solution, *Proceedings of IWOSC 1984 6th International Congress on Soilless Culture*, pp. 633-650. Wageningen, The Netherlands, Apr 29-May 5, 1984.

Steiner, A.A. 1968. Soilless Culture, *Proceedings of the IPI 1968 6th Colloquium of the Internacional Potash Institute*, pp: 324-341. Florence, Italy.

Trejo-Téllez, L. I.; Gómez-Merino, F. C. and Alcántar G. G. 2007. Elementos Benéficos, In: *Nutrición de Cultivos*, G. Alcántar G and L. I. Trejo-Téllez, L. I. (Eds.), pp. 50-91. Mundi-Prensa, México, D. F., México.

Wedemeyer, G.A. 1977. Environmental requirements for fish health. In Proceedings of the International Symposium on Diseases of Cultured Salmonids, pp. 41-55. Tavolek, Inc., Seattle, Washington.

Windsor, G. and Schwarz, M. 1990. Soilless Culture for Horticultural Crop Production. FAO, Plant Production and Protection. Paper 101. Roma, Italia.

Woods, May (1988). Glass houses: history of greenhouses, orangeries and conservatories. Aurum Press, London.

http://www.hightimes.com/read/history-hydroponics-0

http://en.wikipedia.org/wiki/Hydroponics#cite_ref-Douglas1975_1-0

http://ruaf-asia.iwmi.org/Data/Sites/6/PDFs/H_Eng.pdf

Index

Floating Gardens. (p. 2)

An Open Hydroponics System. (p. 10)

NASA Researcher Checking Hydroponic Onions with Bibb Lettuce to his Left and Radishes to the Right. (p. 15)

Lettuce Grown in an
Aeroponics Apparatus
(p. 26)

Roots Suspended in Aeroponics Chamber (p. 26)

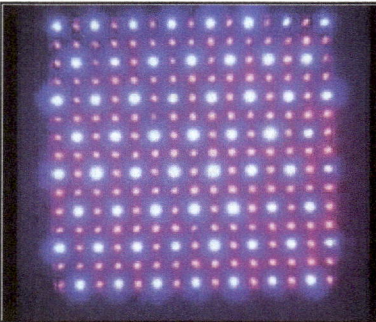

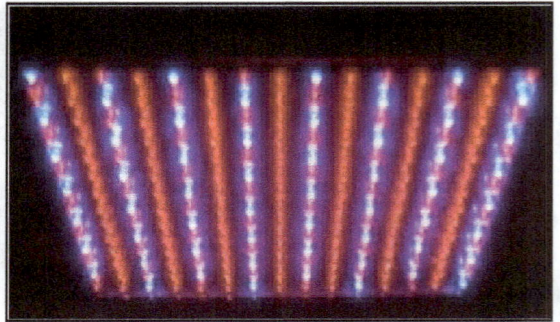

LED Grow Lights (p. 53)

Leaf Burn Due to Excessive
EC Nutrient Solution
Touching the Foliage. (p. 85)

Expanded Clay Aggregate (p. 100)

Rose Plants in Cocopeat (p. 101)

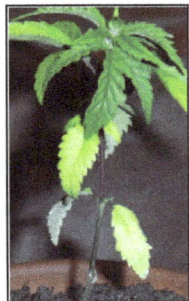

Plants/leaves Showing Nitrogen Deficiency (p. 118)

Leaves Showing Potassium Deficiency (p. 120)

**Symptoms of Magnesium Deficiency
Showing Yellowing in Older Leaves. (p. 121)**

**Symptoms of Magnesium Deficiency Showing
Brown Spots in Leaves. (p. 122)**

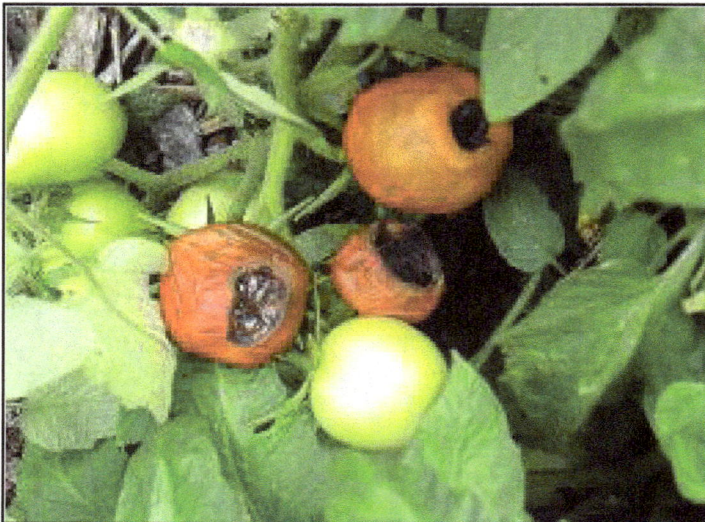

Calcium Deficiency Showing Blossom-End Rot of Tomato (p. 123)

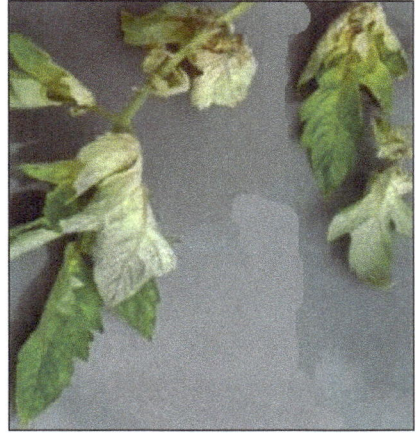

Leaves Showing Calcium Deficiency (p. 123)

Sulphur Deficiency in Corn and Soybean (p. 124)

Leaves Showing Zinc Deficiency (p. 127)

Leaves Showing Copper Deficiency (p. 128)

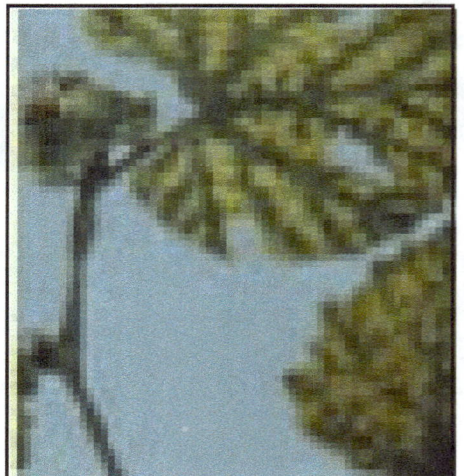

Boron Deficiency Symptoms in Pepper Leaves (p. 129)

Leaves Showing Molybdenum Deficiency (p. 130)

Powdery Mildew (p. 140)

Leaf Mould (p. 143)

Crown Gall Disease Caused by
***Agrobacterium* (p. 144)**

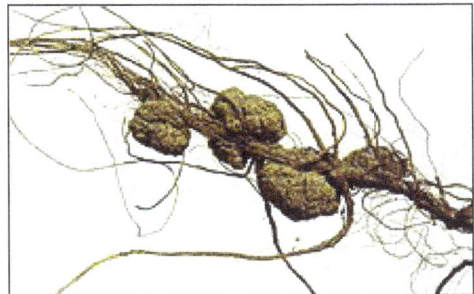

ToMV (p. 146)

Algae (p. 146)

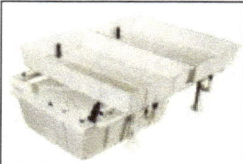

Images of some Types of Hydroponics Systems (p. 176)

www.ingramcontent.com/pod-product-compliance
Lightning Source LLC
Chambersburg PA
CBHW060247230326
41458CB00094B/1471